basic
electricity:
theory
& practice

basic electricity: theory & practice

Milton Kaufman

President, Electronic Writers and Editors, Inc.
Coauthor of Understanding Radio Electronics, *4/e*

J. A. Wilson

Coordinator, Engineering Technologies
Kent State University, Trumbull Regional Campus

McGraw-Hill Book Company

New York	*Kuala Lumpur*	*Panama*
St. Louis	*London*	*Rio de Janeiro*
San Francisco	*Mexico*	*Singapore*
Düsseldorf	*Montreal*	*Sydney*
Johannesburg	*New Delhi*	*Toronto*

Library of Congress Cataloging in Publication Data
Kaufman, Milton.
    Basic electricity.

    1. Electricity. I. Wilson, J.   A
II. Title
QC523.K38   537   72-12835
ISBN 0-07-033402-1

Basic Electricity: Theory and Practice

34567890EBEB798765

*The editors for this book were Alan W. Lowe and Alice V.
Manning, the designer was Marsha Cohen, and its produc-
tion was supervised by James E. Lee. It was set in Basker-
ville by Progressive Typographers.
It was printed and bound by Edwards Brothers Incor-
porated.*

contents

preface

Basic Electricity: Theory and Practice is a beginner's text. It is written particularly for students who have no background in electricity.

The book is planned as a complete course text for beginners, who can use it for self-study or as a class text for school use. The book is written in a simplified but comprehensive manner, to provide the student with a strong background in the subject of electricity and some of its practical uses. The authors recommend that all students read this preface carefully in order to obtain the maximum utilization of this book.

What Topics Are Covered in This Book?

The book begins with a discussion of the structure of the universe. This includes an explanation of the structure of atoms and molecules. The relationship between the atom and electricity is given, and static electricity is explained.

The student is then introduced to the interesting and important principles of magnetism and its relation to electricity. Examples of simple electromagnetic devices are given.

Next, the important principles of conductors, insulators, and resistors are discussed from the viewpoint of electron theory. Included in this overall discussion are examples of wire sizes and the different types of resistors, as well as resistor color codes.

The vital subjects of volts, amperes, ohms, and Ohm's law are the next to be explained. The volt is described in terms of electron theory, and the analogy is made between hydraulic pressure and voltage. The ampere is also described in terms of electron theory, and here, the analogy is made between fluid flow and the ampere. The ohm is described as a unit of opposition to current flow, or a unit of resistance. Here, the analogy is made between electrical resistance and mechanical friction.

Following these topics, the generation of dc electricity is described. The chemical (battery) and mechanical (generator) methods are explained, and brief discussions are given on the methods of generating electricity with heat, light, and pressure.

The means of measuring dc electricity is described, in conjunction with a simple dc meter. Simple methods of measuring direct current, dc voltage, and resistance are shown.

The next three chapters described the all-important principles of the series, parallel, and series-parallel circuits. Current and voltage relationships for these circuits are explained. The characteristics and practical uses of the circuits are described.

The operation of dc generators and motors is covered next. The various parts and their functions are explained. The operating characteristics of these devices are explained, as well as their practical uses and repair.

Alternating current and its measurement are next described. Ac and dc are compared, and methods of generating ac are given. Alternators (ac generators) and dc generators are compared. The purpose of rectifier diodes in changing ac to dc is explained, as is the use of such diodes in an ac measuring instrument. Methods of making ac measurements are given.

The vital principles of inductance and capacitance are described in basic terms and in conjunction with the original experiments regarding them. Practical uses of inductors (coils) and capacitors are shown. The following chapter uses the basic principles of inductance to explain the operation of a simple transformer. Various types of common transformers are shown and their principles discussed. Practical uses are given.

Since almost all students are interested in automobiles, this vehicle offers an interesting means of instruction in various electrical principles. The auto lighting system is described, as are the various fuses and circuit breakers. The battery, its charging system, and the ignition system are simply explained. Troubles in a car's electrical system are described.

The final chapter shows the student the very important principles of house wiring. The types of wiring and fuse panels are shown. Three-wire systems used for appliances are described.

Three appendixes appear at the back of the book and offer much valuable information. Appendix A is devoted to the vital topic of safety practices and first aid. Appendix B summarizes all the symbols and special vocabulary words used in this text. In Appendix C, the circuit board used throughout this text and all its associated parts are completely described.

What Is the Plan of This Book?

Basic Electricity: Theory and Practice is written and arranged so that it can be used either as a self-study book or a classroom textbook. Since it is a beginner's book, mathematics is held to the absolute minimum and never rises above the level of simple arithmetic. Whenever an equation is introduced, a simple, worked-out example is provided to explain the practical meaning of the equation. All possible variations of the equation are given so that the student can solve for any desired unknown, without the need for transposing.

Great emphasis is given to the use of simple and liberally illustrated discussions, which are followed by simple programmed reviews to reinforce the learning of the material. A minor amount of new material is also introduced in the programmed reviews.

The use of long theoretical discussions is avoided in this book. Rather, a more practical, "what can you do with it" approaches is used, which is far more interesting to beginning students. The book is specifically designed to provide the student with a good foundation in electrical theory, in an easy-to-read approach.

The Chapter Format

To simplify study and provide greater text utilization for both students and instructors, the same format is used for each chapter (except Chap. 1). The format has been carefully designed so that the maximum information may be obtained from each chapter with the least effort.

1. Each chapter begins with an introduction which describes the importance

and purpose of the chapter material. A list of the general chapter topics to be covered is given in question form.

2. The next section is called "Instruction." This portion presents the basic information taught in the chapter. It is arranged in relatively short and easy-to-absorb sections. Liberal use is made of simple illustrations. Within this section, a brief summary follows each major topic, so that several such summaries may appear in each chapter.

3. The "Programmed Review Questions" section is next. This is a simplified, programmed approach which reviews the subject matter discussed in the "Instruction" section. In addition, a minor amount of new information related to the chapter subject is presented.

4. The "experiments" section is next. This section contains simple experiments which the student can easily perform to reinforce his understanding of the principles explained in the chapter. Each experiment is completely but simply described and is divided into the following sections: "Purpose," "Test Setup," "Procedure," and "Conclusion."

What is the End-Of-Chapter Format?

Each chapter concludes with a section consisting of a "Self-Test with Answers." This affords the student another opportunity to review the basic material of the chapter. It also permits him to check himself as to how well he has learned the information.

The "Self-Test" is a multiple-choice examination with answers. Most answers are explained to aid understanding. Questions are carefully composed for simplicity and value.

Special Features of This Book

1. *Repetition.* All important concepts are repeated several times to reinforce understanding. They are first introduced in the "Instruction" section; they are then reinforced in the "Programmed Review Questions"; their effects are described in the "Experiments" section; and finally, the student tests his comprehension of the concepts in the "Self-Test with Answers." However, each time the material is repeated, it is with a different technique, so that the student will not become disinterested in the material.

2. *Use of Second Person.* The material is presented as though an instructor is talking directly to the student. Second person and the active voice are used.

3. *Vocabulary.* A complete vocabulary of terms and symbols is included in the appendix. This makes it convenient for the student to refresh his memory of terms that were first introduced in earlier chapters.

A careful and conscientious study of all the material in this book will provide the student with a basic and comprehensive background in the principles and practice of electricity. With the background provided herein, the student will be prepared to begin a career in the field of electricity and electrically operated devices. Electricity is a fascinating and highly lucrative field, and one that will return substantial rewards to anyone who applies himself diligently to its study.

Milton Kaufman
J.A. Wilson

basic electricity: theory & practice

1. What Is the Universe Made Of?

Introduction

Can you name ten ways in which you have used electricity this week? Most people can. Electricity has become so commonplace in our modern society that we take it for granted. Yet, only 100 years ago—a short time in the history of man—electricity was only a laboratory curiosity.

One of the most amazing things about electricity, considering the wide variety of jobs it can do, is the fact that its existence depends on a tiny particle called an *electron*. It is so small that it cannot be seen with the strongest microscope in the world. It would take 1,000 million, million, million, million electrons to make a weight of 1 gram. (A gram is about $1/28$ ounce.) Yet this tiny particle is the basis of electricity that lights cities and operates the machinery of factories.

In order to understand electricity, then, a good place to start is with the study of the electron.

You will be able to answer these questions after studying this chapter:

What is the universe made of?
What are elements made of?
How do electrons, protons, and neutrons combine to make an atom?

What is an electrical charge?
What are electrons and protons?
How big is an electron?
What is static electricity?

Instruction

WHAT IS THE UNIVERSE MADE OF?

In the fourth century B.C. Aristotle tried to answer this question by simple reasoning. In his time, the idea of determining facts by experiment was unheard of. Aristotle concluded that everything in the world was made of four basic elements: fire, earth, water, and air. According to his theory, every type of matter in the universe could be made by the right combination of these four elements. Two thousand years later, alchemists were trying to make gold by combining basic ingredients. Figure 1-1 shows the ingredients they had to work with. Their life's work was based on Aristotle's unproved theory.

Today, we know that Aristotle's elements are *not* the basic ingredients of the universe. To understand what the universe is really made of, let's start with

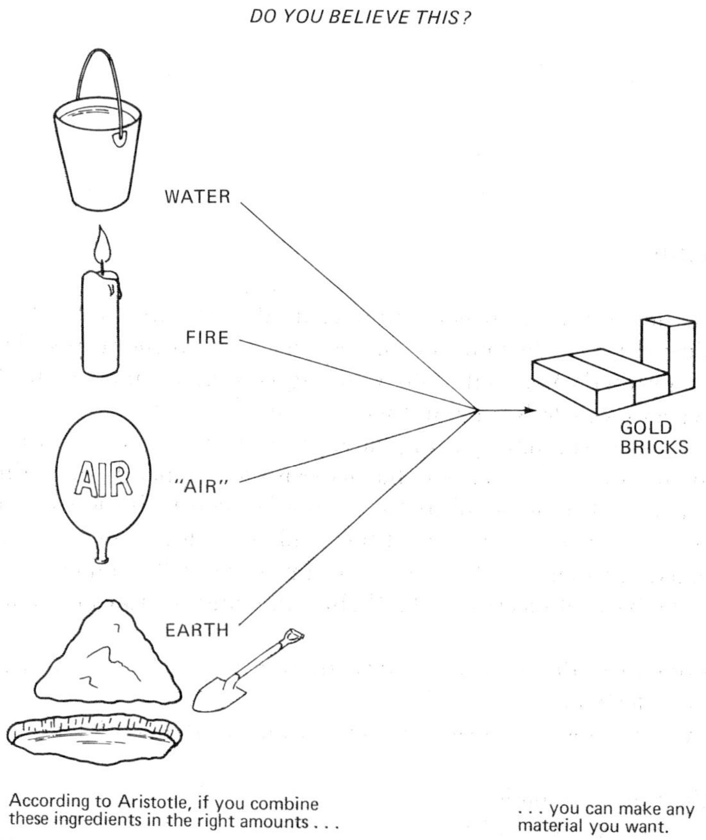
DO YOU BELIEVE THIS?
WATER
FIRE
AIR
"AIR"
EARTH
GOLD BRICKS
According to Aristotle, if you combine these ingredients in the right amounts . . .
. . . you can make any material you want.

Figure 1-1. Early philosophers tried to use their reasoning power to determine what are the basic ingredients of the world.

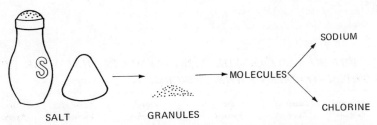

Figure 1-2. Salt can be divided into granules, and the granules can be divided into molecules. If you make any further division, you no longer have salt.

a substance that everyone is familiar with—salt. If you look at a pinch of salt closely, you can see that it is composed of very tiny granules. These granules are not so small that they cannot be divided. In fact, they can be divided again and again.

If you continue to divide the granule of salt into smaller and smaller pieces, you will finally get a piece so small that it cannot be divided again and still be salt. That small piece is called a *molecule*. A molecule is the smallest piece of a substance that can exist by itself and retain all the properties of the substance.

As shown in Fig. 1-2, if you could divide a molecule of salt into smaller pieces, you would find that it is composed of two different materials—*sodium* and *chlorine*. Neither of these two materials is anything like the salt that you started with.

Suppose that instead of dividing salt we had decided to divide a granule of sugar. This is illustrated in Fig. 1-3. Again, we *could* divide it over and over until ultimately we reach a point where we can no longer divide it and still have sugar. This smallest particle of sugar is, of course, a molecule. What happens if we divide the molecule of sugar? We find that it is made of carbon, hydrogen, and oxygen. Once again, these substances do not resemble the sugar that you started with.

If you spent your whole lifetime dividing particles of every material in the universe known to man, you would ultimately come to a strange and fascinating conclusion: there are really only 92 basic ingredients from which everything in the universe is made. These ingredients are called *elements*. Table 1-1 lists all the 92 basic elements from which the materials of the universe are made. (The atomic number listed in Table 1-1 will be explained later.)

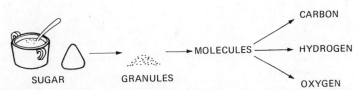

Figure 1-3. Sugar can be divided into granules, and the granules can be divided into molecules. If you make any further division, you no longer have sugar.

TABLE 1-1. LIST OF THE BASIC ELEMENTS THAT MAKE UP
THE MATERIALS OF THE UNIVERSE

Atomic Number	Name of Element	Its Symbol	Atomic Number	Name of Element	Its Symbol
1	Hydrogen	H	47	Silver	Ag
2	Helium	He	48	Cadmium	Cd
3	Lithium	Li	49	Indium	In
4	Beryllium	Be	50	Tin	Sn
5	Boron	B	51	Antimony	Sb
6	Carbon	C	52	Tellurium	Te
7	Nitrogen	N	53	Iodine	I
8	Oxygen	O	54	Xenon	Xe
9	Fluorine	F	55	Caesium	Cs
10	Neon	Ne	56	Barium	Ba
11	Sodium	Na	57	Lanthanum	La
12	Magnesium	Mg	58	Cerium	Ce
13	Aluminium	Al	59	Praseodymium	Pr
14	Silicon	Si	60	Neodymium	Nd
15	Phosphorus	P	61	Promethium	Pm
16	Sulfur	S	62	Samarium	Sm
17	Chlorine	Cl	63	Europium	Eu
18	Argon	A	64	Gadolinium	Gd
19	Potassium	K	65	Terbium	Tb
20	Calcium	Ca	66	Dysprosium	Dy
21	Scandium	Sc	67	Holmium	Ho
22	Titanium	Ti	68	Erbium	Er
23	Vanadium	V	69	Thulium	Tm
24	Chromium	Cr	70	Ytterbium	Yb
25	Manganese	Mn	71	Lutetium	Lu
26	Iron	Fe	72	Hafnium	Hf
27	Cobalt	Co	73	Tantalum	Ta
28	Nickel	Ni	74	Wolfram	W
29	Copper	Cu	75	Rhenium	Re
30	Zinc	Zn	76	Osmium	Os
31	Gallium	Ga	77	Iridium	Ir
32	Germanium	Ge	78	Platinum	Pt
33	Arsenic	As	79	Gold	Au
34	Selenium	Se	80	Mercury	Hg
35	Bromine	Br	81	Thallium	Tl
36	Krypton	Kr	82	Lead	Pb
37	Rubidium	Rb	83	Bismuth	Bi
38	Strontium	Sr	84	Polonium	Po
39	Yttrium	Y	85	Astatine	At
40	Zirconium	Zr	86	Radon	Rn
41	Niobium	Nb	87	Francium	Fr
42	Molybdenum	Mo	88	Radium	Ra
43	Technetium	Tc	89	Actinium	Ac
44	Ruthenium	Ru	90	Thorium	Th
45	Rhodium	Rh	91	Protoactinium	Pa
46	Palladium	Pd	92	Uranium	U

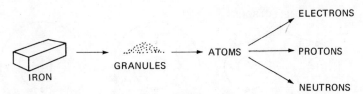

Figure 1-4.   An atom is the smallest particle of an element—such as iron—that retains all the properties of that element. If you divide an atom, you get electrons, protons, and neutrons.

SUMMARY

1. Ancient philosophers tried to determine what the world is made of by pure reasoning. Using this approach, Aristotle concluded that water, fire, air, and earth are the ingredients.
2. Today, we know that there are 92 basic elements from which millions of different kinds of materials can be made. (Additional elements beyond the 92 shown in Table 1-1 have been created artificially in the laboratory, but the 92 are found in nature.)
3. The molecule is the smallest possible particle of a material that has all the properties of that material.

WHAT ARE ELEMENTS MADE OF?

What would happen if you were to take one of the elements of Table 1-1 and divide it into smaller and smaller particles? Would you ultimately arrive at a point where it could no longer be divided? The answer is *yes. The smallest particle that you could obtain by dividing an element that would still contain all the properties of the element is called an* **atom.** Compare the process of dividing an element as shown in Fig. 1-4 with that of dividing a material made by combining elements, such as salt or sugar. As small as the atom is, it can still be divided into smaller particles.

If you could somehow divide the atom into its composite parts as shown in Fig. 1-4, you would find that there are three basic ingredients—*electrons, protons,* and *neutrons.* Electrons and protons are tiny particles that are present in *every* atom. Neutrons are tiny particles, but they are not found in *every* atom.

HOW DO ELECTRONS, PROTONS, AND NEUTRONS COMBINE TO MAKE AN ATOM?

To see how electrons, protons, and neutrons combine to make an atom, we will start with the simplest atom in the universe, that of *hydrogen.* It consists of a proton in its center and an electron moving around the proton at a fantastically high speed. Every hydrogen atom in the universe is identical, with one proton in the center and one electron moving around that proton at a tremendously high speed. Figure 1-5 shows how the hydrogen atom looks in comparison with other atoms.

A logical question is: can we divide the electron and the proton into smaller parts? The answer to that question is *no.* No one has ever been able to do it so far. These particles are considered to be indivisible.

The hydrogen atom is the simplest one in the universe. As shown in Fig.

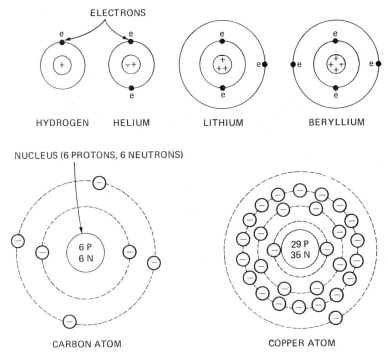

Figure 1-5. *Each element has a different style of atom. The difference between one atom and another is in the number of electrons, protons, and neutrons.*

1-5, the next simplest one is the helium atom. If you were able to look inside a helium atom, you would find that it is comprised of two protons in the center and two electrons moving around the center of the atom—called the *nucleus*—at enormous speeds. The electrons manage to move without colliding with one another. If you could look inside all the atoms of the 92 elements, you would find that they are similar in one way: they all contain protons in the center, or nucleus, and they all contain electrons moving around the center at enormously high speeds. Figure 1-5 shows several atoms for comparison.

In some atoms you would find another particle in the center. It is called the *neutron*. The neutron is very similar to the proton with one major difference: it does not contain an *electrical charge*.

WHAT IS AN ELECTRICAL CHARGE?

We need to talk about electrical charges because they are very important in the study of electricity. If you tied a rock to a string and twirled it around at high speeds, you would eventually reach a speed so high that the string would break. The rock exerts a great pull on the string in an effort to break away, and when its speed is great enough it actually *does* break away. However, the electrons moving around the protons do not fly off despite the fact that they are moving at enormously high speeds. The reason is that the electrons and protons are attracted strongly, electrically.

The proton has a *positive electrical charge*, and the electron has a *negative electrical charge*. A very important basic law in electricity is illustrated in Fig. 1-6. It shows that **unlike charges attract.** The positive charge of the proton attracts

Unlike electrical charges are attracted toward each other.

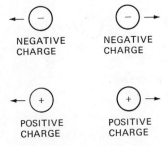

POSITIVE
CHARGE

POSITIVE
CHARGE

Figure 1-6. A very important law is illustrated here: **Unlike charges attract and like charges repel.**

Like charges are repelled by each other.

the negative charge of the electron with a force large enough to prevent the electron from flying off into space. In other words, the electrical force of attraction acts like the string that holds the twirling rock in a circular path.

Another very important basic rule of electricity illustrated in Fig. 1-6 is that *like charges repel.* Thus, two negative electrons will move away from each other if free to do so. Likewise, two positive charges will move away from each other when they are free to do so.

An atom normally has an equal number of electrons and protons. Their electrical charges cancel, and the atom normally does not have an overall electrical charge.

SUMMARY

1. The smallest particle into which an element can be divided is the atom.
2. Atoms are made of electrons, protons, and neutrons. The protons and neutrons are in the center, or *nucleus*. The electrons move around the nucleus at enormous speeds.
3. The atoms of one element are different from the atoms of all other elements because of the number of electrons, protons, and neutrons they contain.
4. Electrons have negative electrical charges, and protons have positive electrical charges. Neutrons do not have electrical charges.
5. An important basic law in electricity is *like charges repel and unlike charges attract.*

WHAT ARE ELECTRONS AND PROTONS?

From our basic study of electrons and protons we can now define what they are. *An electron is a basic particle in an atom, and it has a negative charge. A proton is a basic particle in an atom, and it has a positive charge.* Electrons, protons, and neu-

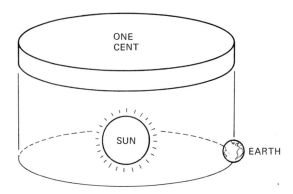

If you could enlarge a penny to this size . . .

. . . its electrons would be the size of baseballs,
and they would be 3 miles apart.

Figure 1-7. This illustration gives you an idea how small electrons are. It also demonstrates that an atom is comprised mostly of empty space.

trons are the basic atom particles which combine to make all the elements shown in Table 1-1. The atomic number of Table 1-1 tells how many protons an atom has.

In our study of electricity we will be most interested in the electron.

HOW BIG IS AN ELECTRON?

The diameter of an electron is approximately 0.000 000 000 0022 inch. To get an idea of how small this really is, suppose you could take a copper penny and enlarge it so that the rim of the penny was as large as the path of the earth around the sun. This is shown in Fig. 1-7. (The earth is about 93,000,000 miles from the sun.) If you could enlarge a penny this much, the electrons within the penny would be only the size of baseballs, and they would be 3 miles apart!

The protons and neutrons in the nucleus are only about $\frac{1}{3}$ of the diameter of electrons, but they weigh much more. In fact, the mass of a proton is about 1,850 times greater than the mass of an electron. (Protons and neutrons have approximately the same mass.) The weight of an electron is about

0.000 000 000 000 000 000 000 000 0009 gram

WHAT IS STATIC ELECTRICITY?

Although the electrons are normally bound to their atoms by the positive charge of the protons in the center, it is possible under some conditions to get an electron away from its atom. One way to do this is to rub electrons from the surface of the material.

If you comb your hair with a hard-rubber comb, some of the electrons on the surface of your hair are rubbed off onto the comb. Your hair is then short of electrons and said to be "positively charged." The comb has too many electrons, and it is said to be "negatively charged." This is an important point to remember: When a body has more than its normal amount of electrons, it is *negatively charged;* when it has less than its normal amount of electrons, it is *positively charged.* A charge caused by an accumulation or deficiency of electrons is called a *static charge.* Nature is always trying to balance the electrical charge. The electrons trapped on the comb will try to return from the comb to your hair. On a dry day you can hear a crackling sound when you comb your hair. This is caused by tiny sparks jumping from the comb back to the hair.

Static charges occur whenever there is an excess or a shortage of electrons at some point. One of the most common demonstrations of static electricity is in a thunderstorm. A large number of electrons can accumulate in a cloud, and beneath the cloud there is a corresponding deficiency of electrons. This charge can become so great that a *spark*—consisting of an enormous number of electrons—moves from the cloud to the earth. The same thing happens when a cloud has a deficiency of electrons. In this case, electrons move from the earth to the cloud to neutralize the charge. To get an idea of how much electricity is involved in a stroke of lightning, consider the fact that your house is usually wired in such a way that the maximum amount of current that should flow in the wires is 15 amperes. (In a later chapter we will define amperes and discuss electric current more thoroughly.) The important point now is to understand that while 15 amperes is sufficient current to run a number of circuits in your household, a stroke of lightning may contain over *150,000 amperes!* In fact, the potential energy in a thunderstorm may be greater than the amount of energy released by the explosion of an atomic bomb.

The problem of static electricity is a serious one and must be dealt with in many branches in industry. For example, in printing presses for newspapers, the paper moving through the press causes a static charge. This charge must be drained off through metal conductors. As another example, the rubber tires on an automobile scraping against the pavement produce a static charge that can cause sparks to jump across wheel bearings. You may have seen a small wire protruding from the pavement in the approach to the toll booth of a turnpike. The purpose of this wire is to bleed off the static charge built up in an automobile so that the booth operator will not receive an uncomfortable shock by touching it.

Static charges can cause sparks which set off explosions in gas-filled rooms. They can also damage sensitive electronic devices such as transistors used for making radios.

You have learned a very important thing about electricity in this chapter: *it can be generated by friction between two bodies.* The bodies must not be made of metal, but metal can be used to drain the charges off. Friction is only one of a number of ways used to generate electricity. Other ways of generating electricity will be discussed in later chapters.

SUMMARY

1. Electrons are negatively-charged particles in atoms. Protons are positively-charged particles in atoms. Neutrons are particles in atoms that have no electrical charge.

2. Static charges result when bodies are rubbed together, but the bodies must not be made of metal.
3. Metal can be used for draining off undesired static charges.

Instructions for Answering Programmed Review Questions

Start with Question 1 in block 1. Answer this question with the choice that you feel is correct. If you think that choice A is correct, proceed to block 7. If, on the other hand, you think that choice B is correct, proceed to block 17.

When you turn to the block indicated by your answer, you will learn if your answer is right or wrong. If your answer is wrong, you will learn why. If your answer is right, you will get another question to answer.

If you have learned the material in the "Instruction" section, you will be able to complete this section easily.

Programmed Review Questions

We will now review the important concepts of this chapter. If you have understood the material, you will progress easily through this section. Do not skip this material, because some additional theory is presented.

1. Static electricity is generated by
   A. rubbing two nonmetallic bodies together. (Proceed to block 7.)
   B. rubbing copper and steel together. (Proceed to block 17.)

2. *Your answer to the question in block 7 is A. This answer is wrong. Mesons and pions happen to be atomic particles, but they are of no interest in this study of electricity.* Proceed to block 12.

3. *Your answer to the question in block 4 is A. This answer is wrong. Study Fig. 1-6.* Then proceed to block 9.

4. *The correct answer to the question in block 11 is A. The electron is considered to be the basic **negative** unit of charge.* Here is your next question.
   Which of the following statements is true?
   A. Like charges attract. (Proceed to block 3.)
   B. Like charges repel. (Proceed to block 9.)

5. *The correct answer to the question in block 16 is A. The center of an atom is called the **nucleus.*** Here is your next question.
   A lead-acid battery is used in the electrical system of an automobile. Some of the plates in the battery are made of pure lead. Which of the following statements is true?
   A. Lead is one of the basic elements of the universe. (Proceed to block 11.)
   B. Lead is a metal made by combining copper and iron. (Proceed to block 18.)

6. *The correct answer to the question in block 19 is* **B***. There is an electrical attraction between the proton in the nucleus and the electron in orbit. This attraction is due to the fact that they are unlike charges.* Here is your next question.

   An accumulation of electrons at a point is called
   A. a negative charge. (Proceed to block 16.)
   B. an ampere of electric current. (Proceed to block 14.)

7. *The correct answer to the question in block 1 is* **A***. Static electricity is made by rubbing nonmetallic bodies together.* Here is your next question.

   Which of the following statements is true?
   A. Electrons are particles made by combining mesons and pions. (Proceed to block 2.)
   B. As far as we know, electrons cannot be divided into smaller particles. (Proceed to block 12.)

8. *Your answer to the question in block 12 is* **A***. This answer is wrong. An atom is the smallest particle of an* **element** *that has all the characteristics of that element. However, we are not dealing with an element here.* Proceed to block 19.

9. *The correct answer to the question in block 4 is* **B***. One of the most important things that you learned in this chapter is that like charges repel and unlike charges attract.* Here is your next question.

   Which weighs more, the electron or the proton?

   _____ (Proceed to block 20.)

10. *Your answer to the question in block 19 is* **A***. This answer is wrong. While it is true that the electron is a very light particle, it would be thrown off into space because of its enormous speed. However, the electrical attraction of the proton holds it in orbit around the nucleus.* Proceed to block 6.

11. *The correct answer to the question in block 5 is* **A***. Table 1-1 shows that lead is one of the basic elements.* Here is your next question.

    An electron is
    A. a negatively-charged particle. (Proceed to block 4.)
    B. a positively-charged particle. (Proceed to block 13.)

12. *The correct answer to the question in block 7 is* **B***. The electron can be considered to be a particle that cannot be divided.* Here is your next question.

    Gasoline is made by combining a number of elements. The smallest particle into which gasoline could be divided and still retain all the characteristics of gasoline is
    A. an atom. (Proceed to block 8.)
    B. a molecule. (Proceed to block 19.)

13. *Your answer to the question in block 11 is* **B***. This answer is wrong. Protons are positively-charged particles.* Proceed to block 4.

14. *Your answer to the question in block 6 is* **B**. *This answer is wrong. As you will learn in a later chapter, an ampere of electric current occurs when there is a motion of charges. However, the question deals with an accumulation of charges at some point, not a motion of charges.* Proceed to block 16.

15. *Your answer to the question in block 16 is* **B**. *This answer is wrong. Neutrons are uncharged particles that may be found in the center of an atom, but the center is not called a* **neutron.** Proceed to block 5.

16. *The correct answer to the question in block 6 is* **A**. *A negative charge may be defined as an accumulation of electrons at some point.* Here is your next question.
     Which of the following statements is true?
     A. The center of an atom is called a nucleus. (Proceed to block 5.)
     B. The center of an atom is called a neutron. (Proceed to block 15.)

17. *Your answer to the question in block 1 is* **B**. *This answer is wrong. Static electricity cannot be generated by rubbing metal bodies together.* Proceed to block 7.

18. *Your answer to the question in block 5 is* **B**. *This answer is wrong. Lead is an element. It cannot be made by combining other metals anymore than gold can be made this way.* Proceed to block 11.

19. *The correct answer to the question in block 12 is* **B**. *Gasoline is a* **compound.** *A compound is made by chemically combining two or more elements. The smallest particle of a compound that has all the properties of that compound is a molecule.* Here is your next question.
     In a hydrogen atom, the electron moves around the proton at an enormous speed. The reason the electron does not fly off into space is that
     A. it does not weigh enough to break away. (Proceed to block 10.)
     B. it is electrically attracted to the proton. (Proceed to block 6.)

20. *The correct answer to the question in block 9 is* **the proton.** *The proton weighs about 1,850 times as much as the electron weighs. As you will learn in a later chapter, it is the lighter electron that moves with an electric current.* You have now completed the programmed questions. The next step is to put some of these ideas to work in laboratory experiments. Proceed to the "Experiment" section that follows.

Experiment

EXPERIMENT 1

Purpose—In this experiment you will learn how to make an instrument to measure the presence of a static charge. You will also learn how to use the instrument to show that like charges repel.

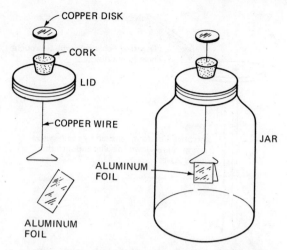

Figure 1-8. Construction details of an electroscope.

Test setup—To construct an electroscope and show how it is used to detect the presence of a static charge.

Procedure—

Step 1—Figure 1-8 shows how the *electroscope* is made. An electroscope is an instrument that can be used for detecting the presence of a static charge.

The electroscope will be more sensitive if the foil is very light and flexible.

Step 2—Using a hard-rubber comb, comb your hair briskly. Hold the comb near the electroscope, as shown in Fig. 1-9. If the comb is charged, and if the electroscope is working properly, the leaves will move apart as the charged comb is brought near the metal top.

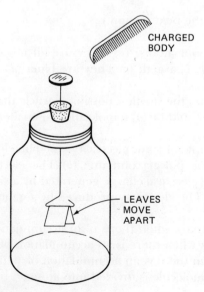

Figure 1-9. Using the electroscope to detect the presence of a negative charge.

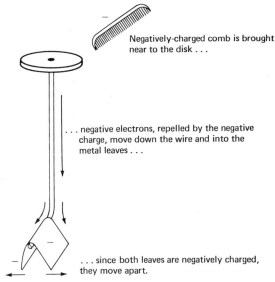

Figure 1-10. Why the leaves of the electroscope move apart when a charged body is nearby.

Conclusion — Figure 1-10 shows why the leaves of the electroscope move apart. As the negative comb is brought near the electroscope, it repels electrons away from the copper disk. These electrons move down the wire and into the metal leaves. Since both leaves have negative electrons, and *like charges repel*, the leaves will move apart.

Self-Test with Answers

(Answers with discussions are given in the next section.)

1. Which of the following is true about atoms? (*a*) They are all negatively charged; (*b*) They are all positively charged; (*c*) They are blue; (*d*) They all have electrons and protons.
2. If you divide the element silicon into the smallest possible particle that still has all the properties of silicon, it would be (*a*) a molecule; (*b*) an atom; (*c*) a neutron; (*d*) an electron.
3. Which of the following is an example of static electricity? (*a*) The electricity delivered to your home by the power company; (*b*) The electricity generated by a car battery; (*c*) The electrical charge generated by a rubber tire scraping against the road; (*d*) The electricity produced by a generator in a car.
4. A negative electrical charge (*a*) occurs when there is an accumulation of electrons at some point; (*b*) occurs when there is an accumulation of protons at some point; (*c*) occurs when there is an accumulation of neutrons at some point; (*d*) occurs when a molecule is divided into atoms.

5. Which of the following statements is *not* correct? (*a*) Unlike charges attract; (*b*) Like charges repel; (*c*) Electrons move around the nucleus of an atom; (*d*) A positive charge occurs when there is an accumulation of neutrons at some point.

6. Which of the following is true about electrons? (*a*) They may be the size of baseballs; (*b*) They are of no interest in the study of electricity; (*c*) They are negatively-charged particles; (*d*) They are red.

7. Electricity can be generated by (*a*) combining sodium and chlorine; (*b*) combining carbon and oxygen; (*c*) friction; (*d*) heating a glass rod.

8. The basic particles that make up the atom (*a*) were first discovered by Aristotle; (*b*) are molecules and granules; (*c*) have no electrical charge; (*d*) are electrons, protons, and neutrons.

9. One of the most important materials used in electrical apparatus is carbon. Which of the following is true about carbon? (*a*) It is one of the 92 elements; (*b*) It is another name for a molecule; (*c*) It has electrons but not protons; (*d*) It is made by combining calcium and iron.

10. You have two large metal spheres. One has an excess of electrons, and the other has a shortage of electrons. Which of the following is true? (*a*) The sphere that has an excess of electrons weighs much more than the one with the shortage of electrons; (*b*) The sphere that has a shortage of electrons weighs much more than the one with the excess of electrons; (*c*) The spheres will be attracted to each other; (*d*) The spheres will move apart if free to do so.

Answers to Self-Test

1. (*d*) — It is possible to take an electron away from an atom. This leaves it with one more proton than electron, and it will have a positive charge. Likewise, an extra electron can be crowded into an atom to give it a negative charge. However, these are special cases. The only choice for this question that is *always* correct is that all atoms have both electrons and protons.

2. (*b*) — Silicon is an element, so its smallest particle is an atom.

3. (*c*) — All the choices are examples of electricity, but only the one concerning friction is an example of static electricity.

4. (*a*) — This is a definition of negative charge.

5. (*d*) — Neutrons do not have an electrical charge, and therefore, an accumulation of neutrons cannot produce a positive charge.

6. (*c*)

7. (*c*) — The only answer you *know* to be true from studying this chapter is (*c*).

8. (*d*)

9. (*a*)

10. (*c*) — Remember that unlike charges attract. The sphere with an excess of electrons has a *negative* charge; the sphere with a shortage of electrons has a *positive* charge. Thus, the spheres with unlike charges are attracted to each other.

2.
How Is Magnetism Related to Electricity?

Introduction

In the introduction of Chap. 1 you were asked if you could name ten ways in which you have used electricity in the past week. This is usually easy to do, considering the broad applications of electricity in our modern life.

Now see if you can name ten different ways in which you have used *magnetism* in the past week. You might be surprised to learn that the number of ways in which magnetism is used in our modern-day living is as endless as the number of uses for electricity. Here is a partial list of components that you may have used within the week. Each of these components depends upon magnetism for its operation.

radios	record players
tape recorders	telephones
television receivers	kitchen appliances
furnaces	air conditioners
electric tools	compasses
automobiles (their electrical system)	

These are just a very few of the many examples of the uses of magnetism.

An application that is much wider in scope is the use of magnetism by power companies. The electricity that is delivered to your home, and to industry, from the power companies is generated by using magnetic fields. Later we will study exactly how this is done, but for the time being it is important to note that if it were not for magnetism, these power companies could not economically generate the large amount of electricity for use in homes and factories.

Because magnetism is so closely related to electricity, it is important to understand the theories of magnetism. Not only is magnetism needed for the generation of electric power by power companies, but it is also a fact that *you cannot have electricity without having magnetism in some form.*

You will be able to answer these questions after studying this chapter:

What are magnetic materials?
What are some different kinds of magnets?
What are some basic rules regarding magnetic fields?
What is magnetism by induction?
What happens when a material is magnetized?
What are some things that can destroy magnetism?
What is the relationship between electricity and magnetism?

Instruction

WHAT ARE MAGNETIC MATERIALS?

One of the things that it is useful to know about magnetism is which types of materials are *magnetic*—that is, which materials are attracted to a magnet. Figure 2-1 shows that there are three different kinds of materials related to the study of magnetism.

You can use a magnet to lift a pair of scissors, nails, tacks, safety pins, and needles. Those materials which are attracted to a magnet are called *magnetic materials.* (Another name for them is *ferromagnetic materials.*)

You cannot use a magnet to pick up a piece of wood or paper, and these materials are called *nonmagnetic.* (Another name for them is *paramagnetic materials.*)

There is a third class of materials, called *diamagnetic,* that act in a very unusual way near a strong magnet. Instead of being attracted to a magnet as are magnetic materials, or indifferent to the magnet as are nonmagnetic materials, the diamagnetic materials actually move away from a magnetic field. So far, diamagnetism has not been put to any large practical use because the effect is relatively small. In other words, it would take a very, very strong magnet to move diamagnetic materials with any noticeable force. However, you should remember that there are three—not two—types of materials related to our study of magnetism. Table 2-1 shows a list of a few materials of each type.

WHAT ARE SOME DIFFERENT KINDS OF MAGNETS?

Early Greeks were familiar with magnetism and experimented with it over 2,000 years ago. They noted that magnetic stones—which they called *magne-*

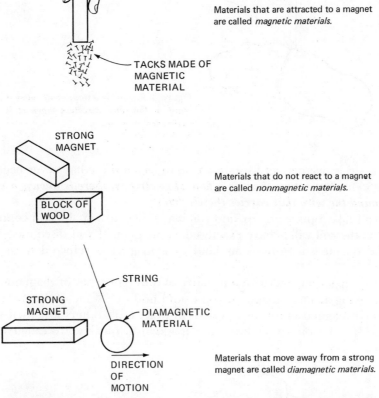

Materials that are attracted to a magnet are called *magnetic materials*.

TACKS MADE OF MAGNETIC MATERIAL

STRONG MAGNET

Materials that do not react to a magnet are called *nonmagnetic materials*.

BLOCK OF WOOD

STRING

STRONG MAGNET

DIAMAGNETIC MATERIAL

DIRECTION OF MOTION

Materials that move away from a strong magnet are called *diamagnetic materials*.

Figure 2-1. All the materials known can be classified as **magnetic, nonmagnetic,** *or* **diamagnetic.**

tite—attracted each other. Later, they found that if they suspended a magnetite stone from a string as shown in Fig. 2-2, it would always point north and south. That is why they also called it a *leading stone* or *lodestone*. A lodestone is an example of a natural magnet, since it is found in nature rather than made by man. Figure 2-3 shows some different kinds of magnets, including a lodestone.

If you take a piece of lodestone and rub a piece of iron with it, the iron will eventually become magnetized. It will retain this magnetism even after the lodestone is removed. Magnets that are made this way are called *artificial magnets*. Figure 2-3 shows an example of an artificial magnet.

An important kind of magnet is very closely related to electricity. Before

TABLE 2-1. EXAMPLES OF MAGNETIC, NON-MAGNETIC, AND DIAMAGNETIC MATERIALS

Magnetic Materials	Nonmagnetic Materials	Diamagnetic Materials
Iron	Wood	Copper
Nickel	Air	Water
Cobalt	Aluminum	Lead

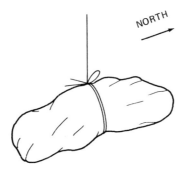

Figure 2-2. A lodestone will always point in the same direction when it is suspended on a string.

we discuss this magnet, let us state a rule to which there has never been found an exception: ***Every time there is a flow of electricity, there is always a magnetic field around the wire that carries the electricity.***

If you take a piece of wire and coil it as shown in Fig. 2-3 and connect it to a battery, the coil will behave exactly like a magnet. If you disconnect the battery, the magnetism stops. This kind of a magnet is referred to as an *electromagnet.*

If you experimented with a number of different types of magnetic materials for making artificial magnets, you would find that some of them are able to retain their magnetism for long periods of time, while others lose their magnetism quickly. The ones that lose their magnetism quickly are called *temporary*

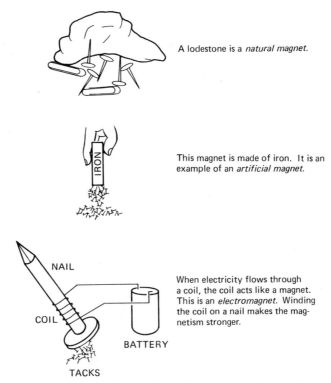

A lodestone is a *natural magnet.*

This magnet is made of iron. It is an example of an *artificial magnet.*

When electricity flows through a coil, the coil acts like a magnet. This is an *electromagnet.* Winding the coil on a nail makes the magnetism stronger.

Figure 2-3. Three different kinds of magnets are illustrated here.

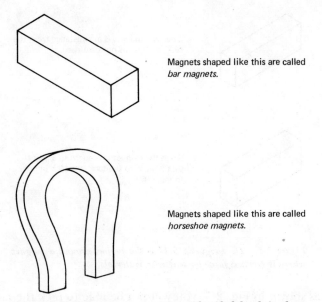

Magnets shaped like this are called *bar magnets.*

Magnets shaped like this are called *horseshoe magnets.*

Figure 2-4. Magnets are sometimes identified by their shape.

magnets, while the ones that can retain their magnetism over a long period of time are called *permanent magnets.* There are applications in electricity in which you need a material to become magnetized for a short period of time (in other words, where you will want temporary magnets), and there are also applications where you will want permanent magnets. *Soft iron* is a material that is used for making temporary magnets. This name is very misleading. There is nothing soft about the iron as far as touching it is concerned. Instead, the name comes from the fact that it cannot retain a magnetic field for any period of time.

Artificial magnets are often identified by their shape. Figure 2-4 shows two examples of artificial magnets that are frequently used. One is a *bar magnet,* and the other is a *horseshoe magnet.* These names obviously come from their shape.

SUMMARY

1. Magnetic materials are attracted to a magnet. Nonmagnetic materials are not affected by magnetism, and diamagnetic materials move away from a magnet.
2. Natural magnets are found in nature. Lodestone is an example of a natural magnet.
3. Artificial magnets are made from magnetic materials.
4. Electromagnets are made by passing electricity through coils of wire.
5. Whenever there is electricity, there is *always* magnetism.

WHAT ARE SOME BASIC RULES REGARDING MAGNETIC FIELDS?

By definition, a *magnetic field* is the space around a magnet where a magnetic material will be attracted by that magnet. If you hold a bar magnet a distance

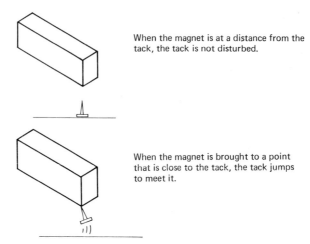

When the magnet is at a distance from the tack, the tack is not disturbed.

When the magnet is brought to a point that is close to the tack, the tack jumps to meet it.

Figure 2-5. The magnetic field is the region around a magnet where it exerts a force on magnetic materials.

from a tack as shown in Fig. 2-5, you will not be able to pick the tack up. However, as you move the bar magnet closer to the tack, you eventually come to a point where the tack is attracted strongly and will actually jump to the magnet. This simple experiment shows that the space around a magnet where attraction occurs is limited.

The magnet cannot attract a material at a great distance. In other words, it cannot exert a force on a magnetic material if the distance is too great. If you take a piece of paper and place it over a magnet, and then sprinkle iron filings on the paper as shown in Fig. 2-6, an interesting thing will happen. The filings arrange themselves around the magnet in lines which are called *flux lines.* (It is usually necessary to tap the paper in order to allow the filings to settle into place.) These lines represent regions of very strong magnetic influence.

You will note that the lines seem to pass from one area in the magnet and go to the other area. For convenience, we call these areas where the lines seem

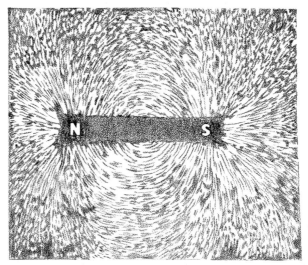

Figure 2-6. Iron filings will produce this pattern of a magnetic field.

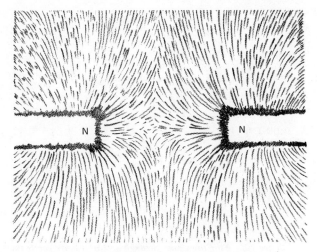

Figure 2-7. Flux pattern produced by two like poles.

to concentrate *magnetic poles*, and identify them as the south and north poles of the magnet. Strictly for definition, we say that the lines leave the north pole and enter the south pole. This simple idea is used extensively, but you should remember that the lines are not actually moving. This is just a way of describing the lines.

Notice that the flux lines are concentrated and close together at the poles, but they are spread apart as the distance from the poles is increased. The attraction of a magnet is stronger in the region of the poles than it is at a distance. Therefore, you can conclude that the concentration of flux lines is related to the strength of the attraction of a magnet.

If two magnets are free to move, they will come together at their north and south poles. This strong attraction of the north to south poles of two magnets is easily demonstrated, and is stated in a very important rule: ***Unlike magnetic poles attract.***

If you were to hold the north poles of two bar magnets close to each other and repeat the experiment with the iron filings, you would get the pattern shown in Fig. 2-7. Notice that the flux lines from each pole never cross, but rather they seem to be pushing each other away. If the magnets are free to move, they will actually move away from each other. (You would get the same result if you held the south poles of two bar magnets close together.) The fact that the like poles oppose each other and tend to move away from each other is stated in a simple and very important rule in magnetism: ***Like poles repel.***

WHAT IS MAGNETISM BY INDUCTION?

The flux lines around a magnet are able to pass through certain materials more easily than others. Soft iron is an example of a material that flux lines can pass through easily. Figure 2-8 shows an experiment that illustrates this. A piece of soft iron is located near the north pole of a magnet. Paper is placed over the magnet and over the soft iron, and iron filings are sprinkled over the paper. Note that the flux lines actually go out of their way in order to pass through the soft iron. This is because the magnetic flux lines can pass through iron more easily than they can pass through air.

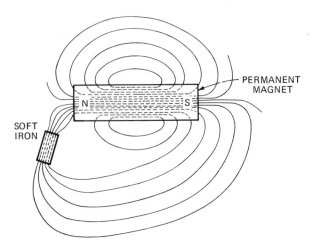

Figure 2-8.   Flux lines pass through soft iron more easily than they pass through air.

As long as the flux lines are passing through the soft iron, it behaves in every way like a magnet. However, if you remove the magnetic field, it no longer retains its magnetism. We can say that we are making a *temporary magnet* out of the soft iron by the method shown in Fig. 2-8. This method of making a magnet is known as *magnetism by induction.*

SUMMARY

1. A magnetic field is the space around a magnet where a magnetic material will be attracted.
2. Unlike magnetic poles attract.
3. Like magnetic poles repel.
4. Magnetic flux lines never cross.
5. Magnetic flux lines can pass through soft iron more easily than they can pass through air.
6. When a piece of soft iron is placed in a magnetic field, it becomes magnetized by induction.

WHAT HAPPENS WHEN A MATERIAL IS MAGNETIZED?

There are many complicated experiments and mathematical equations to demonstrate what happens inside a magnetic material when it becomes magnetized. One theory that has been used for many years is illustrated in Fig. 2-9. The magnetic material is assumed to be made up of thousands and thousands of tiny magnets as shown in Fig. 2-9*a*. When the material is unmagnetized, all these tiny magnets point in different directions. When you magnetize a material, what you are really doing is aligning all the tiny magnets so that the north poles are all in one direction, as shown in Fig. 2-9*b*, and the south poles are all in the other. The flux lines from all the north poles add together to make the north pole of a magnet, and the flux lines of all the south poles combine to form the south pole of the magnet.

To summarize, you have two possible conditions for a magnetic material:

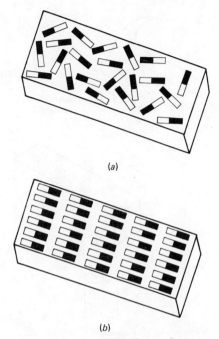

(a)

(b)

Figure 2-9. A popular theory of magnetism. (a) *An unmagnetized piece of iron is made of tiny magnets pointing in all directions.* (b) *When the iron is magnetized, all the magnets point in the same direction.*

unmagnetized and *magnetized*. After the material becomes magnetized it will stay that way until something happens to knock the tiny magnets out of position.

WHAT ARE SOME THINGS THAT CAN DESTROY MAGNETISM?

"Permanent" magnets are only permanent if they are taken care of. As shown in Fig. 2-10, there are three things that can destroy the magnetism. They are vibration, heat, and a strong, rapidly varying magnetic field.

Striking a magnet with a hammer will cause the tiny magnets within to change their directions. When they are no longer lined up so that they point in the same direction, the material will no longer be magnetized.

When you are magnetizing a piece of material, it helps to strike it with a hammer. In this case, the vibration helps the tiny magnets to turn into position.

When a magnet is heated to a high temperature, the tiny magnets get out of alignment. The actual temperature at which magnetism disappears may be the point where the metal turns red hot, or it may be at a lower temperature for some materials.

A varying magnetic field that can destroy a magnet may come from electricity. If you look at the ratings of appliances used in your home, you will often see the expression *USE WITH AC ONLY* or *115V AC*. The initials AC mean that the electricity is flowing back and forth in the wire. The magnetic field around the wire fluctuates rapidly, and this is the varying field that can destroy magnetism.

If a watch becomes magnetized, it will not work right. The same is true of

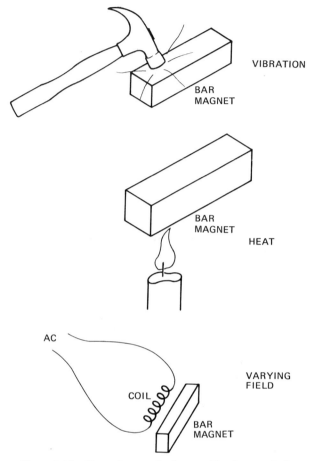

Figure 2-10.   Magnetism can be destroyed by these methods.

a picture tube in a color television set. In practice, the magnetism is removed by using the varying field produced by alternating current. The process of removing magnetism is called *degaussing*.

Magnets should be stored as shown in Fig. 2-11. The *keeper* provides an easy path for the flux from the north to the south magnetic pole. The bar magnets are stored so that their flux lines take an easy path from pole to pole.

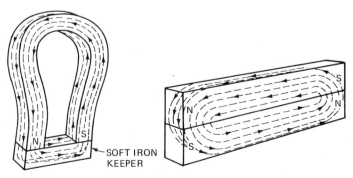

Figure 2-11.   Magnets should be stored like this.

*WHAT IS THE RELATIONSHIP BETWEEN
ELECTRICITY AND MAGNETISM?*

Early experimenters spent a considerable amount of time trying to determine what the relationship is between electricity and magnetism. In an experiment a Danish scientist named Oersted accidentally found that whenever electricity flows, there is a magnetic field around the wire. This experiment is illustrated in Fig. 2-12.

> **NOTE: DO NOT CONNECT A CIRCUIT LIKE THE ONE SHOWN IN FIG. 2-12! YOU COULD RUIN A BATTERY THIS WAY. THIS PICTURE IS ONLY FOR THE PURPOSE OF SHOWING THE THEORY. AN EXPERIMENT THAT YOU *CAN* PERFORM IS DESCRIBED IN THE "EXPERIMENTS" SECTION OF THIS CHAPTER.**

In Fig. 2-12*a* the wire is aligned with the pointer of the compass but not connected to the battery. The compass needle points north and south. In Fig. 2-12*b* the wires are connected to the battery, and this causes electricity to flow through the wire. When electricity is flowing, the needle points at right angles to the wire. Since the needle of the compass is a tiny magnet which is free to move, it is aligning itself so that its north pole points to the north pole of the magnetic field around the wire. Oersted's experiment showed that whenever there is an electric current flowing, there is always a magnetic field associated with the current.

It is reasonable to expect that if there is always a magnetic field around an

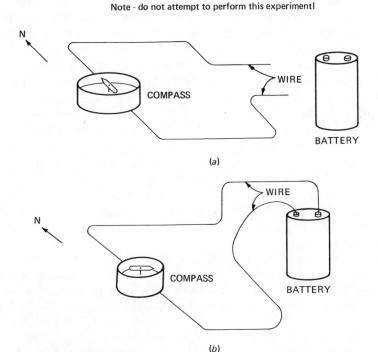

Figure 2-12. Oersted's experiment. (a) *When the compass is laid on the wire, it points north and south.* (b) *When the wire is connected to the battery, the needle of the compass turns at right angles to the wire.*

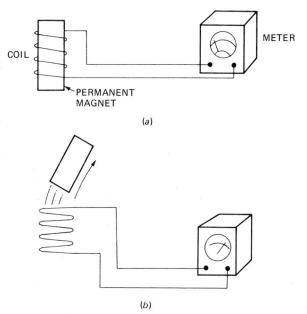

Figure 2-13. Magnetism can be used to generate electricity. (a) *When the permanent magnet is at rest in the center of the coil, no electricity is generated.* (b) *When the magnet is pulled out of the coil, the meter indicates that electricity is generated.*

electric wire, magnetism could somehow be used to create electricity. Michael Faraday and Joseph Henry discovered how to do this. They both made the discovery at the same time, but while working independently. Figure 2-13 shows how electricity is generated with magnetism. In Fig. 2-13*a* a permanent magnet is shown inside of a coil of wire. The ends of the coil are connected to a meter. This meter is able to measure electricity. It is not necessary to know at this time exactly how the meter works, but you should know that whenever there is electricity, the meter needle deflects. As long as the magnet is at rest inside of the coil, the needle of the meter is also at rest. However, if you pull the magnet out of the coil quickly, as shown in Fig. 2-13*b*, the meter needle deflects, indicating that electricity has been generated, or made. The same thing happens if you push the magnet back into the coil. During the time that the magnet is moving, the meter will indicate that electricity is flowing.

If you repeated this experiment a number of times, you would find that there is electricity only during the time that the magnet is moving. However, if you hold the magnet steady and move the coil back and forth, electricity is generated again.

After conducting the experiment a number of times, you would come to the following very important rule: ***Whenever there is motion between a conductor and a magnetic field, electricity is generated.*** This is the principle used by power companies for generating the electricity used in the home and in industry.

SUMMARY

1.   Magnetism can be destroyed by vibration, heat, or a varying magnetic field.
2.   The process of removing magnetism is called *degaussing.*
3.   Whenever there is electricity, there is a magnetic field.

4.   Whenever there is movement between a wire and a magnetic field, electricity is produced.

Programmed Review Questions

(Instructions for using this programmed section are given in Chap. 1.)

We will now review the important concepts of this chapter. If you have understood the material, you will progress easily through this section. Do not skip this material, because some additional theory is presented.

1.   Which of the following is true?
A.   When electricity is flowing in a wire, there is a magnetic field around the wire if, and only if, the wire is lying north and south. (Proceed to block 7.)
B.   When electricity is flowing in a wire, there is always a magnetic field around the wire. (Proceed to block 17.)

2.   *Your answer to the question in block 19 is **B**. This answer is wrong. Wood is paramagnetic—that is, it is nonmagnetic. Flux lines will not pass easily through nonmagnetic materials.* Proceed to block 14.

3.   *Your answer to the question in block 4 is **A**. This answer is wrong. You will learn in a later chapter that heat can be used to generate electricity, but not in the amounts that power companies must generate.* Proceed to block 9.

4.   *The correct answer to the question in block 6 is **B**. Note in this illustration that the flux lines do not cross.* Here is your next question.
Power companies generate electricity by
A.   heating large iron bars. (Proceed to block 3.)
B.   moving wires in magnetic fields. (Proceed to block 9.)

5.   *Your answer to the question in block 10 is **B**. This answer is wrong. Heating an iron bar until it is **red** hot is not a method used for magnetizing.* Proceed to block 19.

6.   *The correct answer for the question in block 15 is **A**. Since the flux lines do not cross, Fig. 2-15a shows how they must be.* Here is your next question. Which of the illustrations in Fig. 2-16 shows the magnetic flux lines correctly?
A.   Figure 2-16*a* shows the flux lines correctly. (Proceed to block 18.)
B.   Figure 2-16*b* shows the flux lines correctly. (Proceed to block 4.)

7.   *Your answer to the question in block 1 is **A**. This answer is wrong. It does not matter which way the wire is pointing. If there is electricity in the wire, there is always a magnetic field around it.* Proceed to block 17.

8.   *The correct answer to the question in block 17 is **A**. Striking the iron while it is being magnetized helps to line up the tiny magnets.* Here is your next question.

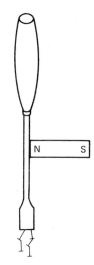

TACKS

Figure 2-14. Which method is being used to magnetize the screwdriver blade?

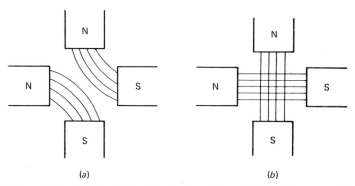

(a)   (b)

Figure 2-15.   Which of these illustrations shows the flux lines correctly?

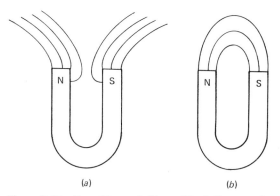

(a)   (b)

Figure 2-16.   Does Figure. 2-16a or Fig. 2-1b correctly show the flux lines around a horseshoe magnet?

A lodestone is an example of
A.   an electromagnet. (Proceed to block 16.)
B.   a natural magnet. (Proceed to block 10.)

9.   *The correct answer to the question in block 4 is* **B**. *Large conductors are moved through magnetic fields to generate the electricity.* Here is your next question.
The fact that there is a magnetic field around a wire that is carrying electricity can be demonstrated with a _____. (Proceed to block 20.)

10.   *The correct answer to the question in block 8 is* **B**. *Lodestones are natural magnets found in Asia Minor.* Here is your next question.
A method commonly used for making magnetism is
A.   by electricity flowing through a coil of wire. (Proceed to block 19.)
B.   by heating an iron bar until it is red hot. (Proceed to block 5.)

11.   *Your answer to the question in block 14 is* **A**. *This answer is wrong. There is no electricity involved, as shown in the illustration.* Proceed to block 15.

12.   *Your answer to the question in block 17 is* **B**. *This answer is wrong. Striking a magnet with a hammer will make it* **weaker,** *not stronger.* Proceed to block 8.

13.   *Your answer to the question in block 15 is* **B**. *This answer is wrong. Flux lines never cross, so they cannot be as shown in Fig. 2-15b.* Proceed to block 6.

14.   *The correct answer to the question in block 19 is* **A**. *Nickel is a magnetic material, and it will allow magnetic flux lines to pass more easily than they will pass through wood.* Here is your next question.
When a screwdriver with a metal blade is not magnetized, it will not pick up tacks. Holding a bar magnet to the shaft as shown in Fig. 2-14 magnetizes it, and as long as the bar magnet is in place, it will hold the tacks. The screwdriver is being magnetized by
A.   electromagnetism. (Proceed to block 11.)
B.   induction. (Proceed to block 15.)

15.   *The correct answer to the question in block 14 is* **B**. *Flux lines from the permanent magnet find an easy path through the screwdriver blade, and the blade is magnetized by induction.* Here is your next question.
Which of the illustrations in Fig. 2-15 correctly shows the flux lines between the poles?
A.   Figure 2-15*a* shows the flux lines correctly. (Proceed to block 6.)
B.   Figure 2-15*b* shows the flux lines correctly. (Proceed to block 13.)

16.   *Your answer to the question in block 8 is* **A**. *This answer is wrong. A lodestone is a natural magnet. It does not require electricity.* Proceed to block 10.

17. *The correct answer to the question in block 1 is **B**. A magnetic field **always** goes with electricity.* Here is your next question.
    Which of these statements is correct?
    A. When you are magnetizing a piece of iron, it helps to strike it with a hammer. (Proceed to block 8.)
    B. If a magnet is becoming weak, it will be made stronger if you strike it with a hammer. (Proceed to block 12.)

18. *Your answer to the question in block 6 is **A**. This answer is wrong. The flux lines of a magnet go between the **north** and **south** magnetic poles.* Proceed to block 4.

19. *The correct answer to the question in block 10 is **A**. When electricity flows through a coil of wire, the coil acts like a magnet. This is one way of making magnetism.* Here is your next question.
    Of the materials listed below, which will pass magnetic flux lines more easily?
    A. Nickel (Proceed to block 14.)
    B. Wood (Proceed to block 2.)

20. *The correct answer to the question in block 9 is **compass**. The needle of a compass is simply a small magnet that is free to turn. It lines up with the magnetic field around a wire that is carrying electricity.*
    You have now completed the programmed questions. The next step is to put some of these ideas to work in laboratory experiments. Proceed to the "Experiments" section that follows.

Experiments

EXPERIMENT 1

Purpose — To demonstrate that there is a magnetic field around a wire that is carrying electricity. The experiment of Oersted shown in Fig. 2-12 will be performed.

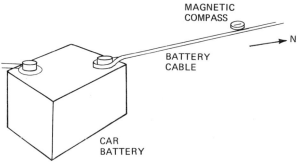

Figure 2-17. An experiment to show that there is a magnetic field around a wire carrying electricity.

Test Setup — For this experiment you will need a small magnetic compass, and you will need to have access to a car battery. Figure 2-17 shows the arrangement.

Procedure —

Step 1 — Lay the compass on the battery cable — that is, one of the heavy wires attached to the battery. Set the compass at some point where the needle of the compass is pointing in the same direction as the cable.

Step 2 — Switch on the lights of the car. This causes electricity to flow in the cable. Note that the compass points at right angles to the wire.

Conclusion — When the lights are switched on, the needle of the compass should turn so that it is at right angles to the cable. This shows that there is a magnetic field around the cable when it is carrying electricity.

EXPERIMENT 2

Purpose — To demonstrate the process of magnetization by induction.

Test setup — For this experiment you will need a paper clip, a magnet, and a nonmagnetized screwdriver with a metal blade. You are going to repeat the setup shown in Fig. 2-14.

Procedure —

Step 1 — Try to lift the paper clip with the screwdriver blade. If the blade is not magnetized, you will not be able to do it.

Step 2 — Hold the magnet onto the shaft as shown in Fig. 2-14, and repeat the experiment. The paper clip should cling to the blade with a magnetic attraction.

Step 3 — Stroke the shaft of the screwdriver with the magnet. Figure 2-18 shows how. Be sure that you do not turn the magnet as you do this. Note that the blade becomes a magnet after you have stroked it a number of times.

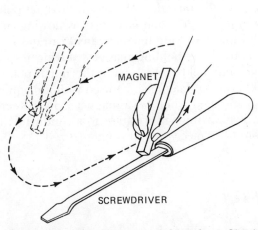

Figure 2-18. Magnetizing a screwdriver by stroking it with a permanent magnet.

Conclusion —

1. The flux lines from the magnet move easily through the shaft, causing it to behave like a magnet.
2. A magnetic material becomes magnetized when you stroke it with a magnet.

Self-Test with Answers

(Answers with discussions are given in the next section.)

1. Which of the following is a ferromagnetic material? (*a*) Copper; (*b*) Aluminum; (*c*) Wood; (*d*) Nickel.
2. The needle of a magnetic compass is made of (*a*) a diamagnetic material; (*b*) a paramagnetic material; (*c*) a magnetic material; (*d*) a nonmagnetic material.
3. A magnetic rock that early Greeks experimented with is (*a*) cobalt; (*b*) nickel; (*c*) lodestone; (*d*) uranium.
4. Whenever electricity flows through a wire, there must always be (*a*) an electric light; (*b*) a magnetic field; (*c*) a sound generated; (*d*) an automobile battery.
5. Which of the following is *not* correct? (*a*) Two north poles repel; (*b*) Two south poles attract; (*c*) A north and a south pole attract; (*d*) All magnets have a north and a south pole.
6. If you suspend a bar magnet from a string, its north pole points to the earth's north pole. This means that (*a*) the earth's north pole is a magnetic north pole; (*b*) the earth's north pole is a magnetic south pole.
7. Which of the following is *not* true? (*a*) Soft iron gets its name from the fact that it feels like soft rubber; (*b*) Magnetic flux lines never cross each other; (*c*) The highest concentration of magnetic flux lines is at the north and south poles of a magnet; (*d*) Excessive heat can destroy magnetism.
8. When a compass is placed over a battery cable in a car, the compass will (*a*) point in the direction that current is flowing; (*b*) point in a direction that is opposite to the direction of current flow; (*c*) point to a direction at right angles to the cable when electricity is flowing in the cable; (*d*) not work, since a rapidly varying magnetic field can destroy magnetism.
9. A *keeper* is (*a*) used for holding magnets during tests; (*b*) used for preventing a magnetic field from leaving the north pole of a magnet; (*c*) a man who takes care of magnets; (*d*) used with a magnet when it is stored.
10. Electricity is generated when (*a*) a magnet is in the center of a coil; (*b*) a magnetic field is moving with relation to a conductor; (*c*) an electric light is switched on; (*d*) a piece of iron is magnetized.

Answers to Self-Test

1. (*d*) — A ferromagnetic material is attracted to a magnet. Only nickel is ferromagnetic in the choices given.

2. (*c*) — The needle of a compass is a small magnet, so it must be made of a magnetic material. Note that choices (*b*) and (*d*) are the same.
3. (*c*)
4. (*b*)
5. (*b*) — Two south poles repel. Remember the rule that *like poles repel.*
6. (*b*) — Since unlike poles attract, the north pole of the earth must be a south magnetic pole.
7. (*a*) — Soft iron is *not* soft to the touch.
8. (*c*) — This was demonstrated in the experiment section.
9. (*d*) — Figure 2-11 shows how a keeper is used.
10. (*b*) — You have not studied about all the ways of generating electricity, so you could not know if the choices are right or not. However, you *know* that choice (*b*) is correct. (See Fig. 2-13.)

3.
What Are Conductors, Insulators, and Resistors?

ntroduction

In studying Chap. 1 you learned about the operation of an electroscope, which is a simple device for determining the presence of static charges. You will remember that when the negative charge is placed near the metal top of the electroscope, the electrons move down through the wire and into the aluminum foil. This causes the aluminum foil to spread apart. The action is based on the fact that *like charges repel*. Since negative electrons move into *both* of the leaves, the leaves repel each other.

An important factor in the operation of the electroscope is the copper wire that connects the top of the electroscope to the aluminum foil. The electrons move through the wire, or rod, when a negative charge is placed near the top.

Suppose the electroscope is made as shown in Fig. 3-1. Instead of using copper wire between the top and the aluminum foil, this one uses a plastic stick. If you bring a negative charge near the top of this instrument, the aluminum foil will be unaffected. This illustrates a very important point in electricity: **There are only certain kinds of materials that the electrons can flow through.**

We know from the experiment described above that electrons *will* flow through a metal conductor and they *will not* flow through a plastic stick. The lightning rod shown in Fig. 3-2 is another example of the use of metal to carry

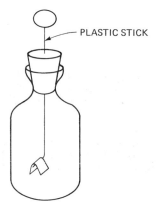

Figure 3-1.   *This electroscope will not work.*

electrons. The lightning rod at the top of the house is connected to a ground rod with a heavy wire. During a thunderstorm, electrons move from the ground through the wire to the lightning rod, or from the lightning rod through the wire to the ground. The actual direction depends upon whether the cloud above has a positive or a negative charge. For example, assume that the cloud has a highly positive charge. Electrons will move from the ground rod through the wire into the lightning rod in an attempt to reach this positive charge. On the other hand, if the cloud has a highly negative charge, the electrons will move out of the lightning rod through the wire into the ground. The electrons move away from the negative cloud because like charges repel. In either case there is a motion of electrons through the wire in an effort to equalize the charges. If lightning strikes, the large amount of electricity associated with the lightning will go through the rod and through the wire and into the ground.

It is important to note that the lightning rod does not prevent lightning from striking the house. Instead, it provides an easy path to ground in case it does strike. What we are interested in is the fact that a *wire* is used to conduct the electricity, and the wire is made of metal. In most cases the metal is either copper or aluminum.

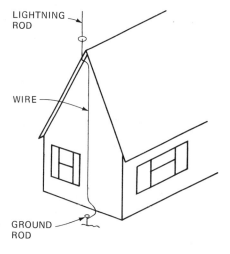

Figure 3-2.   *A conductor is used between a lightning rod and ground.*

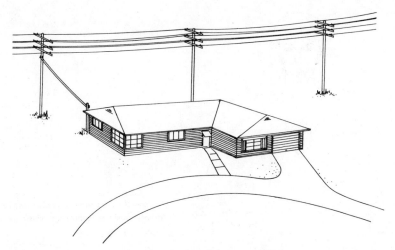

Figure 3-3. Power is delivered to your home through metal wires.

You are no doubt familiar with the power lines, like the ones shown in Fig. 3-3, that carry electricity to your home. These power lines are made of metal wires which are covered with a protective material that will not conduct electricity. The coating on the wire is called *insulation*.

For the purpose of studying electricity, we can say that all materials in the universe can be put into one of three categories. Those which conduct electricity are called *conductors,* and those which will not conduct electricity are called *insulators*. There is a group of materials that are neither good conductors nor good insulators, and they are called *resistors* or *semiconductors*.

In order to study conductors and insulators, it will be necessary to learn more about electricity. In this chapter you will learn about voltage, current, and resistance. These are three basic factors that are always associated with electricity.

You will be able to answer these questions after studying this chapter:

What is electricity?
What causes an electric current to flow in a circuit?
How does an electric circuit work?
What are the symbols used for drawing a circuit?
What are some important characteristics of wires?

Instruction

WHAT IS ELECTRICITY?

You may be surprised to learn that no one has ever come up with a good definition of electricity. Occasionally magazines will hold contests for the best definition of electricity, and the entries are judged by a person who is well known in the industry. Of course, you can go to the dictionary and get a definition of electricity, but scientists tell us that this definition is not satisfactory. They say

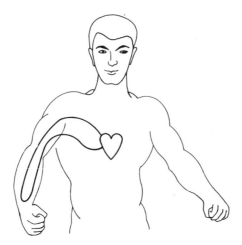

Figure 3-4. Blood circulation in the human body is compared with an electrical system.

that the definition tells what electricity *does*, but not what it *is*. In order to be accurate, then, we should say that no one has ever defined electricity according to the rules that they have for making definitions.

We will not concern ourselves with all the problems of making definitions. Instead, we will define electricity by saying what it does, and then concern ourselves with the more important question: *What can you do with it?*

When you are studying a new subject, it is sometimes convenient to start by comparing it with something you already know. This is called "making an analogy." The study of electricity is sometimes begun by comparing an electrical system to a fluid system. There is a very good reason for this kind of comparison. Early experimenters thought that electricity was some type of invisible fluid. They even tried to store it in jars for future use. Because they were convinced it was a fluid, some of the terms we use in electricity still sound like it is related to a liquid. For example, we talk about the "flow" of current and electrical "pressure," and when we talk about storing electricity, we refer to "capacity."

The human heart supplies a pressure to push the blood through the arteries and veins. This is illustrated in Fig. 3-4. The arteries carry blood away from the heart, and the veins carry it back toward the heart. As the blood moves through the body its flow is opposed by the smaller veins, and by the resistance offered by the arteries.

The heart system is a *closed system*. All the blood that leaves the heart returns to the heart.

Electrical systems are similar in some ways to the heart system. The electric pump is a battery, or generator, or some device that creates an "electrical pressure." This electrical pressure is called the *voltage*.

The electricity that flows in an electrical system can be compared to the blood flowing in the heart system. The flow of electricity is called *current*.

Opposition to the flow of electricity is called *resistance*. The useful part of an electrical system is often the resistive part.

The heart system can be called a *circuit* because the blood goes around it continuously. The same is true in an electrical system. Electrons can move through the wire to the electroscope, or they can move through the wire of the lightning rod, but if you want a continuous flow of electricity, you need a *closed circuit.*

WHAT CAUSES AN ELECTRIC CURRENT TO FLOW IN A CIRCUIT?

You will remember that in Chap. 1 two very important rules were stated: *unlike charges attract* and *like charges repel.* Look at the two metal spheres shown in Fig. 3-5. It is assumed that one has an excess of electrons and one has a shortage of electrons. Thus, the spheres are negatively and positively charged. This is indicated by the minus and plus signs on the spheres.

If you connect a metal wire between the two spheres, electrons will flow away from the negative charge and toward the positive charge. This flow of electrons is called an *electron current,* or more simply, *current.* The flow of electrons is due to the fact that the negative electrons are repelled away from the negative charge (like charges repel) and attracted toward the positive charge (unlike charges attract).

At this point we come to a very odd situation in the study of electricity. Some time ago, before it was understood that electric current is a flow of electrons, someone decided that current must flow from a positive charge toward a negative charge. This was not just a haphazard guess. There are actually good reasons for believing that current flows this way. Of course, electricity that flows from a positive charge to a negative charge could not be a flow of electrons. Instead, it must be a flow of imaginary positive charges.

The flow of electricity from plus to minus is called *conventional current flow,* as compared to the *electron current flow,* which goes from minus to plus. It seems like it would be a simple matter for everyone to get together and decide which way it goes and stick with that decision. However, for some reason or another this has never been done. So, some books say that electricity flows from plus to minus, and other books say that electricity flows from minus to plus. In this book when we talk about an electric current flow we will always mean current flowing from minus to plus. Electric current, then, is as shown in Fig. 3-6, a flow of electrons through a wire.

From Fig. 3-5, you know that electrons will flow through a wire from a minus to a plus charge. Did you ever notice that a battery has a *minus* (−) and *plus* (+) marked on its terminals? The minus terminal of a battery is simply a place where there is an excess of electrons, and the positive terminal is a place where there is a shortage of electrons. We do not need to concern ourselves at this time with how this comes about. Our interest is in the fact that we have a minus and a plus charge just like the charges on the spheres in Fig. 3-5.

If you connected a wire between the negative and positive terminals of the

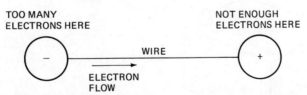

Figure 3-5. Electrons flow from a negative point to a positive point.

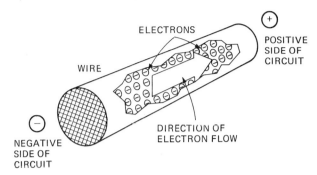

Figure 3-6. When electrons flow through a wire, there is an electric current.

battery in Fig. 3-7, a current would flow between these terminals. However, *you must never, never do this to a battery!* The reason is that there would be such a large current flow—that is, there would be so many electrons flowing from minus to plus—that it would probably ruin the battery. A battery, like so many other electrical and mechanical devices that you use every day, has a limit to what it can provide or endure. In the normal operation of an electric circuit, then, there is always something that opposes the flow of electricity. This "something" is called *resistance*.

SUMMARY

1. Electric current is a flow of electrons through a conductor.
2. Due to the fact that early experimenters thought electricity was an invisible fluid, we have terms in electricity like "current," "flow," and "pressure."
3. Electrical systems, like the blood-circulation system, are closed circuits. Just as blood leaves the heart and returns to the heart, an electric current leaves the battery and returns to the battery.
4. The battery is the *voltage* source that pushes electricity through the wires. The electricity is a flow of electron *current*.
5. Opposition to the flow of current is called *resistance*.
6. Electron current flows from negative to positive. Conventional current flows from positive to negative.
7. Electric current flows through metal *conductors*, but not through *insulators*.

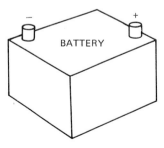

Figure 3-7. A battery has a negative and a positive terminal.

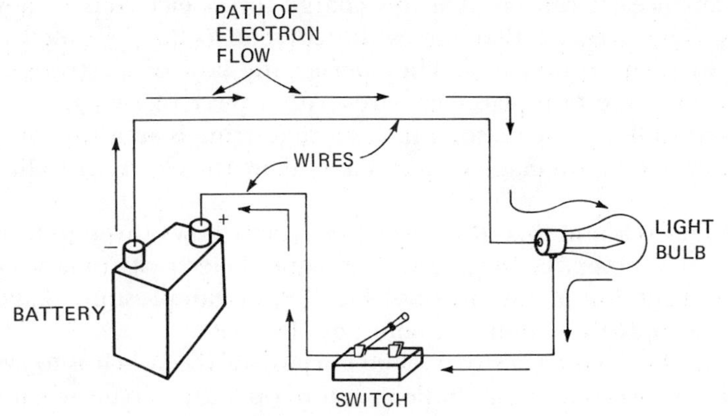

PATH OF
ELECTRON
FLOW

WIRES

LIGHT
BULB

BATTERY

+

SWITCH

Figure 3-8. A simple electric circuit.

HOW DOES AN ELECTRIC CIRCUIT WORK?

Figure 3-8 shows a simple electric circuit. You can easily follow the path of the electrons in this circuit. This procedure is called *tracing the circuit.* Starting at the negative terminal of the battery, electrons flow away from the terminal to the light, through the switch, and back to the positive terminal. If enough electrons flow through the light, it will glow brightly.

We have not discussed the switch yet, but it is simply a method of completing the circuit. It is important to understand that there is no complete circuit, and no electrical accomplishment, unless the electrons can *leave the negative terminal of the battery, flow through the circuit,* and *get back to the positive terminal.* If the electrons cannot make the complete circuit, electricity will not flow. This is one of the most important things you can learn about electricity: *a **complete circuit** is required.*

The purpose of a switch in a circuit is to complete the path or open the path. Figure 3-9 shows the switch in two positions. When the switch is in the position shown in Fig. 3-9a, electrons cannot flow through the switch. If the switch is in this position in the circuit of Fig. 3-8, the light will not light. When the switch is in the closed position of Figure 3-9b, there is a complete path for electrons to flow through the switch, and the light will light.

Now, let's summarize the purpose of each of the parts in the electric circuit of Fig. 3-8. First, there is a battery. It supplies positive and negative charges.

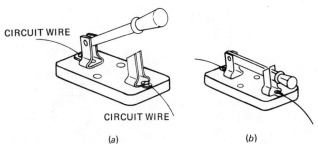

CIRCUIT WIRE

CIRCUIT WIRE

(a)   (b)

Figure 3-9.   Electricity cannot flow through an open switch.
(a) *This switch is* **open.** (b) *This switch is* **closed.**

For convenience we can say that this charge forces electrons to flow through the wires. The pressure that forces the current to flow is called the *voltage*.

The wires are conductors. They permit the flow of electrons around the circuit. If you were to replace the wires with a piece of string, you would not get an electron flow. The reason for this is that string is an insulator. The wires, on the other hand, are made of metal and allow the electrons to flow through them.

The light is a unit that glows when electricity flows through it. As you will study later, it is necessary for a certain amount of electricity to flow in order for the light to light, but in our circuit of Fig. 3-8, we can presume that the battery shown can supply the required amount of electricity.

Next in the circuit is a switch. The purpose of the switch is to close the circuit when you *want* electricity to flow, and to open the circuit when you do *not* want it to flow. Closing the switch in the circuit of Fig. 3-8 will cause the light to light.

SUMMARY

1. To trace an electric circuit, start at the negative terminal of the battery. See if electrons can move completely through the circuit and back to the positive terminal. If they cannot, there is no current flow.
2. A switch is a convenient method of opening and closing a circuit. When the switch is open, current cannot flow through it. When the switch is closed, it completes the circuit to permit electron current to flow.
3. A battery may be used to supply the *voltage* for the circuit. This is the electrical pressure used for pushing the current through the wire.
4. The electron flow around the circuit is called the *current*.

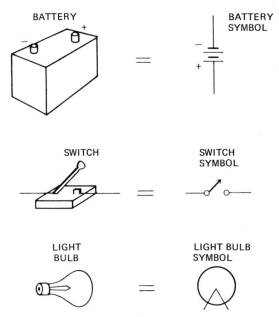

Figure 3-10. Symbols are used for simplifying drawings.

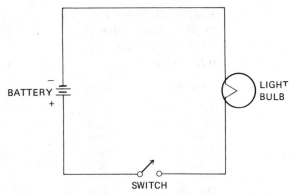

Figure 3-11. The circuit of Fig. 3-8 redrawn with symbols.

WHAT ARE THE SYMBOLS USED FOR DRAWING A CIRCUIT?

There are many electricians who are not good artists, and they would find it very difficult to draw the simple circuit in Fig. 3-8. To draw a more complicated circuit, like the one used in your house wiring system, would be an impossible chore and a great waste of time. To get around this problem we usually draw the circuits with simple symbols like the ones shown in Fig. 3-10.

Instead of drawing the battery, we can use a simple battery symbol. You will notice that there is a minus and plus sign on the battery symbol representing its minus and plus terminals. However, the battery symbol is sometimes used without the minus and plus signs. You should remember that the short bar on the battery symbol is always the negative terminal, and the long bar always represents the positive terminal. If you redraw the circuit of Fig. 3-8 using symbols instead of drawings, the result is as shown in Fig. 3-11. This may not look very much like the original circuit, but to an electrician this drawing (which is sometimes called a *schematic drawing*) is every bit as clear as the one shown in Fig. 3-8.

WHAT ARE SOME IMPORTANT CHARACTERISTICS OF WIRES?

Having studied a basic electric circuit, you are now in a better position to understand the purpose of the conductor. The conductor is used to carry the electrons around the circuit from the voltage source, or battery, through the circuit, and back to the battery. Electricity is generated by a power company in a different way from the way it is generated in a battery. However, the wires still serve the same purpose.

Most of the conductors used are made of copper. The reason for this is that electricity flows through copper more easily than it does through most other materials. In other words, the copper does not resist the flow of electricity as much as aluminum or brass. Copper is not the best conductor known. As a matter of fact, the best one is silver. Table 3-1 lists the materials that could be used for conductors in wiring. You will note that silver is the best conductor, then copper, and then gold. The reason for using copper instead of silver is that copper is much less expensive.

If you use a wire with a larger diameter, electricity will flow through it

TABLE 3-1. LIST OF
CONDUCTORS SHOWING
HOW THEY COMPARE
WITH SILVER

Material	Rating
Silver	best—100%
Copper	98%
Gold	78%
Aluminum	61%
Tungsten	32%
Zinc	30%
Platinum	17%
Iron	16%
Lead	15%
Tin	9%
Nickel	7%
Mercury	1%
Carbon	0.05%

TABLE 3-2. TABLE OF WIRE SIZES

Wire Size (Also Called the Gauge of a Wire)	Diameter of Wire (in Mils)	Area of Cross Section (in Circular Mils)
0000	460.0	212,000.0
000	410.0	168,000.0
00	365.0	133,000.0
0	325.0	106,000.0
1	289.0	83,700.0
2	258.0	66,400.0
3	229.0	52,600.0
4	204.0	41,700.0
5	182.0	33,100.0
6	162.0	26,300.0
7	144.0	20,800.0
8	128.0	16,500.0
9	114.0	13,100.0
10	102.0	10,400.0
11	91.0	8,230.0
12	81.0	6,530.0
13	72.0	5,180.0
14	64.0	4,110.0
15	57.0	3,260.0
16	51.0	2,580.0
17	45.0	2,050.0
18	40.0	1,620.0
19	36.0	1,290.0
20	32.0	1,020.0
21	28.5	810.0
22	25.3	642.0
23	22.6	509.0
24	20.1	404.0
25	17.9	320.0
26	15.9	254.0
27	14.2	202.0
28	12.6	160.0

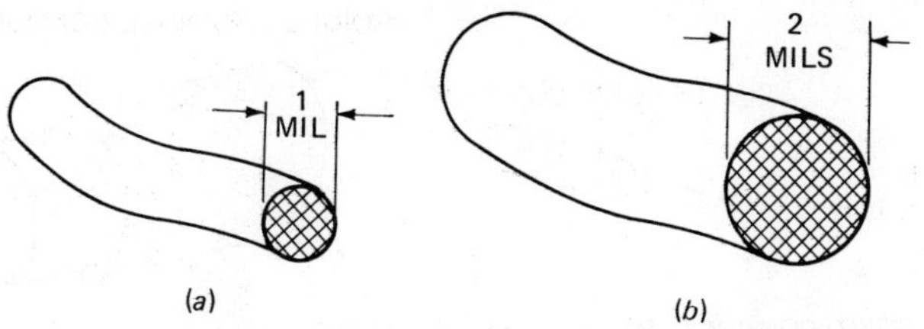

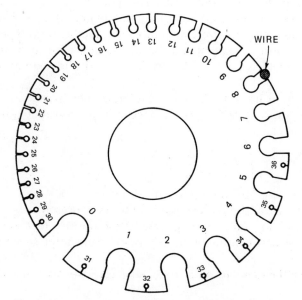

Figure 3-12. *The meaning of the term "circular mil."* (a) **This conductor has an area of 1 circular mil.** (b) **This conductor has an area of 4 circular mils.**

more easily than it will through a smaller wire. This is reasonable to expect, just as you would expect that a large vein can carry more blood than a small vein. Then why not *always* use a wire with a large diameter? The answer is simply that it costs more because more copper is needed.

The size of a piece of wire is identified by a wire-size table such as Table 3-2. Note that the diameters of the different wire sizes are measured in thousandths of an inch, or *mils*. One mil is a thousandth of an inch.

If you cut a piece of wire, the end is usually a circle. This is shown as the two shaded areas in Fig. 3-12. The larger the diameter of this circle, the larger the area. Wire sizes are sometimes identified by this cross-sectional area, rather than by their diameter.

The cross-sectional area of wire is measured in *circular mils*. You can find the area in circular mils by *squaring the diameter*—that is, by multiplying the diameter by itself. This is only true if you know the diameter in mils. Here is the equation:

$$\begin{matrix} AREA\ OF\ WIRE\ IN \\ CIRCULAR\ MILS \end{matrix} = \begin{matrix} DIAMETER\ OF \\ WIRE\ IN\ MILS \end{matrix} \times \begin{matrix} DIAMETER\ OF \\ WIRE\ IN\ MILS \end{matrix}$$

Figure 3-13. *A wire gauge for finding the size of a wire.*

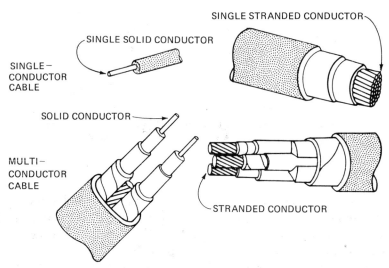

Figure 3-14. Wires may be stranded or solid.

As an example, in Fig. 3-12*a* the diameter of the wire is 1 mil—that is, $\frac{1}{1000}$ inch. Its area in circular mils would simply be 1×1, or 1 circular mil. In Fig. 3-12*b* the conductor has a diameter of 2 mils. In circular mils this is simply 2×2, or 4 circular mils.

A very important thing about Table 3-2 is that it shows the relationship between the wire size and the diameter of a wire. *Note that the smallest wire sizes correspond to the largest diameter, and the largest wire sizes correspond to the smallest diameter.* For example, a wire size 20 has a diameter of 32 mils, but a wire size 14 has a diameter twice as large, or 64 mils. You should remember that for a wire to carry a large current, it must have a smaller wire gauge number than one used to carry a small current.

Figure 3-13 shows a simple gauge for measuring wire size. To use this gauge you set the wire into the slot and read the gauge number from the slot.

Wires are made in two different ways. The conductor can be a solid piece of copper (or other metal used for conducting material), or it may be made of a number of smaller strands combined into a conductor. Both types are shown in Fig. 3-14. The most important reason for making wire in strands is that the wire is more flexible. This is the way lamp-cord and extension-cord wire is made.

The insulation used on the wire may be rubber, plastic, or other nonconducting materials. *Outside wire* uses a special insulation that can withstand the battering of the elements.

SUMMARY

1. Symbols are used to simplify circuit drawings.
2. The short bar on the battery symbol always represents the negative terminal, and the long bar always represents the positive terminal.
3. Silver is the best conductor of electricity. Next after silver is copper, and then gold.
4. The larger the diameter of a wire, the greater the amount of current it can carry.

5. The larger the diameter of a wire, the smaller the wire size.
6. Wire area is measured in circular mils.
7. Stranded wire is more flexible than solid wire.

Programmed Review Questions

(Instructions for using this programmed section are given in Chap. 1.)

We will now review the important concepts of this chapter. If you have understood the material, you will progress easily through this section. Do not skip this material, because some additional theory is presented.

1. Wires are made with stranded conductors
   A. to make them more flexible. (Proceed to block 7.)
   B. because it is cheaper to make them this way. (Proceed to block 17.)

2. *Your answer to the question in block 3 is **A**. This answer is wrong. For the meaning of the word "reluctance," proceed to block 16.*

3. *The correct answer to the question in block 20 is **A**. A mil is 1/1000 inch. Here is your next question.*
   Opposition to the flow of electricity is called
   A. reluctance. (Proceed to block 2.)
   B. resistance. (Proceed to block 16.)

4. *Your answer to the question in block 19 is **B**. This answer is wrong. Electrons are negative, so they would not flow toward a negative point. Remember that **like charges repel**. Proceed to block 8.*

5. *Your answer to the question in block 6 is **A**. This answer is wrong. Gold, like silver, is more expensive than copper. However, unlike silver, gold is not a better conductor than copper. Proceed to block 25.*

6. *The correct answer to the question in block 24 is **A**. It is convenient to think of the voltage as being the pressure that pushes the current through the wire. Here is your next question.*
   Which of the following is a better conductor of electricity?
   A. Gold (Proceed to block 5.)
   B. Copper (Proceed to block 25.)

7. *The correct answer to the question in block 1 is **A**. Conductors used in extension cords, for example, are stranded to make them more flexible. Here is your next question.*
   Electron current flows through a switch when it is
   A. open. (Proceed to block 23.)
   B. closed. (Proceed to block 10.)

8. *The correct answer to the question in block 19 is **A**. Negative electrons flow away from a negative charge and toward a positive charge. Here is your next question.*

Power companies use copper instead of silver for conductors because

A.   copper is a better conductor. (Proceed to block 12.)

B.   copper is less expensive. (Proceed to block 18.)

9.   *Your answer to the question in block 18 is* **B**. *This answer is wrong. Aluminum will oppose the flow of electricity more than copper.* Proceed to block 21.

10.   *The correct answer to the question in block 7 is* **B**. *Remember that a closed circuit is needed for a continuous current to flow, and a closed switch is needed to make a closed circuit.* Here is your next question.

Which of the following wire sizes is for a wire with a larger diameter?

A.   Number 12 wire (Proceed to block 20.)

B.   Number 14 wire (Proceed to block 22.)

11.   *Your answer to the question in block 25 is* **A**. *This answer is wrong. A lightning rod cannot prevent lightning from striking.* Proceed to block 19.

12.   *Your answer to the question in block 8 is* **A**. *This answer is wrong. As shown in Table 3-1, silver is a better conductor than copper.* Proceed to block 18.

13.   *Your answer to the question in block 16 is* **B**. *This answer is wrong. The meaning of power in electric circuits has not been discussed yet.* Proceed to block 24.

14.   *Your answer to the question in block 24 is* **B**. *This answer is wrong. Current is the amount of electricity flowing, and voltage can be considered to be the pressure that pushes the current through the wire.* Proceed to block 6.

15.   *Your answer to the question in block 20 is* **B**. *This answer is wrong. It sounds like a mil would be a millionth of an inch, but it is not.* Proceed to block 3.

16.   *The correct answer to the question in block 3 is* **B**. **Resistance** *is the opposition to current flow.* **Reluctance** *is the opposition that a material offers to magnetic flux lines. This is a term that you will not be using often in your work with electricity. It was not mentioned in Chap. 2 when you studied magnetism.* Here is your next question.

A complete electrical path is called

A.   a circuit. (Proceed to block 24.)

B.   power. (Proceed to block 13.)

17.   *Your answer to the question in block 1 is* **B**. *This answer is wrong. It is easier and cheaper to make the wire with a solid conductor.* Proceed to block 7.

18.   *The correct answer to the question in block 8 is* **B**. *The fact that copper is cheaper is the reason it is used for making wires.* Here is your next question.

If you had two wires of the same diameter and the same length, but one was made of copper and one was made of aluminum, which would conduct electricity more easily?

A.   The copper wire (Proceed to block 21.)

B.   The aluminum wire (Proceed to block 9.)

19.   *The correct answer to the question in block 25 is* **B**. *Lightning rods provide an easy path for the high currents to flow when lightning strikes.* Here is your next question.

Electron current flows from

A.   negative to positive. (Proceed to block 8.)

B.   positive to negative. (Proceed to block 4.)

20.   *The correct answer to the question in block 10 is* **A**. *The smaller the gauge number, the larger the wire.* Here is your next question.

A mil is

A.   a thousandth of an inch. (Proceed to block 3.)

B.   a millionth of an inch. (Proceed to block 15.)

21.   *The correct answer to the question in block 18 is* **A**. *Electricity flows through copper more easily than it flows through aluminum.* Here is your next question.

The area of the conductor in a wire is an important factor in how much current the wire can carry. For wires, the area is measured in _____. (Proceed to block 26.)

22.   *Your answer to the question in block 10 is* **B**. *This answer is wrong. Study the wire gauge shown in Fig. 3-13.* Then proceed to block 20.

23.   *Your answer to the question in block 7 is* **A**. *This answer is wrong. Study the switches in Fig. 3-9.* Then proceed to block 10.

24.   *The correct answer to the question in block 16 is* **A**. *A circuit is a closed electrical path for current flow.* Here is your next question.

To simplify our understanding of electric circuits, we consider the pressure that pushes electricity through the wire to be the

A.   voltage. (Proceed to block 6.)

B.   current. (Proceed to block 14.)

25.   *The correct answer to the question in block 6 is* **B**. *Copper is preferred to gold for two reasons: it is a better conductor, and it is cheaper.* Here is your next question.

"Lightning rods prevent lightning from striking." Is this statement true or false?

A.   True (Proceed to block 11.)

B.   False (Proceed to block 19.)

26.   *The answer to the question in block 21 is* **circular mils.**

You have now completed the programmed questions. The next step is to put some of these ideas to work in laboratory experiments. Proceed to the "Experiments" section that follows.

Experiments

An experiment board is described in Appendix C. This board can be used for the experiments described in this chapter and in the following chapters.

EXPERIMENT 1

Purpose — To make an *electronic battery*.

Theory — In order to perform experiments in electricity, you will need a source of voltage. You could use a battery, but as the battery gets old, an experiment may fail due to the fact that the battery can no longer supply the needed voltage.

Instead of using a battery, you can use a circuit that is called a *power supply*. Since it takes the place of a battery, we can call it an *electronic battery*.

You must remember this very important rule about batteries and power supplies: **Never connect the positive and negative leads together.**

Test Setup — Figure 3-15 shows how the electronic battery is wired. As far as the experiments are concerned, the two *leads* of this supply could come from this equipment, or from a battery.

Procedure —
Step 1 — Wire the electronic battery as shown in Fig. 3-15.
Step 2 — Make sure that the test leads are not connected, and then plug the power lead into a power socket.
Step 3 — Touch the leads to the light terminals as shown in Fig. 3-16. If the battery and the light bulb are both working, the light should glow.

Conclusion — The test board can be wired as a power source for experiments.

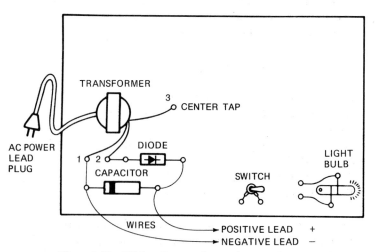

Figure 3-15. Wiring diagram for an electronic battery.

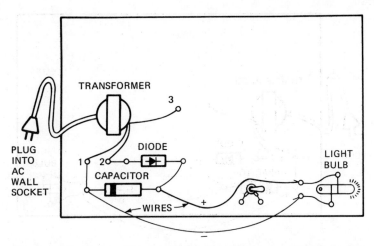

Figure 3-16. *Test to determine if the electronic power supply is working.*

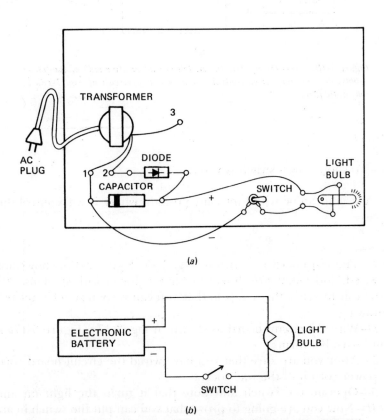

(a)

(b)

Figure 3-17. *This is a circuit that permits the light to be controlled by a switch.* (a) *Circuit for switch control of the light.* (b) *The circuit of Fig. 3-17a in schematic form.*

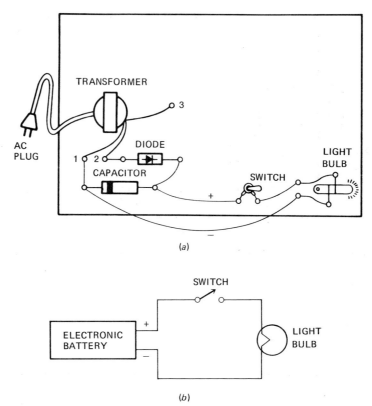

Figure 3-18.   *This circuit has the switch on the positive side of the power supply.* (a) *Another switch-control circuit.* (b) *The circuit of Fig. 3-18a in schematic form.*

EXPERIMENT 2

Purpose — To show how a switch is wired.

Theory — A switch can be wired into any position where it will control the flow of electricity.

Procedure —

Part I — The test board is shown in Fig. 3-17. Figure 3-17a shows how the board is wired, and Fig. 3-17b shows the circuit drawn with symbols. It is important for you to study this setup so that you can recognize a circuit from its symbol drawing.

Step 1 — Wire the circuit board as shown in Fig. 3-17a. Figure 3-17b shows the circuit in symbol form.

Step 2 — After you are sure that you have wired the circuit board correctly, plug the power cord into the wall socket.

Step 3 — Operate the switch and note that it turns the light on and off.

Part II — Now you are going to prove that you can put the switch in another location and it will still control the light.

Step 1 — Wire the circuit board as shown in Fig. 3-18. The wiring plan is shown in Fig. 3-18a, and the schematic drawing is shown in Fig. 3-18b. Note that the switch is now between the positive terminal of the supply and the light, but in the last experiment it was between the negative terminal and the light.

Step 2—After you are sure that you have wired the circuit board correctly, plug the power cord into the wall socket.

Step 3—Operate the switch and note that it turns the light on and off.

Conclusion—A switch can be wired into a circuit in any position as long as it can interrupt the flow of current.

Self-Test with Answers

(Answers with discussions are given in the next section.)

1. Electron current flows from (*a*) top to bottom; (*b*) right to left; (*c*) negative to positive; (*d*) positive to negative.
2. Which of the following conducts electricity with the least opposition? (*a*) Gold; (*b*) Silver; (*c*) Copper; (*d*) Lead.
3. A switch in a circuit (*a*) must be placed in the negative side of the circuit; (*b*) must be placed in the positive side of the circuit; (*c*) must be open for current to flow through it; (*d*) must be connected so that the current flows through it when the switch is closed.
4. A wire with a diameter of 1 mil has an area of (*a*) 1 circular mil; (*b*) 1 square mil; (*c*) 3.14 circular mils; (*d*) 0.01 square inch.
5. Of the following wire sizes, which one represents the largest diameter of wire? (*a*) 14; (*b*) 00; (*c*) 6; (*d*) 22.
6. All useful electric circuits have (*a*) batteries; (*b*) light bulbs; (*c*) resistance; (*d*) switches.
7. Which of these materials could *not* be used as an insulator? (*a*) Glass; (*b*) Plastic; (*c*) Lead; (*d*) Wood.
8. Which of the following statements is true? (*a*) Gold will not conduct electricity; (*b*) Electricity cannot flow uphill; (*c*) There is an excess of electrons at the negative terminal of a battery; (*d*) The current in a circuit is the electrical pressure.
9. Which of the following statements is true? (*a*) You should never connect a wire between the negative terminal and the positive terminal of a battery; (*b*) A battery should not be used for operating light bulbs; (*c*) Power companies supply electricity through large insulators; (*d*) The cross-sectional area of a wire is measured in square mils.
10. To make a wire more flexible, the conductor is (*a*) made of gold; (*b*) made of tungsten; (*c*) stranded; (*d*) inserted into a magnetic field.

Answers to Self-Test

1. (*c*)—Note that the question asks for the direction of electron current flow. If the question had asked for the direction of conventional current flow, then the correct answer would have been (*d*).

2. (*b*) — Electricity flows through silver with less opposition than it gets when flowing through copper. Copper is used more for a number of reasons. The most important is cost. Other factors are its strength, weight, flexibility, and availability.

3. (*d*) — One of the experiments proved that you can put the switch in any part of the circuit as long as it is in a position to stop the current when the switch is open.

4. (*a*) — Figure 3-12 shows this.

5. (*b*) — Table 3-2 shows this.

6. (*c*) — Useful circuits may or may not have batteries, light bulbs, and switches. However, they all have resistance.

7. (*c*) — Remember that most metals — such as lead — will conduct electricity. This means that they make poor insulators.

8. (*c*)

9. (*a*) — This is a very, very important question. One of the most important things you have learned about electricity so far is that you must *not* connect a conductor across the terminals of *any* power source.

10. (*c*)

4.
What Is a Volt?

Introduction

You have no doubt seen the word "voltage" or its abbreviation "V" used on different types of electrical equipment. If you buy a flashlight battery, for example, it is usually rated at "1.5 volts," and this value of voltage is marked on the battery case. Some transistor radios use 9-volt batteries, while automobiles use either 6- or 12-volt batteries. The power company delivers a voltage to your house which is usually rated at some value between 110 and 120 volts.

Figure 4-1 shows two completely different types of electrical devices which are both rated at 12 volts. Obviously, the battery does a different kind of job in an electrical circuit than a light bulb does. Why, then, are they both rated at 12 volts?

To answer this question, you should understand that there are really two kinds of voltages in a circuit. One is called the *applied voltage* or *source voltage*. It comes from the *voltage source* (such as from a battery or generator, or from a power company). The other is a *voltage rating*, which tells what kind of a circuit a device should be connected into. Thus, a 12-volt battery will *deliver* 12 volts, and a 12-volt light bulb must be connected into a circuit which will produce 12 volts across it.

You learned in an earlier chapter that voltage can be considered to be an

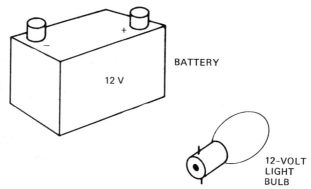

Figure 4-1.   The battery and light bulb are both rated in volts.

electrical pressure that pushes electricity through the circuit. You cannot tell anything about the amount of voltage from the size of the container that the voltage source comes in. For example, the two car batteries shown in Fig. 4-2 are identical in size, and yet one is a 6-volt battery and the other is a 12-volt battery. The 12-volt battery can exert a greater force on electrons to make them flow around a circuit.

There is more to the story than just the voltage alone. Note, for example, that the 6-volt battery shown in Fig. 4-3 is much larger (physically) than the 9-volt battery. This 6-volt car battery can deliver a very large current for starting an automobile, but the 9-volt battery could not begin to start it. Obviously, the amount of voltage is not the only factor that determines how much electricity will flow in a circuit. Another way of saying this is that the amount of current flow in an electrical system is only partially dependent upon the amount of voltage in that circuit. Another important factor is resistance.

The voltage *rating* of a device tells how much electricity must be delivered to it in order for it to operate correctly. As in the case of voltage sources, you cannot tell the voltage rating of a device from its size. Notice that the two light bulbs shown in Fig. 4-4 look identical, but they are rated at 6 volts and 12 volts. If you placed 6 volts across the 6-volt light, it would produce exactly the same amount of light as would be produced if you placed 12 volts across the 12-volt light.

You will be able to answer these questions after studying this chapter:

What does a resistor do in a circuit?
What causes the voltage across a resistor in a circuit?

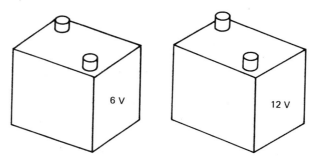

Figure 4-2.   These auto batteries have the same size but not the same voltage.

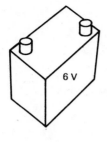

**6-VOLT
BATTERY**

9 V

**9-VOLT
BATTERY**

Figure 4-3. The 6-volt auto battery can supply enough electricity to start a car, but the 9-volt battery cannot start a car.

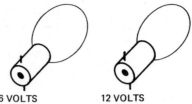

6 VOLTS   12 VOLTS

Figure 4-4. These light bulbs look identical but they have different voltage ratings.

In what way is a light bulb like a resistor?

How is the voltage rating of a device related to its circuit?

What happens when an improper voltage is applied to a device?

Instruction

WHAT DOES A RESISTOR DO IN A CIRCUIT?

You have already learned that every electric circuit has voltage, current, and resistance. In order to better understand about voltages in circuits, you will have to know more about resistors. Figure 4-5 shows a popular type of resistor, and the symbol used for a resistor on circuit diagrams. *A resistor is a part that opposes the flow of electric current in a circuit.*

By way of review, look at the water system shown in Fig. 4-6. Here there are two tanks, with water in tank *A* and no water in tank *B*. The control, or valve, between the two points prevents all the water in tank *A* from rushing to tank *B*. The control may be thought of as being a resistor because it resists the flow of water.

The pressure in tank *A* is created by the depth (or quantity) of the water. This can be compared to the electrical pressure, or voltage, in an electrical system. The water flowing between tank *A* and tank *B* can be compared to the flow of electrons in an electrical system.

COLOR BANDS

This is a resistor. It limits the amount of electricity that can flow in a circuit.

This is the symbol used to represent a resistor in a circuit drawing.

Figure 4-5. A resistor is an important part of an electric circuit.

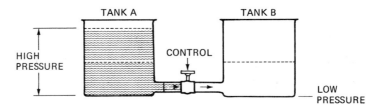

Water will flow until the pressures are equal.

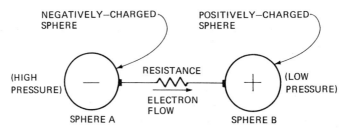

Because there is a difference in electrical pressure (that is, *a voltage*),
electrons will flow until the charges are equal.

Figure 4-6. Comparison of a water system to a system of electrical charges.

The simple electrical system shown in Fig. 4-6 is like the water system in
many ways. Sphere *A* has an excess of electrons, and it is said to be "negatively
charged." Thus, sphere *A* has more electrons than it normally has just as tank *A*
has more water than it normally has. The dotted line on tank *A* shows the
average water level. Sphere *B* has a shortage of electrons, and it is said to be
"positively charged." Tank *B*, which has no water, may be said to have a shortage of water.

Between sphere *A* and sphere *B* there is resistance created by inserting a
resistor into the conductor that connects the two spheres. Thus, electrons
flowing from the negative sphere to the positive sphere will encounter this
resistance, and the presence of the resistance will increase the amount of time
that it takes for the charges of the spheres to become equal. In this way it is
directly related to the control valve which determines the amount of time that it
takes tank *A* to flow into tank *B* and equalize the pressures.

If you allowed the control in the water system to remain open long
enough, the pressure of the two tanks would become equal. That would occur
when the water in tank *B* rose to the dotted line. The same thing will happen
with the charges in the spheres. Eventually, enough of the excess electrons in
sphere *A* will flow into sphere *B* so that the two spheres will have identical
charges. When this happens there is no longer an electrical pressure between
the two, and current will stop flowing.

A battery or an electric generator is a device that can continually add electrons to the negative terminal and take electrons away from the positive side.
This means that if a battery is supplying the charges in Fig. 4-6, the current can
flow continuously. In other words, unlike the system shown in Fig. 4-6 where
one sphere finally becomes discharged, a battery supplies the negative terminal
continually with negative electrons. If there is a complete circuit, the negative
electrons will continuously flow through the circuit.

Since there is a difference in charges in sphere *A* and sphere *B* just as there is a resistance in the terminals of a battery, we say that there is an "electrical pressure," or voltage, which tends to move the charges from one to the other. You know that the moving force is actually due to the fact that the electrons are negatively charged, and they are attracted to the positive charge. This is the basic rule that *unlike charges attract.*

You should understand that a voltage exists between spheres (or between the terminals of a battery) whether or not there is any current flow. This is again similar to the water system shown in Fig. 4-6, where there is a pressure from tank *A* regardless of whether or not there is any water flowing from tank *A* to tank *B*.

SUMMARY

1. A resistor is a device that opposes the flow of an electron current.
2. A resistor can be compared to a valve in a water circuit that controls, or limits, the flow of water.
3. The voltage in an electric circuit can be compared to the pressure in a water-pipe system.
4. Electron current flow in an electric circuit can be compared to water flow in a water system.
5. A voltage may exist without the need of a current flow. For example, when you buy a 6-volt battery there is 6 volts across the battery whether you put it into a circuit or not.
6. Although you can have a voltage without having a circuit, you cannot have a continuous flow of current without a closed circuit.

WHAT CAUSES THE VOLTAGE ACROSS A RESISTOR IN A CIRCUIT?

*Whenever an electric current flows through a resistor, there is **always** a voltage across the resistor.* This is an important rule that you will want to remember. To understand why the voltage exists across a resistor, we will compare the flow of electrons in a wire to the flow of cars in traffic. Figure 4-7 shows a freeway which has two lanes blocked off. Notice that the normal direction of traffic flow is from left to right. However, when the cars get to the road block, a traffic jam develops. This is because they cannot move through one lane as easily as they moved through the three lanes before the road block. It is very important to note the fact that there are more cars on one side of the road block than on the other. You may have seen a road block like this on a freeway or on a highway.

Figure 4-7 also shows what happens when electrons flow through a resistor. The resistor offers a partial road block to the flow of electron current. When the electrons reach the resistor, they tend to accumulate on one side, just like the cars accumulate on one side of the road block. Note that on the one side of the resistor electrons have accumulated so that there are more electrons at this point than there are on the right side of the resistor. You can see the left side of the resistor is going to be negative—that is, it will have an excess of electrons compared to the number of electrons on the right side of the resistor. Since there is a difference in the number of electrons between the two points, there is (by definition) a voltage between those points. This *always* happens when electrons flow through a resistor.

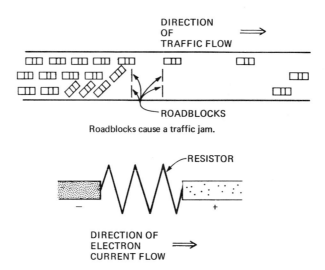

Figure 4-7. *There are more cars on one side of the road block, and there are more electrons on one side of the resistor.*

To distinguish between the voltage that exists across the resistor due to current flowing through it and the voltage that exists across a battery, we use two different terms to describe them. In the case of the battery, we call it a *voltage rise*, or *source voltage*. In the case of a voltage across a resistor, we call it a *voltage drop*. There is an important difference between the two types of voltages. The source voltage will exist whether or not there is any current flow. It is generated by the source, and it is always present at the battery terminals. However, you cannot have a *voltage drop* across a resistor unless there is a current flowing through the resistor. This brings us to another very important rule that you will want to remember: *If there is a voltage drop across a resistor, there must be current flowing through it.*

SUMMARY

1. Whenever there is an electron current flowing through a resistor, there is *always* a voltage drop across that resistor.
2. Whenever there is a voltage across a resistor, there is *always* a current flowing through it.
3. When there is an electron current flowing through a resistor, the side of the resistor where the current is entering is negative compared with the other side. This difference in the number of electrons accounts for the voltage across the resistor.
4. The voltage of a battery is called a *source voltage*. It exists whether there is a circuit or not.
5. The voltage across a resistor when current is flowing through it is called a *voltage drop*. The voltage drop cannot exist without a current flow.

IN WHAT WAY IS A LIGHT BULB LIKE A RESISTOR?

There are other devices besides resistors which have resistance—that is, which have the ability to oppose the flow of electricity. Figure 4-8 shows a closeup of a

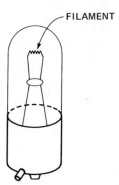

FILAMENT

Figure 4-8. The filament in a light bulb acts like a resistor.

light bulb, which is an example of a component that has resistance. If you look closely at a light bulb (assuming that it has a clear glass), you can see a tiny filament between two wires. Electricity flows through this filament and causes it to get red hot. That is where the light comes from.

Our interest in the light filament at this time is in the fact that it opposes the flow of electric current through it. Therefore, the filament does the same thing as a resistor in a circuit.

HOW IS THE VOLTAGE RATING OF A DEVICE RELATED TO ITS CIRCUIT?

The voltage rating of a device, such as the light bulb, tells what kind of a circuit it *must* be connected into. For example, a 6-volt light *must* be connected into a circuit in which there will be 6 volts across it, or the light cannot operate correctly. Figure 4-9 shows an example of such a circuit. The 6-volt light is connected directly across the 6-volt battery. You will remember that in a previous chapter you were told that you should never connect a wire directly across battery terminals. The circuit of Fig. 4-9 is not such a case, because the light actually has resistance in its filament. Therefore, there will be a limit to the amount of current that can flow through the circuit.

Since the light has resistance, the current flowing through it will cause a voltage drop. The circuit of Fig. 4-9 is too difficult to draw, so one using symbols is used instead. This is shown in Fig. 4-10. Once you get used to the

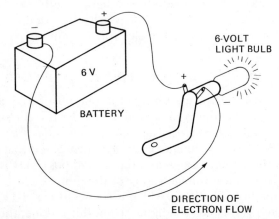

6-VOLT
LIGHT BULB

6 V

BATTERY

DIRECTION OF
ELECTRON FLOW

Figure 4-9. There is a voltage across the light when it is on.

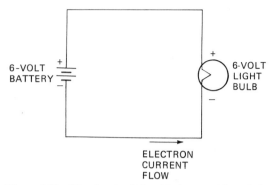

Figure 4-10. The circuit of Fig. 4-9 drawn with symbols.

drawings using symbols, it will no longer be necessary to draw components in the circuit as shown in Fig. 4-9.

There are several important things about the circuit of Fig. 4-10 that you should notice. The *6-volt* battery is connected directly across the *6-volt* light. Notice that *the source voltage is equal to the light voltage.* This is a requirement for all devices that have voltage ratings—that is, *they must always be connected across a circuit that will produce a voltage across them equal to their voltage rating.*

In the circuit of Fig. 4-10, as in all circuits, electron current is flowing *away* from the negative terminal of the battery, through the light, and back to the positive terminal of the battery. Note that since the light has a resistance, there is a voltage drop across the light. The voltage drop is always negative on the side that the electron current enters, and always positive on the side where the current leaves. Therefore, the negative side of the voltage drop across the light is at the negative side of the battery, and the positive side of the voltage across the light is at the positive side of the battery. This is an important point to remember.

SUMMARY

1. A light bulb has a filament that acts like a resistor to oppose current flow.
2. When a light bulb is rated at 6 volts, it must be placed in a circuit so that there is a 6-volt drop across it. Otherwise, it will not operate the way it is designed to.
3. The negative side of the voltage across a light is at the side toward the negative terminal of the source. In other words, the negative side of the voltage drop is at the point where the electron current enters.

WHAT HAPPENS WHEN AN IMPROPER VOLTAGE IS APPLIED TO A DEVICE?

It has been noted that the voltage rating of a unit must be equal to the voltage across it in a circuit if the unit is to operate properly. It is important for you to know what happens if the *source voltage* and *voltage rating* of a unit, such as a light, are not equal. Two possibilities exist. The first one is that the source voltage is *larger* than the voltage rating, and the second possibility is that the source voltage is *not as great* as the voltage rating of the unit.

Let us take the first condition, which is shown in Fig. 4-11. Here a 12-volt battery is connected across a 6-volt light. When this is done, the electrical pres-

DO NOT CONNECT THIS CIRCUIT!

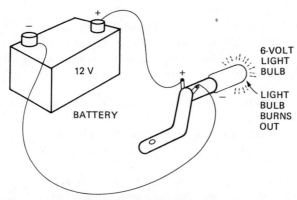

Figure 4-11. If the voltage of the source is greater than the voltage rating of the device, the device may burn out.

sure of the battery is too great for the light, and it forces too many electrons to flow through it. The result is that the device gets overheated and burns out. *You should never connect a circuit like this.*

The second possibility is that the battery voltage may not be as great as the voltage rating of the device. This is shown in Fig. 4-12. In this case, the battery cannot produce enough pressure to force current through the light to make it operate properly. The result will be that the light will not glow as brightly as it would if the full voltage were across it. This will not hurt the light, but it is not a very good way to use it.

SUMMARY

1. If an electrical device is to operate properly, the voltage across it must equal its voltage rating.
2. If the voltage across an electrical device is greater than its voltage rating, it can be ruined.
3. If the voltage across the device is not as great as its voltage rating, the device will not operate properly.

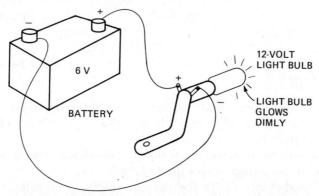

Figure 4-12. If the voltage of the source is not as great as the voltage rating of the device, the device will not operate as it should.

Programmed Review Questions

(Instructions for using this programmed section are given in Chap. 1.)

We will now review the important concepts of this chapter. If you have understood the material, you will progress easily through this section. Do not skip this material, because some additional theory is presented.

1. Is this statement true? "There can be a voltage even though there is no current flowing."
    A. The statement is true. (Proceed to block 7.)
    B. The statement is false. (Proceed to block 17.)

2. *The correct answer to the question in block 8 is **A**. You cannot make a definite statement about the brightness of a light based only on its voltage rating.* Here is your next question.
    Which of the following is correct?
    A. There is a voltage flow across the terminals of a battery. (Proceed to block 24.)
    B. There is a voltage between the terminals of a battery. (Proceed to block 6.)

3. *The correct answer to the question in block 15 is **B**. As shown in Fig. 4-6, the pressure is like the voltage.* Here is your next question.
    Whenever an electron current flows through a resistor, there is a voltage across it. Is it also true that whenever there is a voltage across a resistor, there must be a current flowing through it?
    A. No, this is not true. (Proceed to block 23.)
    B. Yes, this is true. (Proceed to block 9.)

4. *Your answer to the question in block 7 is **A**. This answer is wrong. You cannot make a flat statement that a 9-volt battery can start a car better than a 6-volt battery. As an example of a case where this would not be true, look at the two batteries shown in Fig. 4-3. The 9-volt battery in this illustration could not begin to start a car. Proceed to block 15.*

5. *The correct answer to the question in block 10 is **B**. There is a voltage across a light or across a resistor when current is flowing through these devices. These are both examples of **voltage drops**.* Here is your next question.
    When 50 volts is placed across a 110-volt light,
    A. it will burn out. (Proceed to block 12.)
    B. it will not glow as brightly as it is supposed to. (Proceed to block 22.)

6. *The correct answer to the question in block 2 is **B**. Always say "voltage between" or "voltage across," but never say "voltage flow."* Here is your next question.
    When an electron current is flowing through a resistor, the negative side of the voltage across the resistor is
    A. at the side where the current enters the resistor. (Proceed to block 11.)
    B. at the side where the current leaves the resistor. (Proceed to block 21.)

7. *The correct answer to the question in block 1 is **A**. Just as there is water pressure available at a faucet in your house—whether you use it or not—there is also electrical pressure (voltage) at the terminals of a battery or at the power receptacles of your house.* Here is your next question.
Can a 9-volt battery start a car better than a 6-volt battery?
A. Yes (Proceed to block 4.)
B. No (Proceed to block 15.)

8. *The correct answer to the question in block 22 is **A**. A positive charge means a shortage of electrons, and a negative charge means too many electrons. If a conductor is placed between a positive and negative point, an electron current will flow until the charges are equal. A battery or other voltage source continues to deliver electrons to its negative terminal to replace those that flow in a circuit. Therefore, a continuous current will flow when a source voltage is connected to a circuit.* Here is your next question.
Is this statement true or false? "A 12-volt light is brighter than a 6-volt light."
A. The statement is false. (Proceed to block 2.)
B. The statement is true. (Proceed to block 14.)

9. *The correct answer to the question in block 3 is **B**. A current through a resistor will always produce a voltage across the resistor, and a voltage across a resistor means that there is a current flowing through it. This rule applies to other components, such as light bulbs, which have resistance.* Here is your next question.
A device that limits the flow of electricity is
A. a resistor. (Proceed to block 19.)
B. a limiter. (Proceed to block 16.)

10. *The correct answer to the question in block 19 is **B**. Although the 12-volt light will not burn out, it will not glow as brightly as a 6-volt light across the 6-volt circuit.* Here is your next question.
The voltage across a resistor due to current flowing through it is called
A. a source voltage. (Proceed to block 20.)
B. a voltage drop. (Proceed to block 5.)

11. *The correct answer to the question in block 6 is **A**. One of the things you should remember about the voltage drop across a resistor is that it is always negative at the side where electrons **enter** the resistance.* Here is your next question.
On a separate sheet of paper draw the symbols used for the following devices.
A. A battery
B. A resistor
C. A light bulb
D. A switch
(Proceed to block 26.)

12. *Your answer to the question in block 5 is **A**. This answer is wrong. This is the same as putting 6 volts across a 12-volt light. The light will burn, but not as brightly as it is supposed to.* Proceed to block 22.

13. *Your answer to the question in block 19 is **A**. This answer is wrong. If the source voltage is **lower** than the voltage rating of a device, the device will not burn out. However, this is not a good way to operate circuits.* Proceed to block 10.

14. *Your answer to the question in block 8 is **B**. This answer is wrong. In order to know which light is brighter, you would have to know more about the circuit. For example, if the 12-volt light is connected across a 6-volt source, it will not glow as brightly as a 6-volt light when connected across a 6-volt source.* Proceed to block 2.

15. *The correct answer to the question in block 7 is **B**. You cannot make a flat statement about how much can be accomplished with a battery by just knowing its voltage.* Here is your next question.
    The pressure in a water system can be compared to
    A. the electron current flow in a circuit. (Proceed to block 25.)
    B. the voltage in a circuit. (Proceed to block 3.)

16. *Your answer to the question in block 9 is **B**. This answer is wrong. The name "limiter" **sounds** like it would be used to describe a device that limits the flow of current, but it is **not**.* Proceed to block 19.

17. *Your answer to the question in block 1 is **B**. This answer is wrong. A 12-volt battery has 12 volts across its terminals regardless of whether or not it is connected into a circuit. Likewise, there is about 115 volts delivered to the power outlet in your house regardless of whether or not a circuit is attached.* Proceed to block 7.

18. *Your answer to the question in block 22 is **B**. This answer is wrong. Remember that electrons are negative. If there are too many at some point, the charge at that point is negative.* Proceed to block 8.

19. *The correct answer to the question in block 9 is **A**. In fact, a resistor may be defined as a device that limits the flow of electron current.* Here is your next question.

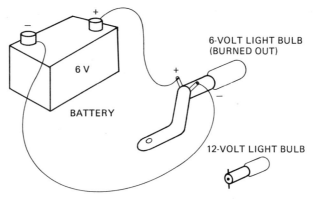

Figure 4-13. The 6-volt light is burned out, and the only one available is rated at 12 volts. It can be used, but it will give a dim light.

A light bulb in a circuit burns out, and you are going to replace it. You note that the light bulb is *rated* at 6 volts, but all you have that fits into the socket is a 12-volt light. The situation is shown in Fig. 4-13. Which of the following is true?

A. If you put the 12-volt light into the socket, it will burn out. (Proceed to block 13.)

B. It is all right to put the 12-volt light into the socket, but it will not burn as brightly as it is supposed to. (Proceed to block 10.)

20. *Your answer to the question in block 10 is **A**. This answer is wrong. The source voltage is the voltage that pushes the electricity through the circuit. A battery and the 115-volt electric power delivered to your house are both examples of source voltages.* Proceed to block 5.

21. *Your answer to the question in block 6 is **B**. This answer is wrong. The side of a resistor at which current is leaving is less negative than the side where it enters.* Proceed to block 11.

22. *The correct answer to the question in block 5 is **B**. Except for the fact that the voltage values are different, this question is very similar to the one given in block 19. Here is your next question.*
When there is a negative charge at some point, it has
A. too many electrons. (Proceed to block 8.)
B. not enough electrons. (Proceed to block 18.)

23. *Your answer to the question in block 3 is **A**. This answer is wrong. You have missed one of the most important rules in this chapter. For a restatement of this rule, proceed to block 9.*

24. *Your answer to the question in block 2 is **A**. This answer is wrong. You should **never, never, never** say "voltage flow." A voltage is a difference in electrical pressure **between** two points. Refer to the water tanks in Fig. 4-6. There is water pressure in tank A, but not in tank B. In other words, there is*

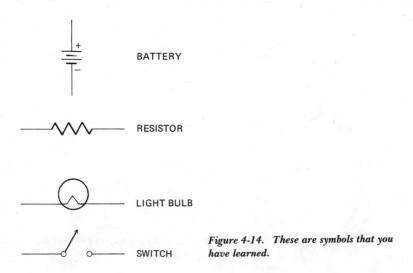

Figure 4-14. These are symbols that you have learned.

BATTERY

RESISTOR

LIGHT BULB

SWITCH

a *difference* in pressure *between* them. *That is why the water flows away from tank A and into tank B. Likewise, the electrons flow out of sphere A and into sphere B because there is a* **difference** *in electrical pressure between them. The difference exists between two points, but in no case will the pressure "flow." In an electric circuit it is the current that* **flows.** Proceed to block 6.

25.   *Your answer to the question in block 15 is* **A**. *This answer is wrong. The flow of water can be compared to the flow of electron current, not the voltage.* Proceed to block 3.

26.   *The correct symbols are shown in Fig. 4-14.*
      You have now completed the programmed questions. The next step is to put some of these ideas to work in laboratory experiments. Proceed to the "Experiments" section that follows.

Experiments

(The experiments described in this section may be performed on the circuit board described in Appendix C or on a similar laboratory setup.)

EXPERIMENT 1

Purpose—To learn how to use a trouble light.

Theory—One way to determine if there is a voltage between two points is to use an instrument called a *voltmeter*. When the two leads of the voltmeter are touched to the two points, the meter will tell if there is a voltage difference, and how much the voltage difference is.

   When you are working on house wiring you are most likely to encounter only three voltage values: 240 volts, 120 volts, and 0 volt. In an automobile that has a 12-volt battery, the voltage will be either 12 volts or 0 volt, while in a car with a 6-volt battery you will find 6 volts or 0 volt. (This does not include wiring for the spark plugs, where the voltage will be many thousands of volts.) Thus,

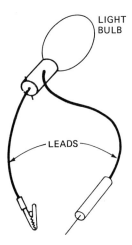

Figure 4-15.   A light bulb can be used to test for voltage.

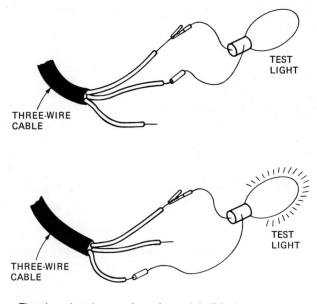

There is a voltage between these wires, and the light *is on.*

Figure 4-16. The test light can be used to determine if there is a voltage between two points.

you can have an indicator that will tell if the two or three voltages are present. That is the purpose of the trouble light.

Figure 4-15 shows how a trouble light can be made. The test leads are soldered to the two points on the base where electrical contact is made. When these leads are placed across a place where there is a voltage, the light will be on.

Figure 4-16 shows the test light being used to determine if there is a voltage between the wires. Only two tests are shown. A third test would normally be made between the center wire and the bottom wire.

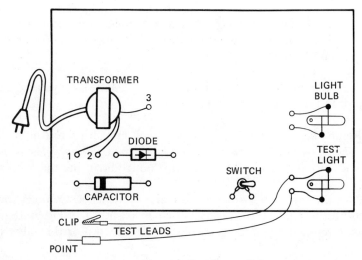

Figure 4-17. The test light.

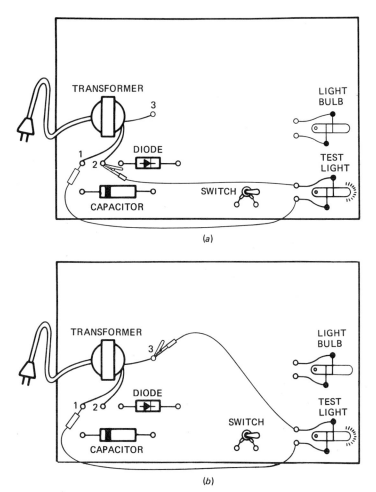

Figure 4-18.   How a test light can be used to check voltages. (a) *Measurement of voltage between terminals 1 and 2.* (b) *Measurement of voltage between terminals 1 and 3.*

In this experiment you are going to make a trouble light and test for voltage. It will not be necessary to solder the wires, since you have convenient connectors on the circuit board.

Test Setup—Wire the test light as shown in Fig. 4-17. Make sure your test leads are long enough to reach across the board.

Procedure—Remember that you are using a 12-volt light for a test light. *It will glow at full brightness when it is across a 12-volt source, and will only glow at partial brightness when it is across a 6-volt source.* The procedure for using the test light is illustrated in Fig. 4-18.

Step 1—Plug the power transformer into the wall socket. Connect the test leads across terminals 1 and 2 of the transformer as shown in Fig. 4-18a. Observe the glow to see if the voltage is 6 volts or 12 volts. Enter the value here.

_____ volts

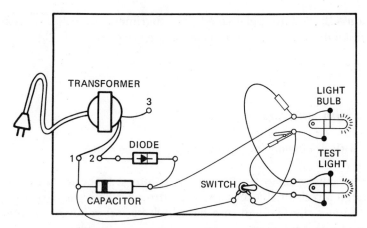

Figure 4-19. Using the test light to see if there is a voltage across a glowing light.

Step 2 — With the power transformer still connected to the wall socket, connect the test leads across terminals 1 and 3. See Fig. 4-18*b*. Observe the light glow to see if the voltage is 6 volts or 12 volts. Enter the value here.

_____ volts

Step 3 — With the power transformer still connected to the wall socket, connect the test leads across terminals 2 and 3. Observe the light glow to see if the voltage is 6 volts or 12 volts. Enter the value here.

_____ volts

Conclusion — Your results should show that the voltage across terminals 1 and 2 is 12 volts. The voltage across terminals 1 and 3 and also across terminals 2 and 3 is 6 volts.

EXPERIMENT 2

Purpose — To show that there is a voltage across a light when it is glowing.

Theory — As stated in the chapter, there *must* be a voltage across a resistor when there is current flowing through it. If a light is glowing, you should be able to measure the voltage across it.

Test Setup — Wire the circuit as shown in Fig. 4-19.

Procedure —
Step 1 — With power applied to the transformer, close the switch so that the light is glowing.
Step 2 — Place your test light across the light to see if there is a voltage across it.
Step 3 — Open the switch and see if there is voltage across the light when it is not glowing.

Conclusion — If you have performed the experiment correctly, you have shown that when a light is glowing there is a voltage across it.

Self-Test with Answers

(Answers with discussions are given in the next section.)

1. There is always a current flow (a) in a battery; (b) between two spheres; (c) through a resistor that has a voltage drop across it; (d) from left to right.
2. At all times the filament in a light bulb has (a) voltage; (b) resistance; (c) current; (d) insulation.
3. The source voltage in a circuit is most like (a) current flowing in a stream; (b) cars moving along a highway; (c) pressure in a water system; (d) the filament in a light bulb.
4. Which of the following statements is true about batteries? (a) They do not have voltage unless they are connected into a circuit; (b) The larger the battery, the larger its voltage; (c) Most batteries have a positive and a negative terminal, but some batteries may have two negative terminals or two positive terminals; (d) When a battery is connected into a circuit, electrons leave its negative terminal and return to its positive terminal.
5. You should never (a) connect a light bulb across a battery; (b) connect a resistor in a circuit where there is a current flow; (c) refer to a current "flow"; (d) connect a wire directly across battery terminals.
6. Electrical pressure is measured in (a) volts; (b) amperes; (c) watts; (d) pounds.
7. A test light is used for (a) determining if there is resistance in a circuit; (b) determining if there is a light bulb in a circuit; (c) determining if a voltage is positive or negative; (d) determining if there is a voltage between two points.
8. You should never connect a 12-volt test light (a) across a circuit that has no voltage; (b) where there is no resistance; (c) across a 115-volt source of voltage; (d) across a 12-volt light.
9. One similarity between a resistor and a light bulb is that (a) they both have resistance; (b) they do exactly the same job; (c) they are both made of silver; (d) they are both rated at 6 volts.
10. When you connect a switch into a circuit, (a) it must be connected to the side of the circuit that goes to the negative battery terminal; (b) it must be connected to the side of the circuit that goes to the positive battery terminal; (c) it must be connected into a circuit in such a way that current flows through it when the switch is closed; (d) it should not be connected in such a way that it can interrupt the flow of current.

Answers to Self-Test

1. (c) — There *may* be a current flow between spheres if they have a difference in voltage and a conductor between them.

2. (*b*) – There *may* be a voltage across a filament, and a current through it, or there may not be. However, the filament always has resistance.

3. (*c*) – It is easy to think of the source voltage as being the pressure that forces current to flow through a circuit.

4. (*d*) – All the statements except (*d*) are very definitely wrong.

5. (*d*) – When a wire is connected directly across the terminals of a battery, it is called a *short circuit*. This will ruin a battery.

6. (*a*)

7. (*d*) – The method of using a test light was described in the "Experiments" section.

8. (*c*) – If you connect a 12-volt test light across a 115-volt source, you will burn out the bulb.

9. (*a*) – Resistors and light bulbs do *not* do exactly the same job in a circuit. You could not replace a light bulb with a resistor. It is true, however, that they both have resistance.

10. (*c*)

5.
What Is an Ampere?

Introduction

Since the electron is such an extremely small charge of electricity, a very large number must flow in a circuit if any useful work is to be accomplished. To give you an idea of how many electrons are needed, consider the problem of starting an automobile engine. It is not uncommon for a starter to require 100 amperes of current flow. One ampere of current flows when

$$6,240,000,000,000,000,000 \text{ electrons}$$

flow past a point in a circuit *every* second. The starter requires 100 times this many electrons per second.

The number of electrons per second required for a current of 1 ampere is too difficult to write, so there is a special name for it. That number of electrons is called *1 coulomb*. The correct value is 6.24×10^{18} but it is sometimes incorrectly given as 6.28×10^{18}. Apparently this comes from the fact that the value of 2π is 6.28, and this number is used frequently in electricity and electronics.

Now the definition of current can be called a flow of one coulomb per second. Here is another way of saying this:

$$1 \text{ ampere} = 1 \text{ coulomb per second}$$

Obviously, you could not count this many electrons to find the amount of current flow. If you could count 100 electrons a minute, and worked at it 8 hours a day, 7 days a week, it would take you over 356 billion years to count all the electrons in 1 coulomb. One of the things you will learn in this chapter is how current can be measured indirectly by using the effects of current.

You will be able to answer these questions after studying this chapter:

What are the effects of current flow in a circuit?
What factors determine how much current will flow in a circuit?
What is a milliampere?
What is a microampere?
How does a fuse work?

Instruction

WHAT ARE THE EFFECTS OF CURRENT FLOW IN A CIRCUIT?

The practical uses of electricity in modern applications are based on the fact that there are *effects* of electric current flow which can be used to accomplish certain tasks. In order to understand how the jobs of electricity are accomplished, you must have a clear understanding of what these effects are.

One important effect of an electric current is that it *always* produces a magnetic field. You have already studied about the magnetic field around a wire,

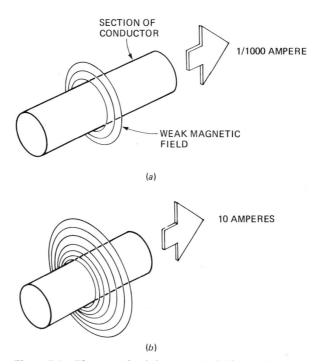

(a)

(b)

Figure 5-1.   The strength of the magnetic field around a wire depends upon the amount of current flow. (a) A weak magnetic field is produced by a small current. (b) A large current produces a strong magnetic field.

and around a coil, when current is flowing. It is important for you to remember that there is a magnetic field around *every* current that flows, regardless of whether the current is flowing through a resistor, a lamp, or a conductor.

The strength of the magnetic field around a wire depends on the *amount* of current. This is shown in Fig. 5-1. As illustrated in Fig. 5-1*a*, a small current results in a weak magnetic field, while Fig. 5-1*b* shows that a larger current produces a stronger magnetic field.

Another effect of a current flow in a circuit is that there is *always* a voltage drop when the current flows through a resistance. No matter how large or how small the resistance of the circuit, if a current flows through it, there must be a voltage drop across it.

The amount of voltage across a given resistor depends directly upon the amount of current flowing through it. This is shown in Fig. 5-2. As illustrated in Fig. 5-2*a*, a current of 1 ampere through a certain resistor will cause a 10-volt drop across the resistor. In Fig. 5-2*b* the current through the same resistor has been increased to 2 amperes, and the voltage across the resistor has increased to 20 volts. Thus, the amount of voltage drop depends directly upon the amount of current. The reason for this voltage drop was discussed in the last chapter. You will remember that the electrons pile up on one side of the resistor (where the current is entering the resistor), and this causes a negative charge at that point. The voltage across the resistor is based on the fact that one side is more negative than the other.

The only way you could avoid having a voltage drop when a current is

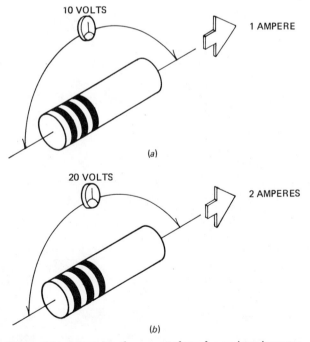

10 VOLTS

1 AMPERE

(a)

20 VOLTS

2 AMPERES

(b)

Figure 5-2. Increasing the current through a resistor increases the voltage drop across it. (a) Current flowing through a resistor causes a voltage drop. (b) Increasing the current through the same resistor increases the voltage drop.

flowing is to have a circuit that has absolutely no resistance. For all practical purposes the resistance of a piece of wire is so small that it may be ignored in *most* (but not all) cases. But you should not be misled into believing that there is *no* resistance in a wire. *All conductors have some resistance.* When an electrician is laying out a wiring job for a large building, he has to take the resistance of the wire into consideration. If a wire is going to be very long, then he must know how much voltage drop will occur across that wire. If the voltage drop is too large, of course, there will not be enough electricity to operate the light bulbs (or whatever is connected to the wire terminal). As a general rule, electricians allow from 2 to 10 percent of source voltage to be dropped across the wire when they are determining the size and the length of wire to be used.

Another effect of electricity flowing through a resistance is that *heat is always generated.* The amount of heat generated depends on the amount of electricity flowing. The heat also depends upon the amount of resistance in the circuit. In other words, you can increase the amount of heat by increasing either the resistance or the current or both. The important thing here is that the *current* determines the heat. Figure 5-3 shows how the current and heat are related. Light bulbs are used in Fig. 5-3, but you know that the filament of a light bulb has resistance. Figure 5-3a shows that a small amount of current produces a small amount of heat. As shown in Fig. 5-3b, increasing the current through the same amount of resistance causes more heat to be generated.

If you have an electric toaster, or an electric stove, or an electric heater in your house, it operates on the principles of heat produced by an electric current flowing through a resistance. The heating element in the appliance is actually nothing more than a resistor that is designed to produce heat when a current flows through it.

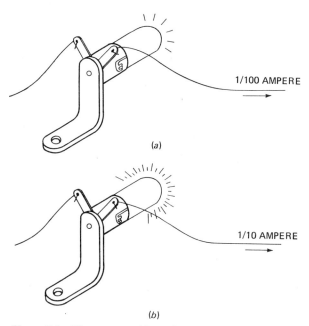

Figure 5-3. The amount of heat depends upon the amount of current flow. (a) *When a small current flows through a light bulb, it becomes warm.* (b) *A higher current through the same light bulb results in a greater amount of heat.*

The three effects of electricity are so important that they are summarized here.

1. Whenever electric current flows, whether or not it is flowing through a resistance, there is a magnetic field surrounding the current.
2. Whenever an electric current flows through a resistance, however small, there is always a voltage drop.
3. Whenever an electric current flows through a resistance, there is always heat generated.

WHAT FACTORS DETERMINE HOW MUCH CURRENT WILL FLOW IN A CIRCUIT?

The amount of current flow in a circuit is dependent only upon two factors:

1. The amount of opposition, or resistance, to the current flowing in the circuit
2. The amount of voltage applied to the circuit

A 12-volt battery, for example, will cause more current to flow through a light bulb than a 6-volt battery. You can make a rule about the relationship between voltage and current, which can be stated as follows: *The amount of current flowing in a circuit depends directly upon the amount of voltage across the circuit.*

The other factor that determines how much current flows is the opposition or resistance in the circuit. If the resistance of a circuit is large, then the amount of current flow is small. If the resistance of a circuit is small, then the amount of current flow will be large. (Always keep in mind that the resistance is only one factor. The other is the amount of voltage applied.)

Here is a rule that relates the amount of current flow in a circuit to the resistance of that circuit. *The greater the amount of resistance in a circuit, the less the current flow.*

If you compare the two things that determine current flow (voltage and resistance), you will notice a very important fact: **Increasing the voltage IN-CREASES the amount of current flow, but increasing the resistance DECREASES the amount of current flow.**

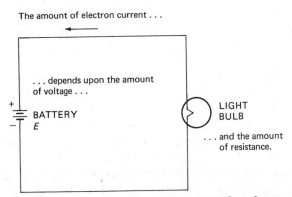

Figure 5-4. The amount of current in a circuit depends upon the amount of voltage and upon the amount of resistance.

The relationship between current, voltage, and resistance is shown in Fig. 5-4.

SUMMARY

1. There is always a magnetic field around an electron current. The strength of the magnetic field depends upon the amount of current flow.
2. When an electron current flows through a resistance, there is *always* a voltage drop. The amount of voltage drop across a resistor depends upon the amount of current flow.
3. When an electron current flows through a resistance, there is *always* heat generated. The amount of heat depends on the amount of current.
4. The amount of current flowing in a circuit depends upon how much applied voltage is used and how much resistance there is in the circuit.

WHAT IS A MILLIAMPERE?

It is an interesting fact that the greatest practical uses of theories in science occur only after it is possible to make measurements. Electricity is no different than other fields of science in this respect. Once the scientists learned how to measure electricity, the door was opened to the design of many practical applications.

In the introduction of this chapter it was explained that electric current is measured in amperes—that is, in coulombs per second. This is the standard *basic* unit of measurement for electric current. Unfortunately, it is not a very convenient unit in many types of circuits. There are many units of measurement that have the same disadvantage. Consider the unit of measurement for time, the *second*. In all the methods of measuring used in science, the second is the *basic* unit of time. However, it is inconvenient to use a second in some applications. For example, suppose a businessman is told it will take 4 hours to fly from New York to Los Angeles. Expressing the time in hours is convenient in this case. If, instead, he was told it would take 14,400 seconds to fly from New York to Los Angeles, the number is inconvenient to work with. In other words, the second is not a convenient unit of time for measuring long periods.

In the example just cited the second is too short as a unit of time. There are also periods of time where the second is much too long for convenient use. In a radar system like the one shown in Fig. 5-5 a small burst of radio wave is sent from the antenna toward the airplane. It bounces off the airplane and returns to the antenna. One of the most basic equations is that

$$DISTANCE = RATE \times TIME$$

Therefore, if you know the speed of the radio wave (186,000 miles per second) and the time that it takes the radio pulse to get from the antenna to the airplane and back, it is possible to calculate the round-trip distance. Of course, you would have to divide this round-trip distance by 2 to determine the distance between the antenna and the airplane. Now remember that the radio wave is traveling at a speed of *186,000 miles per second!* Suppose the airplane is only a mile away. The amount of time that it takes the radio wave to go to the plane and back is practically instantaneous. Of course, this time is too short to be measured by any kind of a watch, so special equipment had to be designed to measure such a short time interval. The use of the *second* as a period of time

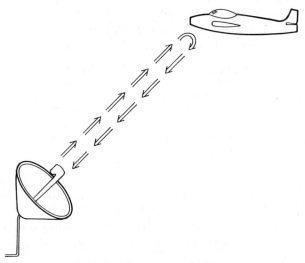

Figure 5-5. The idea behind radar is easy to understand. A short burst (or pulse) of radio wave is bounced off an airplane, and the time for the round trip of the pulse is measured. The round-trip distance equals the speed of the pulse multiplied by the time required for the pulse to make the round trip (distance = rate × time). A second is a very long time in this type of system.

in such an experiment measuring such a short time is very inconvenient, and so the millionth of a second (or millionth of a millionth of a second) is used.

To summarize, the unit measurement of time is the second, but in some cases the second is too short for convenience, and in other cases the second is too long for convenience.

The same situation exists with an electric current. The basic unit of measurement is the ampere, which in many cases is much too large for convenience, and so a much smaller unit is used. The smaller unit is called the *milliampere*. By definition a milliampere is one-thousandth of an ampere.

It is sometimes necessary to change a current given in amperes into milliamperes, and to change the measurement in milliamperes into amperes. A simple way to change from one unit to another is to use Table 5-1. This table shows that if you want to change a value in amperes into milliamperes, simply move the decimal place to the *right* three places. If you want to change a value in milliamperes into amperes, simply move the decimal place to the *left* three places. Two problems will be worked to show how this is done.

Example 5-1 — The current in a certain circuit is measured and found to be 5 milliamperes. How many amperes is this?

Solution — According to Table 5-1, to change the value in milliamperes to amperes, you move the decimal point to the *left* three places.

decimal point goes here **x** 0 0 5.

5 milliamperes = 0.005 ampere

Thus, 5 milliamperes is another way of saying the current is 0.005 ampere.

The current in a circuit may be expressed in amperes when milliamperes may be more convenient. This is shown in the following example.

TABLE 5-1. CONVERSION FROM AMPERES TO MILLIAMPERES AND MILLIAMPERES TO AMPERES

To Change from	To	Move Decimal Point
Amperes	milliamperes	to the right three places
Milliamperes	amperes	to the left three places

Example 5-2 — The current in a certain circuit is known to be 0.01 ampere. What is this current as stated in milliamperes?

Solution — Table 5-1 states that amperes can be changed to milliamperes by moving the decimal point to the right three places. Therefore,

$$0.0\,1\,0\,\mathbf{x} \text{ decimal point goes here}$$

$$0.01 \text{ ampere} = 10 \text{ milliamperes}$$

You can say that the amount of current flow in the circuit is 0.01 ampere, or you can say it is 10 milliamperes. Either way you say it, it means the same thing.

WHAT IS A MICROAMPERE?

There are cases where the milliampere is too large, and another, more convenient method is to express the current in *microamperes*. A microampere is a millionth of an ampere. Table 5-2 shows how to convert from amperes to microamperes and from microamperes to amperes. Note this is similar to the changes in Table 5-1 except that the decimal point is moved six places instead of three. Two examples will be worked to show how this is done.

Example 5-3 — A certain instrument is so delicate that it cannot withstand a current greater than fifty-millionths of an ampere (0.000050 ampere). Express this number as amperes and microamperes.

Solution —

$$0.0\,0\,0\,0\,5\,0\,\mathbf{x} \text{ decimal point goes here}$$

Fifty-millionths of an ampere equals 50 microamperes. Thus, you can say

TABLE 5-2. CONVERSION FROM AMPERES TO MICROAMPERES AND MICROAMPERES TO AMPERES

To Change from	To	Move Decimal Point
Amperes	microamperes	to the right six places
Microamperes	amperes	to the left six places

that the maximum current that is allowed to flow through the instrument is 50 microamperes, or you can say it is fifty-millionths of an ampere. Both ways mean exactly the same thing. Both methods express exactly the same amount of current.

Example 5-4 — The amount of current flowing through a certain antenna wire is 100 microamperes. How may this current be expressed in amperes?
Solution —

decimal point goes here **x** 0 0 0 1 0 0.

Thus, you can say that the current flowing in the antenna wire is 100 microamperes, or you can say it is 0.0001 ampere. In either case, you are saying the same thing.

Tables 5-1 and 5-2 show you how to convert from milliamperes or microamperes to amperes, and from amperes to milliamperes or microamperes. However, these tables can also be used in other cases. Occasionally, a voltage may be expressed in millivolts or microvolts, and you might have to convert these values into volts. You can use Tables 5-1 and 5-2 for this purpose by just substituting the word "volts" for "amperes." In fact, these figures can be used *any time* you desire to convert any unit from micro and milli units into regular units.

SUMMARY

1. Current is measured in amperes, but this unit may be too large for some applications.
2. The milliampere is a smaller unit of measurement than the ampere. It takes 1,000 milliamperes to make 1 ampere.
3. The microampere is a smaller unit than a milliampere. It takes 1,000,000 microamperes to make 1 ampere.
4. It is an easy matter to convert from one unit to another by moving the decimal point as explained in Tables 5-1 and 5-2.

HOW DOES A FUSE WORK?

Whenever an electric current flows through a resistance, heat is generated. It does not make any difference if this heat is desired (as in the case of electric stoves) or undesired (as in the case of many pieces of electrical equipment where the heat is simply troublesome).

Whether it is wanted or not, heat will *always* be generated. Sometimes it is possible to make use of this heat. One example is for fuses. A fuse is simply a wire or strip of metal made of a material that will melt if it becomes too hot. Figure 5-6 shows a closeup view of one type of fuse. When the current flowing through the fuse is below a certain value, the conductor in the fuse becomes warm, but not hot enough to melt. However, if the current becomes excessive — that is, above the ampere rating of the fuse — the resulting heat will melt the wire and open the circuit. When the circuit is open, of course, the current will stop flowing.

Figure 5-7 shows some different types of fuses that are used in electrical

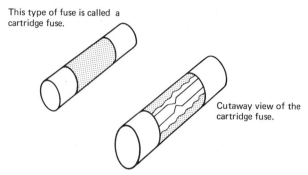

This type of fuse is called a cartridge fuse.

Cutaway view of the cartridge fuse.

Figure 5-6. One type of fuse used for protecting a circuit from an overload.

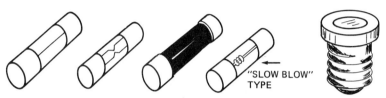

"SLOW BLOW" TYPE

Figure 5-7. Some different types of fuses. All work on the same principle. An overload heats the element, melts it, and opens the circuit.

FUSE

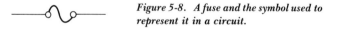

Figure 5-8. A fuse and the symbol used to represent it in a circuit.

work. Most of these are designed to open the circuit immediately when the heat becomes larger than its rating. In some circuits, however, the fuse is designed to operate with an excessive current for a short period of time without opening the circuit. This type of fuse is called a *slow blow*, and it is also shown in Fig. 5-7.

Figure 5-8 shows the symbol used for a fuse in circuits. This same symbol is used for all the types shown in Fig. 5-7.

Figure 5-9 shows how a fuse is connected into a circuit. The circuit is shown in Fig. 5-9*a*, and Fig. 5-9*b* shows the circuit drawn with symbols. Note that the circuit current must flow *through* the fuse if it is to be able to stop the current when it burns out.

An overload will occur when the resistance of the circuit is lowered. One way that this can happen is by a *short circuit*. A short circuit is a very low resistance path. It is usually not intended to be in the circuit. The short circuit removes opposition to current flow and causes the current to increase above a value that the battery (or other source) can safely deliver.

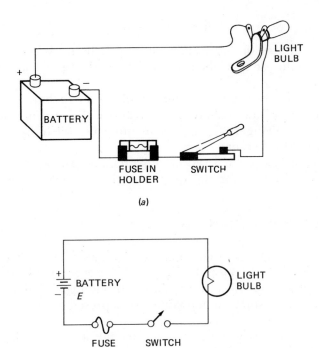

(a)

(b)

Figure 5-9. How a fuse is connected in a circuit. (a) *The fuse must be located in a circuit so that the current flows through it.* (b) *Schematic drawing of the circuit in Fig. 5-9a.*

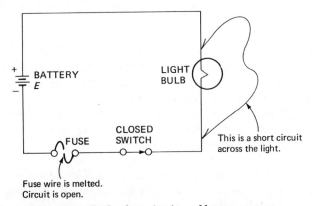

Figure 5-10. Normally the short circuit would cause an excessive current to flow from the battery. There is so little resistance in the short-circuit wire that all the current flows through it rather than through the resistance of the light. However, the fuse burns out and opens the circuit.

Figure 5-10 shows how a short circuit across the light will cause an excessive current to burn out the fuse. Note that when the fuse is burned out, the circuit path is open and no current can flow. This is the same situation you get when the switch is open, so you might think of the fuse as being an automatic switch that protects the circuit.

SUMMARY

1. A fuse is a device that is used to protect a circuit. When the current is greater than the rating of the fuse, the wire in the fuse melts and opens the circuit.
2. A 1-ampere fuse can conduct a current of 1 ampere, but not more.
3. A fuse must be connected into a circuit so that the current flows *through* it.
4. A fuse may be thought of as being an automatic switch that opens the circuit if an overload occurs.
5. A short circuit is a very low resistance path that causes an overload current. Fuses are used for protecting equipment when a short circuit occurs.

Programmed Review Questions

(Instructions for using this programmed section are given in Chap. 1.)

We will review the important concepts of this chapter. If you have understood the material, you will progress easily through this section. Do not skip this material, because some additional theory is presented.

1. A friend tells you he is saving to buy a car, and he already has 5,000,000 microdollars. Can he buy a car that costs $5,000? Which of the following is true? (*Hint:* Use Table 5-2 to convert microdollars to dollars.)

   A. He can buy a $5,000 car with the money he has saved. (Proceed to block 7.)

   B. He is not even close to having enough money saved. (Proceed to block 17.)

2. *The correct answer to the question in block 25 is **B**. If the fuse is across the resistance, it will be like the short circuit of Fig. 5-10. Remember that it must have the circuit current flowing **through** it if it is going to operate properly.* Here is your next question.

   Is this statement true or false? "A fuse in a circuit acts like an automatic switch to stop current from flowing in case there is an overload." (Proceed to block 26.)

3. *Your answer to the question in block 15 is **A**. This answer is wrong. It is very important for you to know the effects of current flowing through a resistor.* Proceed to block 22.

4. *Your answer to the question in block 5 is **B**. This answer is wrong. A flow of current equal to 1 coulomb per hour would be very small compared to the coulomb per second that is required to produce 1 ampere.* Proceed to block 11.

5. *The correct answer to the question in block 20 is **A**. An electric current flows through coils of wire in the heater. This produces the heat required for heating the water.* Here is your next question.

   A current of 1 ampere is

A. a flow of current past a point equal to 1 coulomb per second. (Proceed to block 11.)

B. a flow of current past a point equal to 1 coulomb per hour. (Proceed to block 4.)

6. *The correct answer to the question in block 11 is **B**. A milliampere is a thousandth of an ampere, and this is larger than a microampere, which is only a millionth of an ampere. In fact, the milliampere is 1,000 times larger than the microampere.* Here is your next question.

Which of the following statements is true?

A. Wires have no resistance unless they are connected into battery circuits. (Proceed to block 13.)

B. The resistance in a piece of wire is so small that it may often (but not always) be ignored. (Proceed to block 24.)

7. *Your answer to the question in block 1 is **A**. This answer is wrong. Convert 5,000,000 microdollars to dollars by moving the decimal place to the left six places. Then decide if that amount of money will be enough to buy a $5,000 car.* Proceed to block 17.

8. *The correct answer to the question in block 22 is **B**. Moving the decimal place to the right six places as required shows that 0.000003 ampere is the same thing as 3 microamperes.* Here is your next question.

If you increase the amount of voltage across a light, the current through the light will

A. increase. (Proceed to block 25.)

B. decrease. (Proceed to block 18.)

9. *Your answer to the question in block 25 is **A**. This answer is wrong. When you connect a fuse across a component, it means that one side of the fuse goes to one side of the component and the other side of the fuse goes to the other side of the component. The fuse would be in the same position as the short circuit in Fig. 5-10. The resistance of a fuse is very low, and it would be a short circuit itself. However, the high current through the fuse would quickly burn it out.* Proceed to block 2.

10. *Your answer to the question in block 24 is **B**. This answer is wrong. If you move the decimal point to the left six places, you will find that 11,000,000 microamperes is the same as 11 amperes. An 11-ampere current will surely burn out a 10-ampere fuse.* Proceed to block 15.

11. *The correct answer to the question in block 5 is **A**. Remember that the second is the unit of time most often used in science. It is very important to note that an ampere is a coulomb per second regardless of the size of wire in which the current is flowing.* Here is your next question.

Which of the following is a larger amount of current flow?

A. 1 microampere (Proceed to block 21.)

B. 1 milliampere (Proceed to block 6.)

12. *Your answer to the question in block 16 is **B**. This answer is wrong.*

Remember that resistance is the opposition to current flow. If you increase the opposition, you will decrease the current. Proceed to block 20.

13. *Your answer to the question in block 6 is **A**. This answer is wrong. In practical wiring circuits the resistance of the wires is disregarded because it is so small compared with the resistance of other components in the circuit. However, this does **not** mean that there is no resistance.* Proceed to block 24.

14. *Your answer to the question in block 22 is **A**. This answer is wrong. If you move the decimal place to the right six places as required, you will find that 0.003 ampere is the same as 3,000 microamperes.* Proceed to block 8.

15. *The correct answer to the question in block 24 is **A**. Moving the decimal place to the left three places shows that 9,000 milliamperes is equal to 9 amperes. This amount of current will not burn out a 10-ampere fuse.* Here is your next question.
     Three effects of current flowing through resistance are a magnetic field, a voltage drop, and
     A. a tension on a spring. (Proceed to block 3.)
     B. heat. (Proceed to block 22.)

16. *The correct answer to the question in block 17 is **A**. The two things that determine the amount of current are voltage and resistance.* Here is your next question.
     If you increase the resistance in a circuit with a 6-volt battery,
     A. you will decrease the amount of current flow. (Proceed to block 20.)
     B. you will increase the amount of current flow. (Proceed to block 12.)

17. *The correct answer to the question in block 1 is **B**. Five million microdollars is the same as $5, which is not enough to buy a $5,000 car.* Here is your next question.
     Two things that determine how much electron current will flow in a circuit are the voltage and
     A. resistance. (Proceed to block 16.)
     B. amount of time. (Proceed to block 23.)

18. *Your answer to the question in block 8 is **B**. This answer is wrong. The amount of voltage across **any** resistance—including the light resistance—**directly** determines the amount of current flow.* Proceed to block 25.

19. *Your answer to the question in block 20 is **B**. This answer is wrong. An electric-clock motor does get warm when it is running, but this heat is not desired.* Proceed to block 5.

20. *The correct answer to the question in block 16 is **A**. Increasing the resistance causes a greater opposition to current flow. This results in a decrease of current.* Here is your next question.

Which of these appliances operates due to the fact that an electric current flowing through a resistor always produces heat?
A. An electric hot-water heater (Proceed to block 5.)
B. An electric clock (Proceed to block 19.)

21. *Your answer to the question in block 11 is* **A**. *This answer is wrong. A microampere is only one-millionth of an ampere. It is one of the smallest units of measurement used for current.* Proceed to block 6.

22. *The correct answer to the question in block 15 is* **B**. *The heat generated is not always desired, but it is always there.* Here is your next question.
Saying that there is a current of 3 microamperes in a circuit is the same as saying that there is a current of (*Hint:* Use the directions in Table 5-2 to convert the values in choices A and B to microamperes. Then compare with 3 microamperes.)
A. 0.003 ampere. (Proceed to block 14.)
B. 0.000003 ampere. (Proceed to block 8.)

23. *Your answer to the question in block 17 is* **B**. *This answer is wrong. The time would be important if you wanted to know the total number of electrons that flow in a circuit. However, electron current is a measure of the number of electrons* **per second,** *not the total number of electrons.* Proceed to block 16.

24. *The correct answer to the question in block 6 is* **B**. *Remember that there is always resistance, even though you can usually ignore it because it is so small.* Here is your next question.
The maximum current that could flow through a 10-ampere fuse continuously is (*Hint:* Use Tables 5-1 and 5-2 to convert these values to amperes.)
A. 9,000 milliamperes. (Proceed to block 15.)
B. 11,000,000 microamperes. (Proceed to block 10.)

25. *The correct answer to the question in block 8 is* **A**. *It is easy to remember that* **decreasing** *the voltage across a light to 0 volt* **reduces** *the current to 0.* **Increasing** *the voltage* **increases** *the current.* Here is your next question.
A fuse is connected into a circuit so that
A. it is across the resistance of the circuit. (Proceed to block 9.)
B. the circuit current flows *through* it. (Proceed to block 2.)

26. **True.** *When the fuse is burned out, it is like an automatic switch that stops current from flowing in the circuit.*
You have now completed the programmed questions. The next step is to put some of these ideas to work in laboratory experiments. Proceed to the "Experiments" section that follows.

Experiments

(The experiments described in this section may be performed on the circuit board described in Appendix C or on a similar laboratory setup.)

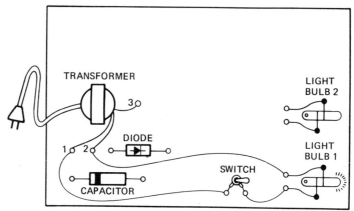

Figure 5-11. This basic test setup allows you to test the components to see if they are working.

EXPERIMENT 1

Purpose — To show that the amount of current flowing through a circuit depends upon the amount of voltage applied and also upon the amount of resistance in the circuit. You will also demonstrate that a short circuit causes an increase in circuit current.

Theory — You will remember that a light has a filament that offers resistance to current flow. Current flowing through the filament resistance causes it to heat — in fact, to become red hot. This is what causes the light bulb to glow.

Up to a point, the greater the amount of current flow, the greater the heat and light. However, if the current becomes too great, the light will burn out. (Even when the right amount of current flows through the light, it will eventually burn out because it ages.)

In the last chapter you used the light bulb to determine the amount of voltage across the light. Now you will use it to determine how much current is flowing.

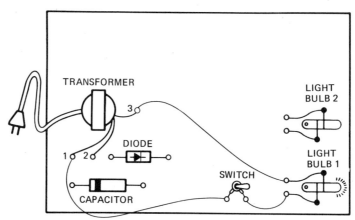

Figure 5-12. When the voltage across a lamp is reduced, less current flows through it, and it does not glow as brightly.

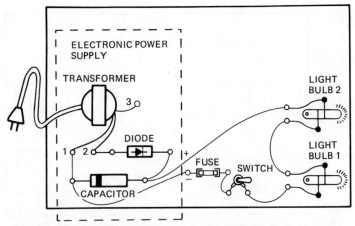

Figure 5-13. When the two lights are connected this way, it is necessary for the electron current to flow through both of them. This means that the opposition is greater than it would be if the current had to flow through only one light.

Test Setup—Wire the circuit as shown in Fig. 5-11. Insert the power transformer plug into the wall socket. Note that the light glows when the switch is closed. The reason for this setup procedure is that you must be sure that your voltage is present and that the light is working.

Procedure—

Step 1—Close the switch and observe the brightness of the light. There is 12 volts across the transformer when it is connected this way.

Step 2—Connect the light across the 6-volt source as shown in Fig. 5-12, and note that the light does not glow as brightly. This shows that lowering the voltage across a resistance reduces the current through it.

Step 3—Connect the circuit as shown in Fig. 5-13. The schematic drawing of this circuit is shown in Fig. 5-14.

Note that the lights do not glow as brightly. The arrows on Fig. 5-14 show the path of current through the circuit when the switch is closed. Note that the current must flow through two resistances instead of just one. The decrease in brightness is due to the fact that the current through the light circuit is lower. You can conclude, then, that when the current has to flow through two resistances, the resistance is greater and the current is less.

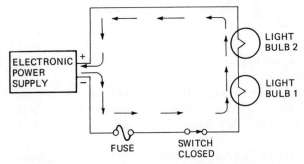

Figure 5-14. The schematic drawing of the circuit shown in Fig. 5-13.

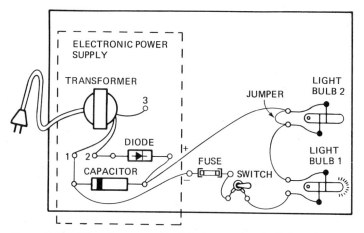

Figure 5-15. The wire across the light is sometimes called a "jumper" because it makes a temporary connection. In this case, it makes a "short circuit"—that is, a very low resistance path around a component.

Step 4 — Place a jumper across *one* of the lights as shown in Fig. 5-15.

CAUTION: BE SURE TO CONNECT THE JUMPER ACROSS *ONLY ONE* LIGHT AS SHOWN IN THE ILLUS-TRATION! IF YOU CONNECT IT ACROSS BOTH LIGHTS, YOU WILL CAUSE A SHORT CIRCUIT ACROSS THE POWER SUPPLY. AS YOU KNOW, THIS SHOULD *NEVER* BE DONE!

When the short circuit is across one of the lights, it does not glow, but the other one shines at full brightness. This shows that the short circuit has lowered the circuit resistance and raised the circuit current.

Conclusion—
1. The greater the voltage across a resistance, such as the light resistance, the greater the current flow.
2. The greater the resistance in the circuit, the lower the current flow.
3. Connecting two lights into a circuit so that the current has to flow through both makes the circuit resistance higher.
4. When there is a short circuit across a resistor, current flows around it. The circuit current increases.

EXPERIMENT 2

Purpose— To show that there is a magnetic field around a coil of wire.

Theory— In an earlier experiment you demonstrated that there is a magnetic field around a wire with current flowing in it. Now you are going to show that a coil of wire has a magnetic field as long as there is a current flowing through it.

Test Setup— Wire the circuit as shown in Fig. 5-16. The purpose of the two lights is to add resistance to the coil circuit. This limits the amount of current flowing through the coil. Too much current will cause the coil to overheat.

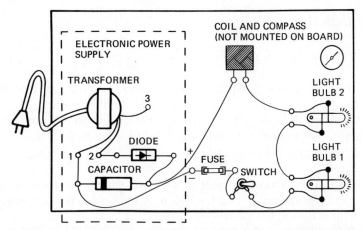

Figure 5-16. In this circuit the lights serve to limit the current flow. This is necessary because the delicate wires in the coil can be burned out when too much current flows through them. The compass will indicate that there is a magnetic field around the coil.

Procedure—

Step 1—With the switch closed move the compass around the sides and ends of the coil. Note that the compass deflects. This shows that there is a magnetic field around the coil when there is a current flowing through it.

Step 2—Stop the current in the circuit by opening the switch. Note that there is no magnetic field when the current stops.

Step 3—Insert a large iron nail (do not use an aluminum nail) through the coil as shown in Fig. 5-17. Note that when current flows through the coil, the nail is magnetized. Use it to pick up paper clips.

Conclusion—A coil becomes a magnet when there is current flowing through it. A nail through the center becomes magnetized by the coil when current is flowing through it. This is called an *electromagnet*.

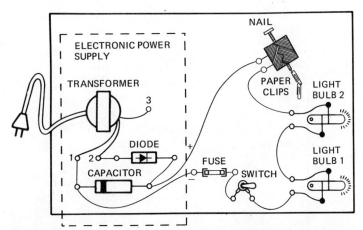

Figure 5-17. When the iron nail is inserted into the center of the coil, the flux lines flow easily through it. This makes the nail a temporary magnet. Removing the current from the coil reduces the flux to zero. (A small amount of magnetism may remain in the nail, but it will not be as strong as it is when the current is flowing.)

Self-Test with Answers

(Answers with discussions are given in the next section.)

1. A current of 1 ampere equals (*a*) 1 coulomb per hour; (*b*) 1 coulomb per minute; (*c*) 1 coulomb per second; (*d*) 1 coulomb per microsecond.
2. Which of the following is not necessarily the result of current flowing through a resistance? (*a*) Light must be produced; (*b*) There must be a voltage drop; (*c*) There must be a magnetic field; (*d*) There must be heat generated.
3. Increasing the current through a wire will always (*a*) increase its resistance; (*b*) increase the strength of its magnetic field; (*c*) decrease the voltage drop across it; (*d*) cause it to shrink.
4. Which of the following is the smallest amount of current? (*a*) 1 ampere; (*b*) 10 milliamperes; (*c*) 100 milliamperes; (*d*) 1,000 microamperes.
5. Which of the following is the largest amount of current? (*a*) 1 ampere; (*b*) 10 milliamperes; (*c*) 100 milliamperes; (*d*) 1,000 microamperes.
6. Which of these components is most like a fuse? (*a*) Coil; (*b*) Switch; (*c*) Light; (*d*) Resistor.
7. The amount of current that flows in a circuit depends upon the amount of resistance in the circuit and also upon (*a*) the distance of the resistors from the switch; (*b*) whether the switch is on the positive or negative line of the battery; (*c*) the time of day that the measurement is made; (*d*) the amount of voltage across the circuit.
8. The basic unit of measurement for current is the (*a*) ampere; (*b*) volt; (*c*) milliampere; (*d*) microampere.
9. Another way of expressing 90 millivolts is (*a*) 900 volts; (*b*) 9 volts; (*c*) 0.09 volt; (*d*) 0.00009 volt.
10. For the circuit of Fig. 5-18, which of these statements is true? (*a*) There is a short circuit across light *B*; (*b*) There will be no current flowing when the switch is closed; (*c*) Light *A* will burn out; (*d*) Light *A* will not glow.

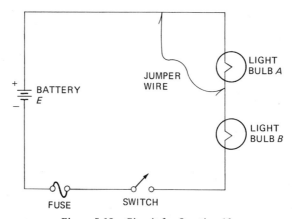

Figure 5-18. Circuit for Question 10.

Answers to Self-Test

1. (c) — As with most technical fields of study, the *second* is used as a standard unit of time in electricity.

2. (a) — There are many circuits that do not have light bulbs.

3. (b) — Figure 5-1 shows how the magnetic field increases with an increase in current. If you thought the *resistance* should increase, as in choice (a), you should know that resistance is a property of the *wire*, not the circuit. In other words, when you buy a length of wire you get the resistance with it.

4. (d) — In order to work a problem like this, convert all the choices to amperes using the procedure described in Tables 5-1 and 5-2.

5. (a) — See the comment for Question 4.

6. (b) — The switch is operated manually to open a circuit, and a fuse automatically opens a circuit when there is an overload.

7. (d)

8. (a) — Milliamperes and microamperes are both used for current, but the *ampere* is the *basic unit* of measuring current.

9. (c) — To answer a question like this, change 90 millivolts to volts and see which of the answers is correct. Use Table 5-1 for converting the millivolts to volts.

10. (d) — Study this question carefully and be sure you understand why each of the other choices is wrong.

6.
What Is an Ohm?

Introduction

You have studied about *voltage* and learned that it is a measure of the electrical pressure that forces the electricity through the circuit. You have also studied about the measurement of *current,* which is the flow of electrons in a circuit, and learned that the ampere is the basic unit of current measurement. In this chapter you will learn about another very important measurement in an electric circuit—that of *resistance.* The basic unit for measuring resistance is the *ohm.*

There was a time in history when scientists and inventors spent much effort trying to make a *perpetual-motion machine.* (A perpetual-motion machine is one that will run forever without the need for fuel and without the need for repairs.) Although many very ingenious devices were actually made, none could produce perpetual motion. The reason is that *all* machinery has friction. Friction may be defined as resistance to motion. When the parts of a machine rub together, they oppose the motion and produce a mechanical loss. This loss, which is usually called a *power loss,* cannot be recovered, and so power must *always* be supplied to a machine to replace the loss due to friction.

You know that heat is always produced when an electric current flows through a resistance. Resistance is sometimes called *electrical friction* because it causes heat just as mechanical friction does. Heat represents a power loss from the circuit. In some cases, such as electric stoves, the heat is desired, but it still is a loss of power in the circuit. The heat generated when there is current flowing through a resistance means that the lost power within the circuit must be replaced. In other words, electric power must *always* be delivered to an electric circuit to make up for the power lost in the form of heat. That is the purpose of the power supply.

Components in electric circuits are sometimes marked with a color-code system. By knowing how to read the color code, you can tell how many ohms of resistance a resistor has, that is, you can tell just how much opposition it will offer to the flow of current. The greater the resistance in ohms, the greater the amount of opposition. It is very important to know how much resistance there is in a circuit, because this is one of the two factors that determine how much circuit current is flowing. (The other factor is the voltage.)

You will be able to answer these questions after studying this chapter:

How much resistance is 1 ohm?
How are resistors made?
How can you tell the resistance value of a resistor?
What are variable resistors?
How are variable resistors connected into circuits?
Why are resistors made in different sizes?

Instruction

HOW MUCH RESISTANCE IS 1 OHM?

The *ohm* is the basic unit of resistance measurement. A unit of measurement is useless unless you know exactly how much that unit is. For example, if you want to measure length in terms of feet, you can obtain a yardstick or a ruler which can be used as a standard. If you want to measure electric current, you can (theoretically) count the number of electrons flowing in a circuit in 1 second. (More practically, you can measure the effects of the current flow and interpret them against some standard.)

In order to be useful, the unit of resistance, like any other unit of measurement, must have some realistic physical meaning. Figure 6-1 shows how the unit of resistance is related to other things.

As shown in Fig. 6-1*a*, a column of mercury 106.3 centimeters long and 1 square millimeter in cross section has a resistance of 1 ohm. As a matter of fact, this used to be the only standard used for obtaining the unit of resistance. By using this as a standard, one could construct a resistance of 1 ohm anyplace in the world. Once you have a standard unit for making 1 ohm, you can design meters for reading larger values.

Figure 6-1*b* shows a roll of number 10 copper wire. This roll of wire is 1,000 feet long and has a resistance of 1 ohm. You will notice that both ends of the wire are available on this spool. The reason for this is that you may want to know for sure that this is a full spool of wire. In order to find out, you could measure the resistance of the wire to determine its length. Remember this important point: *The resistance of wire is directly related to its length.* If you measured the resistance of the number 10 wire on the spool of Fig. 6-1*b* and found it to

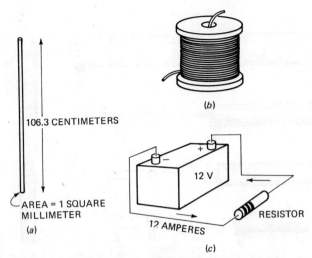

106.3 CENTIMETERS

AREA = 1 SQUARE MILLIMETER

(a)

(b)

12 V

12 AMPERES

RESISTOR

(c)

Figure 6-1. *Different ways of defining 1 ohm.* (a) *A column of mercury with these dimensions has a resistance of 1 ohm.* (b) *A roll of number 10 copper wire that is 1,000 feet long has a resistance of 1 ohm.* (c) *If 12 amperes of current flows when a 12-volt battery is connected across a resistor, the resistance is 1 ohm.*

be ½ ohm, you would know that it is not 1,000 feet long. If the size of the wire is different than number 10, you would have to consult a wire table to find out how much resistance there is per foot or per 1,000 feet before you could determine its length by this process. However, the important fact is that a length of wire could be used as a standard of resistance.

Figure 6-1*c* shows another way of interpreting 1 ohm of resistance. If a 12-volt battery is connected across a resistor and the resulting current flow is 12 amperes, then the resistor has a resistance of exactly 1 ohm. If the resistance is larger than 1 ohm, the current will be less than 1 ampere, and if the resistance is lower than 1 ohm, the current will be greater than 1 ampere. Thus, a known voltage and a known current can be used as a standard for 1 ohm.

It has been shown that the unit of resistance can be established in any of three ways. A physical tube of mercury can be made to certain specific dimensions and its resistance will be exactly 1 ohm. A length of wire can be used for establishing a resistance of 1 ohm. Finally, a known voltage will produce an exactly known current if the resistance in the circuit is 1 ohm.

Note that each of the methods used for establishing a resistance of 1 ohm is a physical thing. It is something that you could make anyplace in the world, and if you followed instructions carefully, you would get an exact value. All the units of measurement that you will use in electricity have a physically related meaning just as the ohm has.

HOW ARE RESISTORS MADE?

Resistors are components that are designed to introduce a desired amount of resistance into a circuit. The value of resistance is measured in ohms.

Resistors are used in circuits to *oppose current, produce a voltage drop,* or

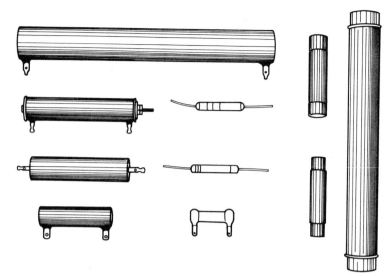

Figure 6-2. Some different types of resistors used in electric circuits.

produce heat. The application in which the resistor is to be used determines, to some extent, the way it is made. Most of the resistors in use today are made in one of two ways. One way is to pack some carbon granules together and attach metal connectors at each end. This type of resistor is heated and pressed so that it will hold its shape. Resistors that are made this way are called *carbon-composition resistors.*

Another way of making resistors is to wind a special type of resistance wire around an insulator form. Resistance wire is simply a wire that has a large amount of resistance in a short length through a nonconducting form. Resistors made this way are called *wire-wound resistors.* Most of the resistors that you will encounter in your work in electricity are either carbon-composition or wire-wound types. Figure 6-2 shows some of the different types of resistors used in electric circuits.

SUMMARY

1. The unit of resistance measurement is the ohm.
2. As with all units of measurement used in electricity, the ohm can be described in terms of physical things.
3. The resistance of a certain column of mercury, or the resistance of a certain length of wire, can be used to define an ohm.
4. A known voltage and current can also be used to define an ohm.
5. Resistors are used for introducing voltage drops, limiting current, and radiating heat.
6. One popular type of resistor is made with carbon granules. This type is called a *carbon-composition resistor.*
7. Another type of resistor is made by winding high-resistance wire on an insulator form. This type is called a *wire-wound resistor.*

HOW CAN YOU TELL THE RESISTANCE VALUE OF A RESISTOR?

Although all the resistors in Fig. 6-2 are different sizes, you cannot tell anything about the amount of resistance they have from this fact. In other words, the larger resistor does not have the most resistance.

One method of determining the value of resistance of a resistor is to measure it. An instrument used for measuring resistance is the *ohmmeter*.

Usually it is possible to determine the value of resistance by inspection. For example, some resistors have their resistance values stamped on their case. This is especially true of the larger types of resistors which are wire-wound types with porcelain coverings. Some smaller resistors have a catalog number printed on their case, rather than a value, and if you want to know their resistance, you can either look it up in a catalog or measure it with an ohmmeter.

The resistance of carbon-composition resistors is very often marked on the case with a color code. Figure 6-3 shows how the colors are marked in bands around the resistor. Here you see three bands of color, each color representing a different number in the color code. Before you can use the color code to determine the resistance for resistors, you must know what number each color stands for. This information is given in the first column of Table 6-1. Note that each color stands for a different number. For example, *brown* stands for the number *1*, and *green* stands for the number *5*. You may want to memorize this color code, because it is used in many areas of electricity other than for determining the resistance of resistors.

The first color band on the resistor, as shown in Fig. 6-3, tells the first digit of the resistance value—that is, the first number of the resistance value. The color code is always read with the bands on the left side of the resistor. Note that the three color bands (sometimes there are four) are crowded on the one side of the resistor. You should always turn the resistor so that it looks like the one in Fig. 6-3, so that the band on the left will be the first band.

The second color band stands for the second number of the resistance value, and the third band tells you how many zeroes you should add to the first two numbers to get the correct value. A few examples will show how this works.

> *Example 6-1*—What is the resistance value of a resistor that is color-coded as follows: *first band*—yellow; *second band*—violet; and *third band*—brown?
> *Solution*—The first number of the resistance value is indicated by the color for yellow, or the number 4. The second color band is indicated by the color violet, or the number 7. Therefore, we know that the first two digits of the resistance are 47. A third color band tells us how many zeroes to add to the two digits. The color code tells us that the third color band, which is

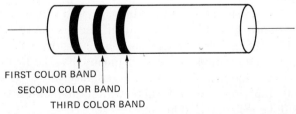

FIRST COLOR BAND
SECOND COLOR BAND
THIRD COLOR BAND

Figure 6-3. The resistance of a resistor can be determined by its color code.

TABLE 6-1. COLOR CODE CHART USED FOR DETERMINING RESISTANCE VALUES

Color Code	First Color Band *First Significant Figure*	Second Color Band *Second Significant Figure*	Third Color Band *Number of Zeros to Add*	Fourth Color Band *Tolerance*
0 Black	0	0	0	No color 20%
1 Brown	1	1	1	Silver 10%
2 Red	2	2	2	Gold 5%
3 Orange	3	3	3	
4 Yellow	4	4	4	
5 Green	5	5	5	
6 Blue	6	6	6	
7 Violet	7	7	7	
8 Gray	8	8	8	
9 White	9	9	9	

brown, stands for 1, which means that we must add *1 zero* to the first two digits. The result is 470, and the resistance of this resistor is 470 ohms.

*Example 6-2—*What is the resistance of a resistor that is color-coded as follows: *first band—*red; *second band—*red; *third band—*red?
*Solution—*All three of the colors on the resistor are red, and its resistance is determined as follows: Noting that the number 2 is indicated by red, we write 2 for the first number and 2 for the second number, giving us the number 22. The third color band tells us how many zeroes should be added to the number 22 to determine the resistance. The third color band is red, indicating that 2 zeroes should be added. The final result is 2,200 ohms, and this is the resistance of the resistor.

*Example 6-3—*What is the resistance of a resistor that is color-coded as follows: *first band—*brown; *second band—*green; *third band—*orange?
*Solution—*The first band is brown, which indicates the number 1, and the second color band is green, which indicates the number 5. We now have the first two numbers of the resistance value: 15. The third color band is orange, which stands for the number 3. This indicates that 3 zeroes are added to the 15 that was obtained for the first two digits. The result is a resistance of 15,000 ohms.

With a little practice you can read the resistance of a resistor from its color code very rapidly.

Sometimes the resistors have a fourth color band which indicates the tolerance of the resistance. It tells you how far off the resistance value can be and still be acceptable. When you buy a 15,000-ohm resistor, it does not mean that its value is *exactly* 15,000 ohms. Instead, it means that the manufacturer made that resistor so that it has *approximately* that resistance value. Its actual resistance could be 15,100 ohms, and it would still be within the manufacturer's tolerance, and it would still be called a 15,000-ohm resistor.

Some manufacturers make resistors with a fifth color band. This band is not useful to technicians. It tells the reliability of the resistor and is of importance primarily in military applications.

You might be surprised to learn that resistors are not highly accurate electrical components. You should understand, however, that it would be very difficult and very costly to manufacture resistors with exactly the resistance value indicated on the color code.

When there is no fourth band of color, the manufacturer says that the resistance value can be as much as 20 percent more or 20 percent less than the resistance indicated by the color code. If there is a fourth band that is silver-colored, the manufacturer says it is a better resistor and its value will be within 10 percent above and 10 percent below its resistance value as indicated by the color code. If the fourth color is gold, it is what is known as a *close-tolerance resistor,* and its value will not vary more than 5 percent above or 5 percent below the resistance value indicated by the color code.

The higher the accuracy of the resistor, the more money you have to pay for it. If you want to spend the money, you can buy a *precision-tolerance resistor* that will not vary more than 1 percent above or below the rated value. You can even buy resistors with more accurate values than that. But remember that when you purchase a resistor with a smaller tolerance, you have to pay more money for it than you pay for one that is not manufactured to accurate values. Many of the electric circuits that you will work with are made so that the resistance need not be highly accurate. In other words, if the resistance values are approximately correct, the circuit will still function the way it is supposed to. Tolerances of 1 percent or closer are usually written on the resistor case.

SUMMARY

1. There are different ways of determining the resistance of a resistor. One way is to measure its resistance with an ohmmeter.
2. For some types of resistors the resistance value is marked on the case, or a catalog number may be given so that you can look it up.
3. Some resistors have their resistance value marked in a color code. Each color of the color code represents a different number.
4. The first three bands of color on a resistor are used for identifying its value of resistance. A fourth band may be used for its *tolerance*—that is, for determining how close to its rated value you can expect its resistance to be.
5. In most electric circuits the value of resistance does not have to be exact. A variation of as much as 20 percent will not affect the operation of most circuits.
6. You can purchase resistors with very accurate resistance values. These are close-tolerance resistors or precision resistors. They cost more than resistors that do not have close tolerance values.

WHAT ARE VARIABLE RESISTORS?

All the resistors illustrated in Fig. 6-2 have a single value of resistance. They are called *fixed resistors* because their value cannot be changed. There are also *variable resistors.* When a resistor is variable, you can choose a resistance value according to the amount of rotation of its shaft, or according to some other method of varying its resistance. A volume control in a radio or a television set

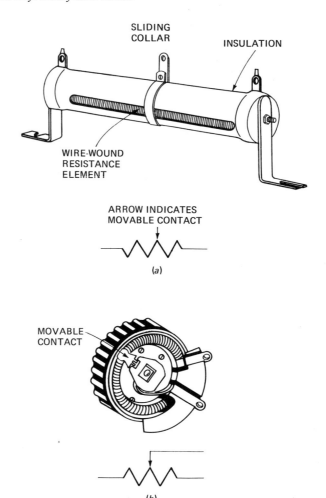

Figure 6-4. Two types of variable resistors. (a) **This type of variable resistor is sometimes called an adjustable resistor.** (b) **A wire-wound variable resistor.**

is actually a variable resistor. Also, the contrast and brightness controls of a television set are variable resistors. In industrial electricity, variable resistors are used for controlling motor speeds, the brightness of lights, and a number of other functions.

Figure 6-4 shows two types of variable resistors. The one in Fig. 6-4a is a heavy-duty wire-wound type. It is sometimes called an *adjustable resistor,* because it is necessary to use a screwdriver to vary its resistance value. The variable resistor in Fig. 6-4b is sometimes called a *rheostat.* (This term is also used to mean any variable resistor that controls current flow.) The resistor in Fig. 6-4b is specifically designed for use in circuits where it is necessary to vary the resistance in order to control or vary the amount of current flow. Other types of variable resistors can be used for this same purpose.

The variable resistor in Fig. 6-5 is a different type. Both of the resistors in Fig. 6-4 are wire-wound types, but the one in Fig. 6-5 is a *carbon-composition variable resistor.* In other words, the resistance element in the variable resistor is

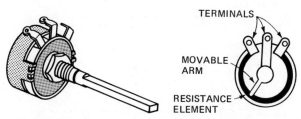

Figure 6-5. *This type of variable resistor is used in circuits with lower current.*

made of carbon granules. This type of resistor is often used in circuits where there is not as much current flowing as in the wire-wound types of circuits. Small wire-wound resistors in cases like the one in Fig. 6-5 are also available. They are usually somewhat more expensive than the carbon-composition types.

HOW ARE VARIABLE RESISTORS CONNECTED INTO CIRCUITS?

There are two different ways variable resistors are connected into circuits. One way is to use the variable resistor to control the amount of *current* through the circuit. In this type of connection, it is called a *rheostat.* A second way is to use a variable resistor to control the amount of *voltage* across some component, and in this connection, it is called a *potentiometer.*

Figure 6-6 shows how a variable resistor is connected into a circuit as a

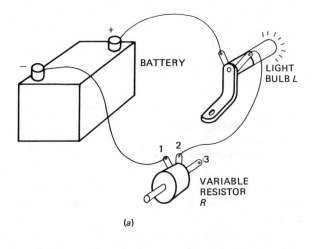

(a)

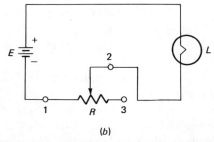

(b)

Figure 6-6. *A variable resistor used as a rheostat.* (a) *In this circuit the variable resistor controls the amount of current in the light.* (b) *The same circuit drawn with symbols.*

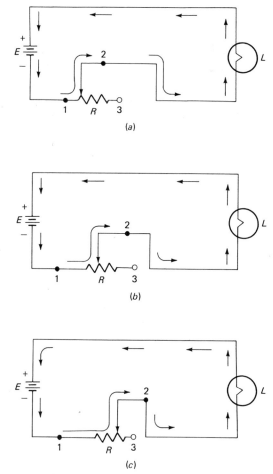

Figure 6-7. *How adjustment of the rheostat controls the current through the light.* (a) *The light is glowing at full brightness.* (b) *The light is dim.* (c) *The light is glowing at minimum brightness.*

rheostat. Figure 6-6*a* is a pictorial drawing of the circuit. Note that the variable resistor shown has three terminals, but only two of the terminals are used for the circuit connection.

It is somewhat easier to understand the circuit operation from the schematic drawing of Fig. 6-6*b*. Here the variable resistor, with terminals marked 1, 2, and 3, is placed in series with the light and the battery circuit. The arrow on the variable resistor symbol indicates the moving element, called the *arm*, for changing the resistance value. This moving element can be clearly seen in Fig. 6-6*b*.

The circuit operation of the rheostat is shown in Fig. 6-7. In Fig. 6-7*a* the arm of the rheostat is placed closest to terminal 1. The arrows show the flow of electron current. Note that it is not necessary for electron current to flow through any part of the resistor in order to go through the light. Therefore, the only opposition to the flow of current in this circuit is the light itself.

In Fig. 6-7*b* the arm of the variable resistor is moved down toward terminal 2 on the symbol. Now the arrows show that the electron current must flow through part of the variable resistance in order to go to the light. In other words, resistance has been added to the circuit, and this means that the amount of current will decrease. The overall result is that the light will not burn as brightly.

In the circuit of Fig. 6-7c, the arm of the variable resistor has been moved all the way toward terminal 3. Now, the electron current must flow through the complete resistor to flow through the light. The maximum amount of resistance has been added to the circuit when the arm of the variable resistor is in this position. Since there is a large amount of resistance added to the circuit, the minimum amount of current will flow, and the light will glow very dimly. In some cases, the light may be extinguished if the electron current is not sufficient to cause a glow with the variable resistor in this position.

The very simple circuit of Fig. 6-7 is used in many applications in electricity. For example, the dash lights in an automobile usually have some provision for changing their brightness. The connection for the dash light circuit is similar to the one shown in Fig. 6-6. When you turn the dash light control, you are actually changing the variable resistor in the circuit.

Figure 6-8 shows a different connection for a variable resistor. This circuit will be discussed only briefly at this time, but applications will be described as you increase your knowledge of electricity. Note that the variable resistor is connected across the battery so that both terminals 1 and 3 are connected in the circuit. This means that there is an electron current path, as shown by the arrows, and there will be a voltage drop between terminals 1 and 3 of the resistor. The voltage drop actually occurs all along the length of the resistor. Therefore, point *a* is more negative than point *b*, and point *b* is more negative than point *c*, and point *c* is more negative than point *d*. In other words, at any place along this variable resistor the voltage is slightly different than at any other point. Moving the arm of the variable resistor up and down between terminals 1 and 3 will cause the arm to pick off different voltage values. The voltage values at terminal 2 can be read with a *voltmeter*—an instrument for measuring voltage.

If the arm of the variable resistor in Fig. 6-8 is moved to terminal 1, then the voltmeter will read 0 volt. This is because the electron current does not have to flow through any part of the resistance at that point, and the voltage drop is zero. If the arm of the variable resistor is moved to terminal 3, then

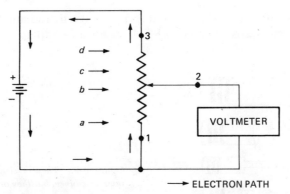

→ ELECTRON PATH

Figure 6-8. In this circuit the variable resistor is connected as a potentiometer.

the electron current must flow through all the resistance to reach that point, and the voltage drop will be maximum.

WHY ARE RESISTORS MADE IN DIFFERENT SIZES?

It has already been noted that you cannot tell the resistance value of a resistor from its size. In other words, the largest resistors do not necessarily have the largest resistance values. If this is true, then why make the resistors different sizes? Why not make them all a simple, single standard size?

The answer to this question is related to the fact that whenever an electron current flows through a resistor, heat is radiated. The amount of heat is directly related to the amount of current flow and also to the amount of resistance. If the resistor of a certain value—say 1,000 ohms—is placed in a circuit where very small current will flow through it, the amount of heat radiated will be very small. On the other hand, if you put the same resistor in a circuit with a large current flow, then a large amount of heat will be radiated.

The purpose of making the resistors different sizes, then, is related to the amount of heat that the resistor must radiate when it is connected into a circuit. Large resistors are made to radiate large amounts of heat, and small resistors are used in circuits where the amount of heat they are required to radiate is small.

All the resistors shown in Fig. 6-9 have the same resistance value in ohms. The only difference between them is the fact that the larger ones are designed to radiate more heat. You can use a large resistor in a circuit that has to dissipate a small amount of heat, but you cannot use a small resistor in a circuit that must dissipate a large amount of heat.

SUMMARY

1. Variable resistors are used when you want to be able to change the resistance value.
2. When a variable resistor is used to control the *current* in a circuit, it is called a *rheostat*.
3. When a variable resistor is used to control the *voltage* in a circuit, it is called a *potentiometer*. (It is also popularly called a "pot.")
4. It is possible for different sizes of resistors to have the same resistance value.
5. Larger resistors are used when it is necessary to radiate more heat, and smaller resistors are used when less heat must be radiated.

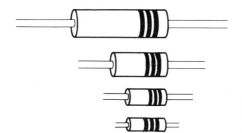

Figure 6-9. These resistors have the same resistance, but they are different in size.

Programmed Review Questions

(Instructions for using this programmed section are given in Chap. 1.)

We will review the important concepts of this chapter. If you have understood the material, you will progress easily through this section. Do not skip this material, because some additional theory is presented.

1. A variable resistor connected so that it can be used to control the circuit current is called a
   A. rheostat. (Proceed to block 7.)
   B. potentiometer. (Proceed to block 17.)

2. *Your answer to the question in block 14 is **B**. This answer is wrong. Resistance is opposition in an electric circuit, and this is not like fuel for an engine.* Proceed to block 9.

3. *Your answer to the question in block 16 is **A**. This answer is wrong. It is the physical size of the resistor that is related to how much heat it can radiate.* Proceed to block 11.

4. *Your answer to the question in block 11 is **A**. This answer is wrong. Consult Table 6-1 for the correct number for blue. Then proceed to block 6.*

5. *The correct answer to the question in block 12 is **A**. A gold-colored fourth band means that the actual resistance value will be within 5 percent of its rated value. Such a resistor would normally be more expensive than one that is only 20 percent accurate.* Here is your next question.
   One way of determining the resistance of a resistor is to measure it with
   A. an ohmmeter. (Proceed to block 14.)
   B. a test light. (Proceed to block 8.)

6. *The correct answer to the question in block 11 is **B**. Remember that color codes are used for other components besides resistors. However, the numbers represented by each color are the same as shown in Table 6-1.* Here is your next question.
   Most of the resistors used in electric circuits are either carbon-composition or
   A. plastic. (Proceed to block 15.)
   B. wire-wound. (Proceed to block 10.)

7. *The correct answer to the question in block 1 is **A**. A rheostat is a variable resistor used for controlling circuit current.* Here is your next question.
   The physical size of a resistor is an indication of
   A. the amount of heat that the resistor can safely radiate. (Proceed to block 12.)
   B. the accuracy (tolerance) to which the resistor is made. (Proceed to block 19.)

8.   *Your answer to the question in block 5 is* **B**. *This answer is wrong. A test light can be used to determine if there is a voltage present, but it cannot be used for measuring resistance values.* Proceed to block 14.

9.   *The correct answer to the question in block 14 is* **A**. *Resistance is similar to friction. In fact, resistance is sometimes called "electrical friction."* Here is your next question.
     A certain resistor is color-coded as follows: the first band is green, the second band is black, and the third band is brown. What is its resistance value?
     A.   500 ohms (Proceed to block 16.)
     B.   50,000 ohms (Proceed to block 18.)

10. *The correct answer to the question in block 6 is* **B**. *Wire-wound resistors are often used in electric circuits when the resistor is required to radiate a large amount of heat.* Here is your next question.
     For a 33,000-ohm resistor that is color-coded, what color should the third band be? _____ (Proceed to block 20.)

11. *The correct answer to the question in block 16 is* **B**. *The fourth band is used for the tolerance of the resistor. The lower the tolerance number, the more accurately the resistor is made.* Here is your next question.
     In the color code for resistors, blue represents the number
     A.   7. (Proceed to block 4.)
     B.   6. (Proceed to block 6.)

12. *The correct answer to the question in block 7 is* **A**. *The larger the size of the resistor, the greater the amount of heat it can radiate.* Here is your next question.
     Which would you expect to pay more for?
     A.   A resistor with a fourth color band of gold (Proceed to block 5.)
     B.   A resistor with no fourth color band (Proceed to block 13.)

13. *Your answer to the question in block 12 is* **B**. *This answer is wrong. When there is no color in the fourth band, the manufacturer says that its rated resistance value is only 20 percent accurate. This is not very accurate, and you would not expect to pay much for such a resistor.* Proceed to block 5.

14. *The correct answer to the question in block 5 is* **A**. *An ohmmeter is an instrument that measures the value of resistance in ohms.* Here is your next question.
     Resistance in an electric circuit is very much like
     A.   the friction in an engine. (Proceed to block 9.)
     B.   the fuel used for an engine. (Proceed to block 2.)

15. *Your answer to the question in block 6 is* **A**. *This answer is wrong. Plastic is an insulator, not a resistor material.* Proceed to block 10.

16. *The correct answer to the question in block 9 is* **A**. *The resistance value is obtained as shown in Fig. 6-10.* Here is your next question.

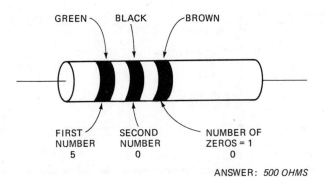

ANSWER: *500 OHMS*

Figure 6-10. How the resistance value for the problem in block 9 is determined.

The fourth band of a resistor color code is used for

A. indicating the amount of heat that the resistor can safely radiate. (Proceed to block 3.)
B. indicating the accuracy (tolerance) to which the resistor is made. (Proceed to block 11.)

17. *Your answer to the question in block 1 is* **B**. *This answer is wrong. A potentiometer is a variable resistor connected so that it controls voltage.* Proceed to block 7.

18. *Your answer to the question in block 9 is* **B**. *This answer is wrong. You have obtained the correct numbers in the first and second bands, but the third band is colored* **brown**. *This means that* **1** *zero is to be added to the first two numbers.* Proceed to block 16.

19. *Your answer to the question in block 7 is* **B**. *This answer is wrong. The physical size of a resistor does not tell you anything about its resistance value or its tolerance.* Proceed to block 12.

20. *The correct answer to the question in block 10 is* **orange**.
You have now completed the programmed questions. The next step is to put some of these ideas to work in laboratory experiments. Proceed to the "Experiments" section that follows.

Experiments

(The experiments described in this section may be performed on the circuit board described in Appendix C or on a similar laboratory setup.)

EXPERIMENT 1

Purpose—To show how a variable resistor can be connected as a rheostat for controlling light intensity.

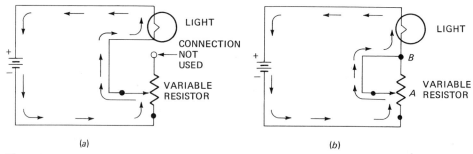

Figure 6-11. Two ways of connecting a three-terminal variable resistor as a rheostat. (a) *In this connection only two of the resistor terminals are used.* (b) *In this connection all three terminals of the resistor are used, but the action is the same.*

Theory — A variable resistor can be connected so that it controls the amount of current in the circuit. In this type of connection it is called a *rheostat*. Some variable resistors are specifically designed to be used as rheostats. An example is shown in Fig. 6-4*b*. Note that there are only two terminals on this variable resistor.

If a variable resistor has three terminals, it can be used as a rheostat by using only two of its terminals. There are two ways of doing this, and they are shown in Fig. 6-11. In Fig. 6-11*a* only two of the three terminals are used. In this connection the variable resistor is used in the same way as any two-terminal rheostat. The arrows show the path of electron current.

In Fig. 6-11*b* all three terminals are connected in the circuit. The arrows show the flow of electron current to be identical to the path taken in the circuit of Fig. 6-11*a*. The arm of the variable resistor is connected directly to point *b* by a piece of wire that has practically no resistance. The electrons will take this easy path rather than flow through the resistive path between points *a* and *b*.

The arm of a variable resistor is the part that is most likely to go bad. The arm usually becomes *open-circuited*. This means that the electrical connection between the arm and the resistor is no longer complete. This may happen at some positions of adjustment, or it may happen over the complete range of adjustment.

Note that in the connection of Fig. 6-11*a*, if the arm is open, then all the cir-

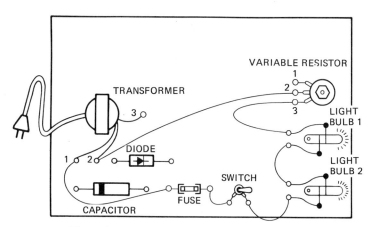

Figure 6-12. Circuit connections for Experiment 1.

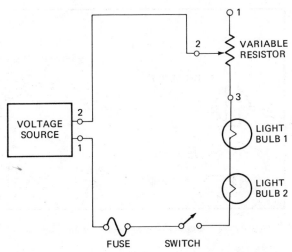

Figure 6-13.   The circuit of Fig. 6-12 shown with symbols.

cuit current stops. On the other hand, if the arm is open in the circuit of Fig. 6-11*b*, there is still a path for current to flow (through the resistance between *a* and *b*). From this, you can conclude that the difference between the two connections is that current stops in one when the arm is open, and it does not stop in the other when the arm is open.

Test Setup—Wire the circuit as shown in Fig. 6-12. Figure 6-13 shows this circuit connection in symbols. Always make sure that you can make the connections from the drawing in symbols as well as from the picture. Remember that in electrical work you will usually get a schematic like the one in Fig. 6-13, and you will be expected to trace the current paths and connections of the circuit from its schematic rather than from a picture drawing.

Procedure—

Step 1—Close the switch and adjust the variable resistor. If you have connected the circuit properly, turning the resistor shaft should change the brightness of the lights.

Step 2—Open the switch and rewire the variable resistor as shown in Fig. 6-14. Notice that in this connection all three terminals of the variable resistor are used. Figure 6-15 shows the circuit in schematic form.

Close the switch and note that the brightness of the lights can be controlled by turning the shaft.

Step 3—If you disconnect the wire leading to terminal 2 of the variable resistor, the lights do not glow. This is because the resistance in the circuit is so large that only a small amount of current will flow. This is not enough current to make the lights glow. However, you should understand that current may be flowing in a circuit whether or not you can see the evidence.

Conclusion—A variable resistor can be used to control the brightness of the lights by using it to control the amount of light current.

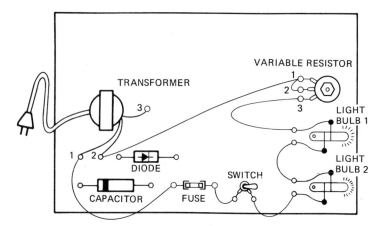

Figure 6-14. A second way of connecting the rheostat.

EXPERIMENT 2

Purpose – To show how a variable resistor can be connected as a potentiometer for controlling light intensity.

Theory – If you change the amount of current through a light, you change its brightness. This was demonstrated in the last experiment. Another way to change its brightness is to change the voltage across it.

If you change the amount of voltage across a light, you will change the amount of current through it. Remember that the amount of current that flows in a resistor depends upon two things: the amount of voltage across it, and the resistance of the resistor. The filament of a light is a specially made resistor that gets red hot when current flows through it.

In this experiment you will connect the variable resistor as a poten-

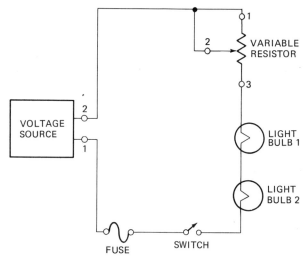

Figure 6-15. Schematic drawing for the test setup in Fig. 6-14.

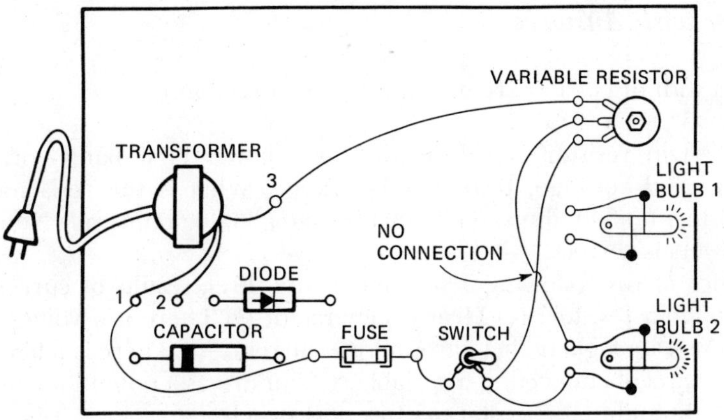

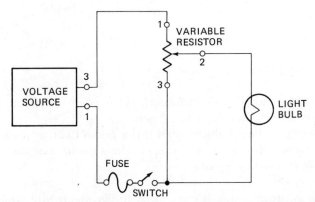

Figure 6-16. Connection of the variable resistor as a potentiometer.

tiometer. You will see that varying the resistor changes the brightness of the light as it did in Experiment 1.

Test Setup—Wire the circuit as shown in Fig. 6-16. Figure 6-17 shows the schematic drawing of the test setup. The parts and connections on the schematic are not in the same position as on the board, but the electrical connections are the same. This is a very important point, and you should study the two illustrations carefully until you are sure that the connections of the wires are the same.

Procedure—

Step 1—Turn the switch on and turn the variable resistor shaft. Note that the brightness of the lights can be changed by changing the resistance.

Conclusion—When the variable resistor is used as a potentiometer, it controls the amount of light.

Figure 6-17. Schematic drawing of the circuit in Fig. 6-16.

Self-Test with Answers

(Answers with discussion are given in the next section.)

1. A certain resistor is color-coded as follows: First band—orange; second band—orange; third band—orange. What is the resistance of this resistor? (*a*) 330 ohms; (*b*) 3,300 ohms; (*c*) 33,000 ohms; (*d*) None of these answers is correct.

2. Which of the following is not one of the three results of current flowing through a resistor? (*a*) Heat is generated; (*b*) There is a voltage drop; (*c*) There is a magnetic field around the current; (*d*) There is a light emitted.

3. Two ways of connecting a variable resistor are as a potentiometer and as a (*a*) light bulb; (*b*) rheostat; (*c*) fuse; (*d*) switch.

4. The movable part of a variable resistor is called the (*a*) eye; (*b*) arm; (*c*) leg; (*d*) font.

5. The resistance of a resistor can sometimes be determined by its color code. Another way is to measure it with (*a*) an ohmmeter; (*b*) a voltmeter; (*c*) a wattmeter; (*d*) an ammeter.

6. The color codes of two different resistors indicate that they have the same resistance value, but they are different sizes. Which of the following is true? (*a*) The larger one will limit the current flow to a smaller value; (*b*) The larger one can carry more current safely without being destroyed by heat; (*c*) The smaller one is much less expensive; (*d*) The color code is wrong.

7. Resistance in electric circuits is like (*a*) speed in an engine; (*b*) friction in an engine; (*c*) fuel in an engine; (*d*) the battery in a flashlight.

8. One practical and indirect way of telling how much wire of a given wire size is on a large spool is to (*a*) read its color code; (*b*) measure how much heat it radiates when current flows through it; (*c*) measure its diameter; (*d*) measure the total resistance of the wire and then determine its length by using a wire table.

9. Two popular types of resistors are the carbon-composition and the (*a*) copper wire; (*b*) wire-wound; (*c*) plastic; (*d*) silver.

10. The fourth band of the color code on a resistor tells (*a*) its cost; (*b*) the type of material it is made of; (*c*) how accurately its resistance value is made; (*d*) how much heat it can radiate.

Answers to Self-Test

1. (*c*)—Orange stands for the number 3 in the color code. The first and second digits are 33, and there are 3 zeroes.

2. (*d*)—There may be a light emitted if the resistor is the filament of a light bulb. However, this is a special case. Light is not *always* emitted when current flows through a resistor.

3. (*b*)

4. (*b*)—This is the part that is most likely to become troublesome.

5. (*a*) — The ohmmeter may be in the same case as a voltmeter and ammeter. The complete instrument is called a *multimeter*.

6. (*b*) — The size of a resistor is not an indication of how much resistance it has, but it does give an idea of how much heat it can radiate safely.

7. (*b*) — Resistance is sometimes called *electrical friction*.

8. (*d*) — A wire table will tell how many ohms a wire has for each 1,000 feet. You can measure the resistance of the wire and then determine how many thousand feet it has.

9. (*b*) — There are other types of resistors, but the carbon-composition and wire-wound types are the most popular.

10. (*c*) — The fourth band is called the *tolerance band*.

7. What Is Ohm's Law?

Introduction

One of the things that you have learned in your studies is that the amount of current flowing through a resistor depends upon the amount of voltage applied across that resistor. If you place a 3-volt battery across a 6-volt light (which is a resistor), there is not a sufficient amount of current flow to cause it to glow at full brightness. On the other hand, if you place a 12-volt battery across a 6-volt light, there will be an excessive amount of current flow, and the light will burn out. Thus, the voltage and the current are directly related in a circuit.

Another important relationship that you have learned is that the amount of current flowing in the circuit depends upon the amount of resistance in that circuit. If you place one light in a circuit across a battery, a certain amount of current flows. If you place two lights in such a way that the current must flow through both of them, then the resistance of the circuit is increased, and this causes a decrease in the amount of current flow. Thus, the amount of current flow is dependent upon both the amount of voltage and the amount of resistance in a circuit.

The voltage, current, and resistance of a circuit are related to each other by a simple relationship known as *Ohm's law*. One of the things that you will learn by studying this chapter is Ohm's law and how it can be used to determine the unknown quantity in a circuit.

You will be able to answer these questions after studying this chapter:

How is Ohm's law used for determining circuit current?
How is Ohm's law used for determining voltage?
How is Ohm's law used for determining resistance?
What is an easy way for remembering the equations for Ohm's law?
How do you get the values of voltage, current, and resistance into the basic units of volts, amperes, and ohms?
What is the effect of temperature on resistance?

Instruction

HOW IS OHM'S LAW USED FOR DETERMINING CIRCUIT CURRENT?

The amount of current that flows in a circuit is dependent upon two things: the amount of *voltage* across it and the amount of *resistance* in the circuit. Ohm's law expresses the relationship between voltage, current, and resistance as follows: *The amount of current (in amperes) that flows in a circuit is equal to the amount of voltage across the circuit (in volts) divided by the resistance of the circuit (in ohms).* Instead of expressing this relationship in words, it is much simpler to use symbols. Ohm's law for finding current in a circuit as expressed in symbols is

$$I = \frac{E}{R}, \text{ where} \tag{7-1}$$

$I =$ the current in *amperes*
$E =$ the voltage in *volts*
$R =$ the resistance in *ohms*

Figure 7-1 shows a simple circuit in which the current is unknown, but the voltage and the resistance are both given. Figure 7-1*a* shows how the circuit actually looks, and Fig. 7-1*b* shows the circuit in schematic symbols. In order to find the current in this circuit, a measurement *could* be taken with an ammeter. However, it can also be found by using Ohm's law [Eq. (7-1)]. The current (*I*) is unknown, but it can be found by dividing the voltage by the resistance as follows:

$$I = \frac{E}{R} \tag{7-1}$$

Substituting 10 volts for *E*, and 20 ohms for *R*:

$$I = \frac{10}{20}$$

$$= 0.5 \text{ ampere} \qquad \qquad \textit{Answer}$$

Thus, the amount of current flowing in the circuit of Fig. 7-1 is 0.5 ampere, or ½ ampere.

It is important to note that whenever the current is to be determined, two of the values—that is, the voltage and resistance—must already be known.

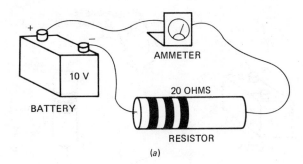

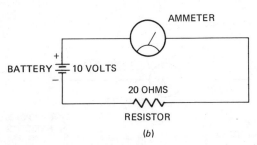

Figure 7-1. In this circuit the voltage and the resistor values are known. The problem is to find the current value. (a) ***Pictorial drawing of the circuit.*** (b) ***Schematic drawing of the circuit.***

HOW IS OHM'S LAW USED FOR DETERMINING VOLTAGE?

It is possible that you may know the amount of current flowing in a circuit as measured by an ammeter, and you may know the amount of resistance in a circuit. You wish to find the amount of voltage. Ohm's law can be restated for determining the voltage. *The amount of voltage across a circuit is equal to the amount of current flowing in the circuit (in amperes) multiplied by the resistance of the circuit (in ohms).* This can be more conveniently expressed in symbols as follows:

$$E = I \times R, \text{ where} \qquad (7\text{-}2)$$

E = the voltage applied across the circuit in volts
I = the amount of current flowing in the circuit in amperes
R = the resistance of the circuit in ohms

The multiplication sign ($\times$) is usually not written, but it is assumed to be present. Thus,

$$E = I \times R$$

means the same thing as

$$E = IR$$

Figure 7-2 shows a simple circuit in which the voltage is unknown. This circuit is shown pictorially in Fig. 7-2a, and Fig. 7-2b shows the circuit in schematic form. Note that the ammeter indicates that the current flow is 2 amperes, and the resistance of the circuit is known to be 10 ohms. The problem is to find the amount of voltage of the battery. Remember that there is only one value of

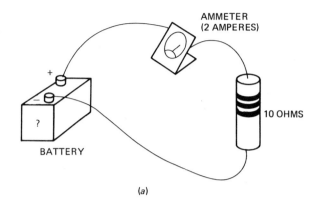

(a)

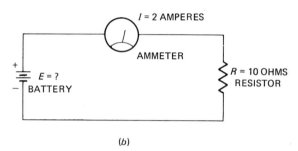

(b)

Figure 7-2. In this circuit the current and resistance values are known. What is the battery voltage? (a) *Pictorial drawing of the circuit.* (b) *Schematic drawing of the circuit.*

voltage that will cause 2 amperes to flow through a 10-ohm resistor. Ohm's law tells us that the voltage value can be determined simply by multiplying the current and the resistance values as follows:

$$E = I \times R \qquad (7\text{-}2)$$

Substituting 2 amperes for *I*, and 10 ohms for *R*:

$$E = 2 \times 10$$
$$= 20 \text{ volts} \qquad \qquad Answer$$

According to Ohm's law, then, if there is 2 amperes flowing through a 10-ohm resistor, the applied voltage must be 20 volts.

In the circuit of Fig. 7-2 we calculated the amount of applied voltage to the circuit, but Ohm's law can also be used to determine a voltage drop across a resistor. Consider the situation shown in Fig. 7-3. In this illustration the current through a known resistance value is given, but we do not know anything about the rest of the circuit. Our only interest here is to *find the amount of voltage across the resistance.*

Figure 7-3*a* shows a pictorial drawing of the circuit with an ammeter measuring the current flow through the resistor. Note in this illustration that the current must flow through the ammeter in order for the ammeter to make the measurements. Figure 7-3*b* shows the circuit in schematic form. We are not interested in any other part of the circuit. We only want to find the voltage drop across the resistor. According to Ohm's law, we can find this voltage drop

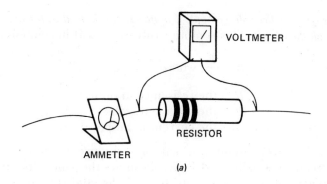

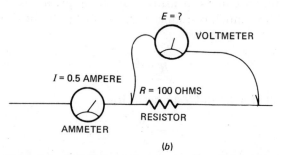

Figure 7-3. *In this circuit the current and resistance values are known. The problem is to find the voltage drop.* (a) **Pictorial drawing of the circuit.** (b) **Schematic drawing of the circuit.**

by simply multiplying the current flow times the resistance value as follows:

$$E = I \times R \qquad\qquad (7\text{-}2)$$

Substituting 0.5 ampere for *I,* and 100 ohms for *R:*

$$E = 0.5 \times 100$$
$$= 50 \text{ volts} \qquad\qquad Answer$$

Thus, in the circuit of Fig. 7-3 the amount of voltage drop across the resistor must be 50 volts.

SUMMARY

1. The voltage, current, and resistance in a circuit are related by Ohm's law.
2. When the voltage and resistance are known, the current can be determined by dividing the resistance value into the voltage value ($I = E/R$).
3. When the current and resistance values are both known, the voltage can be determined by multiplying the amperes times the ohms ($E = IR$).
4. The equation $E = IR$ can be used for finding the applied voltage in a circuit, or it may be used for finding the voltage drop across a resistor.

HOW IS OHM'S LAW USED FOR DETERMINING RESISTANCE?

If the current and the voltage in a circuit are both known, the resistance can be found by applying Ohm's law. Stated in words, Ohm's law says that *the resistance in a circuit is equal to the voltage applied to the circuit divided by the amount of current flowing through the circuit.* This is more easily expressed in symbols as

$$R = \frac{E}{I}, \text{ where} \tag{7-3}$$

$R =$ the resistance in ohms
$E =$ the voltage in volts
$I =$ the current in amperes

Figure 7-4 shows a circuit in which the resistance is unknown. Figure 7-4*a* shows the circuit pictorially, and Fig. 7-4*b* shows the schematic drawing. Note that the applied voltage is known to be 12 volts, and the amount of current flowing through the circuit as measured by the ammeter is 2 amperes. The problem is to find how much resistance there is in the circuit. This can be done by applying Ohm's law as follows:

$$R = \frac{E}{I} \tag{7-3}$$

Substituting 12 volts for *E*, and 2 amperes for *I*:

$$R = \frac{12}{2}$$

$$= 6 \text{ ohms} \hspace{3cm} Answer$$

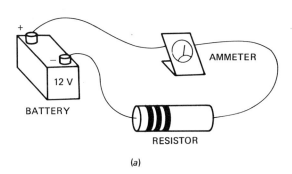

(a)

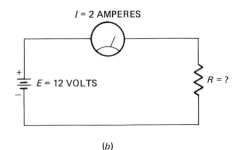

(b)

Figure 7-4. **In this circuit the voltage and current are known. The problem is to find the resistance.** (a) **Pictorial drawing of the circuit.** (b) **Schematic drawing of the circuit.**

Thus, the amount of resistance in a circuit can be determined when its voltage and current are both known.

WHAT IS AN EASY WAY FOR REMEMBERING THE EQUATIONS FOR OHM'S LAW?

There are three different factors to be considered in a circuit: *voltage, current,* and *resistance.* If you know any two of these, the third can be determined by Ohm's law. A different equation is needed for finding each of the factors when the other two are known. Figure 7-5 shows a convenient way of being able to write the Ohm's law equations. If you memorize the simple symbol diagram shown in Fig. 7-5a, it will not be necessary to memorize the three different versions of Ohm's law.

The symbol diagram of Fig. 7-5a indicates the three factors which must be determined in the circuit. *When you cover the one you wish to find, the other two are in the proper relationship for the formula.* For example, suppose you know the voltage and the resistance, and you wish to find the current in the circuit. Figure 7-5b shows that when you cover the current (I) in the symbol, the other two values are arranged for division: E/R. Thus you can write that I (the quantity covered by your finger) is equal to E/R.

Suppose you know the current and the resistance in the circuit, and you wish to find the voltage. Using the symbol as shown in Fig. 7-5c, you cover the voltage (E) with your finger, because that is the quantity which is unknown. Note that I times R remains. Thus, you can write that the voltage (the unknown) is equal to I times R.

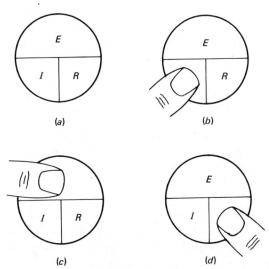

Figure 7-5. *The symbol in Fig. 7-5* (a) *is useful for working Ohm's law problems. Use of the symbol is illustrated in Figs. 7-5* (b), (c), *and* (d). (a) *This symbol will be helpful for working Ohm's law problems.* (b) *Covering* **I** *shows that it can be calculated by dividing* **R** *into* **E.** (c) *Covering* **E** *shows that it is equal to* **I** *times* **R.** (d) *Covering* **R** *shows that it is equal to* **E/I.**

Suppose you wish to know the resistance of the circuit when you already know the voltage and the current. Figure 7-5d shows how this symbol can be used. Covering the resistance (R) with your finger because that is the quantity you want to know, you see that the voltage is divided by the current. You can then write the equation $R = E/I$.

If you will take a few seconds to memorize the symbol diagram of Fig. 7-5a, it will be helpful in working Ohm's law problems. After you have used the simple Ohm's law equations a number of times, you will find that you have memorized all three forms, and the symbol diagram will no longer be required.

SUMMARY

1. When the voltage and the current are both known, the resistance can be determined by dividing the volts by the amperes ($R = E/I$).
2. There are three different ways of writing Ohm's law:

$$I = \frac{E}{R}$$

$$E = I \times R$$

(This may be written without the multiplication sign. IR means $I \times R$.)

$$R = \frac{E}{I}$$

3. The choice of which version to use depends upon which two quantities are known and which is to be determined.
4. The symbol diagram shown in Fig. 7-5a makes it an easy matter to determine one of the circuit factors when the other two are known. In order to use this symbol diagram, cover the value you wish to find, and the other two factors are shown in the proper relationship for finding that value.

HOW DO YOU GET THE VALUES OF VOLTAGE, CURRENT, AND RESISTANCE INTO THE BASIC UNITS OF VOLTS, AMPERES, AND OHMS?

In the simple equations you have been given for Ohm's law, it is important to note that the voltage must always be in volts, not in kilovolts or megavolts. The current in the circuit must always be given in amperes, not milliamperes or microamperes. Finally, the resistance in the circuit must always be given in ohms, not kilohms or megohms.

You will remember in your study of electric current that the ampere is much too large a unit of measure for many applications, and milliamperes and microamperes are used instead. A milliampere is a thousandth of an ampere, and a microampere is a millionth of an ampere.

In the case of resistors, the ohm is generally too small a unit, and larger units are more convenient. The term "kilo" (or its abbreviation k) is a convenient way of saying a thousand. Thus, instead of saying that a resistor has a value of 50,000 ohms, it is generally referred to as a 50k or 50-kilohm resistor. The word "mega" is a convenient way of writing a million. It is sometimes abbreviated M. Instead of saying that a resistor has a value of 5,000,000 ohms, the value is generally expressed as 5 megohms or 5 M.

TABLE 7-1. CONVENIENT REFERENCE FOR CHANGING UNITS OF MEASUREMENT

To Change from	To	Move the Decimal Point to the
Mega units	units	right 6 places
Kilo units	units	right 3 places
Milli units	units	left 3 places
Micro units	units	left 6 places
Units	mega units	left 6 places
Units	kilo units	left 3 places
Units	milli units	right 3 places
Units	micro units	right 6 places

For use with Ohm's law, it is necessary to express resistors in *units* (ohms) of measurement rather than kilo or mega units. Table 7-1 shows how to convert between the units. Some examples will show how this is done.

> *Example 7-1* — A certain resistor has a value of "50k" stamped on its case. How many ohms of resistance does this resistor have?
> *Solution* — The resistor has a value of 50 kilohms. According to Table 7-1, in order to convert kilohms to ohms, move the decimal place to the right three places. This is shown in the first row of the table, but the word "units" is used instead of ohms. This is so the table can also be used for kilovolts, kilowatts, etc. Therefore, 50 kilohms equals 50,000 ohms.

> *Example 7-2* — A certain resistor has a resistance value of 5,000,000 ohms. How many megohms resistance does it have?
> *Solution* — Table 7-1 tells you that in order to convert from ohms to megohms, move the decimal point to the left six places. (This information is given in the fifth row of the table.) Therefore, 5,000,000 ohms equals 5 megohms.

The words "kilo" and "mega" are used in other units besides resistance. For example, a very high voltage circuit may be rated as having a voltage of 5 kilovolts. To convert kilovolts to volts, the procedure is exactly the same as converting kilohms to ohms. According to Table 7-1, kilovolts can be converted to volts by moving the decimal place to the right three places. Thus, 5 kilovolts = 5,000 volts.

The first step in solving an Ohm's law problem should be to get the known values into volts, amperes, and ohms. Table 7-1 summarizes the procedure for converting from the so-called simplified units to the basic units of measurement for use with Ohm's law. You will need to refer to this table whenever you are working a problem in which the quantities are not given in basic units. As with the equations for Ohm's law, after a while you will get so that you can make the conversions without referring to the table.

WHAT IS THE EFFECT OF TEMPERATURE ON RESISTANCE?

For most of the problems that you will work in Ohm's law, you can assume that the value of resistance is unchanged, regardless of the value of current that flows through it. However, you should understand that this is an assumption which is made for convenience in working in Ohm's law problems, and is not necessarily true.

You know that whenever a current flows through a resistance value, heat is always generated. The amount of heat depends upon the amount of current flow and the amount of resistance. When a metal or other material becomes heated, *its electrical resistance may change!* There are two different ways in which it can change. One way is that a temperature rise can cause an increase in resistance. When a metal or other material has this characteristic, it is said to have a "positive temperature coefficient." The second way occurs for some types of materials when an increase in temperature causes a *decrease* in the resistance of the material. When this happens the material is said to have a "negative temperature coefficient."

Suppose you are working with a circuit in which the resistance is said to be 10 ohms. If you place a 10-ohm resistor in a circuit with current flowing through it, the current will heat the resistor and change its resistance value very slightly. For most applications, the amount of change in resistance is so small that it can be ignored. However, you should always remember that there is a possibility that the temperature will cause a resistance change which cannot be neglected, and *when this happens you cannot use Ohm's law for solving the problem!* A resistor is said to be "nonlinear" if a small change in temperature causes a large change in the resistance value. Remember this: *You cannot work a problem with Ohm's law if the resistance is not linear—that is, if the resistance value changes widely with relatively small changes in temperature.*

None of the resistances that you work with in this book will be nonlinear. This precaution is only given for future reference.

SUMMARY

1. In order to use Ohm's law, the circuit quantities must be expressed in basic units of volts, amperes, and ohms.
2. Table 7-1 can be used for converting from convenient units to the basic units required for use with Ohm's law.
3. Resistances are conveniently expressed in kilohms or megohms. They must be converted to ohms if they are to be used in Ohm's law equations.
4. If an increase in temperature causes the resistance of a metal or other material to *increase*, it is said to have a "positive temperature coefficient."
5. If an increase in temperature causes the resistance of a material to *decrease*, it is said to have a "negative temperature coefficient."
6. When a small change in temperature causes a large change in the resistance of a resistor, it is said to be "nonlinear."
7. Ohm's law does not work for circuits that have nonlinear resistors.

Programmed Review Questions

(Instructions for using this programmed section are given in Chap. 1.)

We will review the important concepts of this chapter. If you have understood the material, you will progress easily through this section. Do not skip this material, because some additional theory is presented.

1. Which of these equations is correct?

   A. $R = \dfrac{E}{I}$ (Proceed to block 7.)

   B. $R = \dfrac{I}{E}$ (Proceed to block 17.)

2. *The correct answer to the problem in block 23 is* **B**.

$$I = \frac{E}{R} = \frac{12}{30} = 0.4 \ ampere$$

   Here is your next question.
   When a current of 1 ampere flows through a 10-ohm resistor, the voltage drop across the resistor is
   A.   0.1 volt. (Proceed to block 24.)
   B.   10 volts. (Proceed to block 10.)

3. *The correct answer to the question in block 10 is* **A**. *In fact, this is the definition of a positive temperature coefficient.* Here is your next question. A resistor is marked "4.7 M" on its case. How many ohms of resistance does it have?
   A.   4,700,000 ohms (Proceed to block 18.)
   B.   4,700 ohms (Proceed to block 11.)

4. *The correct answer to the question in block 18 is* **B**. *A resistance of 20 ohms will result in a larger current flow than a resistance of 30 ohms. Always remember that* **decreasing the circuit resistance results in increasing the circuit current provided the battery voltage is the same.** Here is your next question.
   If there is a 10-volt drop across a 250-ohm resistor, how much current must be flowing through that resistor?
   A.   0.04 ampere (Proceed to block 19.)
   B.   25 amperes (Proceed to block 20.)

5. *The correct answer to the question in block 19 is* **A**.

$$R = \frac{E}{I} = \frac{6}{0.1} = 60 \ ohms$$

   Here is your next question.
   When the resistance of a resistor varies widely with small changes in temperature, it is said to be
   A.   nonlinear. (Proceed to block 22.)
   B.   linear. (Proceed to block 15.)

6. *Your answer to the question in block 22 is* **B**. *This answer is wrong. If you converted 5 kilohms to 5,000 ohms, then divided this value into the voltage, you obtained a value of 0.001. This is the circuit current in* **amperes,** *but the question asks for the current in milliamperes. Consult Table 7-1 for the procedure for converting amperes to milliamperes.* Then proceed to block 8.

7. *The correct answer to the question in block 1 is **A**. This equation shows that circuit resistance can be found by dividing the circuit current into the applied voltage.* Here is your next question.
Ohm's law cannot be used for finding the current through a resistor if the resistor is
A. linear. (Proceed to block 21.)
B. nonlinear. (Proceed to block 14.)

8. *The correct answer to the question in block 22 is **A**.*

$$I = \frac{E}{R} = \frac{5}{5,000} = 0.001 \ ampere = 1.0 \ milliampere$$

Here is your next question.
What is the Ohm's law equation for finding the voltage across a resistor when you know the resistance value and the current through the resistor? _____ (Proceed to block 26.)

9. *Your answer to the question in block 23 is **A**. This answer is wrong. The current in a circuit is found by dividing the voltage (12 volts) by the resistance (30 ohms). Rework the problem.* Then proceed to block 2.

10. *The correct answer to the question in block 2 is **B**. E = I × R = 1 ampere × 10 ohms = 10 volts.* Here is your next question.
If the resistance of a material increases when its temperature increases, it is said to have
A. a positive temperature coefficient. (Proceed to block 3.)
B. a negative temperature coefficient. (Proceed to block 25.)

11. *Your answer to the question in block 3 is **B**. This answer is wrong. According to Table 7-1, you convert megohms to ohms by moving the decimal point to the right six places. Rework the problem.* Then proceed to block 18.

12. *Your answer to the question in block 18 is **A**. This answer is wrong. **The larger the resistance, the lower the current.** Remember that the larger resistor offers greater opposition to current flow.* Proceed to block 4.

13. *Your answer to the question in block 19 is **B**. This answer is wrong. Did you use 100 milliamperes in the equation for Ohm's law? Remember that you must only use volts, **amperes**, and ohms in Ohm's law equations. Table 7-1 shows that you can convert 100 milliamperes to amperes by moving the decimal point to the left three places. Thus, 100 milliamperes = 0.1 ampere. Use this value in the equation*

$$R = \frac{E}{I}$$

and rework the problem. Proceed to block 5.

14. *The correct answer to the question in block 7 is **B**. When a resistor is nonlinear, it is not possible to determine how much current flows through it by*

using Ohm's law. For the problems that you will work in this book, only linear resistors will be considered. Here is your next question.
Which of the following memory aids is correctly drawn?
A.   The one in Fig. 7-6 (Proceed to block 23.)
B.   The one in Fig. 7-7 (Proceed to block 16.)

15.   *Your answer to the question in block 5 is **B**. This answer is wrong. If the resistance value varies widely with small changes in temperature, then the resistor is nonlinear.* Proceed to block 22.

16.   *Your answer to the question in block 14 is **B**. This answer is wrong. If you cover I with your finger to indicate the quantity to be found, the symbol in Fig. 7-7 indicates that the procedure is to multiply E by R. But this is wrong, since I = E/R. Thus, the symbol is not correct.* Proceed to block 23.

17.   *Your answer to the question in block 1 is **B**. This answer is wrong. The resistance of a circuit is equal to the voltage divided by the current. Review the memory symbol shown in Fig. 7-5a.* Then proceed to block 7.

18.   *The correct answer to the question in block 3 is **A**. The value 4.7 megohms is converted to ohms by moving the decimal point to the right six places. This procedure is described in Table 7-1.* Here is your next question.
Consider the circuits of Figs. 7-9 and 7-10. Which circuit will have the larger current flow?
A.   The circuit in Fig. 7-9 will have the larger current. (Proceed to block 12.)
B.   The circuit in Fig. 7-10 will have the larger current flow. (Proceed to block 4.)

19.   *The correct answer to the question in block 4 is **A**.*

$$I = \frac{E}{R} = \frac{10}{250} = 0.04 \; ampere$$

Here is your next question.
When a voltage of 6 volts is applied across a light bulb, the current flow is 100 milliamperes. What is the resistance of the light bulb?
A.   60 ohms (Proceed to block 5.)
B.   0.06 ohm (Proceed to block 13.)

20.   *Your answer to the question in block 4 is **B**. This answer is wrong. The current is found by dividing the voltage by the resistance (I = E/R). Rework the problem.* Then proceed to block 19.

21.   *Your answer to the question in block 7 is **A**. This answer is wrong. A linear resistor is the only type of resistor that **can** be used in Ohm's law problems.* Proceed to block 14.

22.   *The correct answer to the question in block 5 is **A**. In fact, this is a definition of a nonlinear resistor.* Here is your next question.

Figure 7-6. Is this memory aid correctly drawn?

Figure 7-7. Is this memory aid correctly drawn?

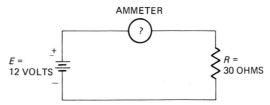

Figure 7-8. Determine the current in this circuit.

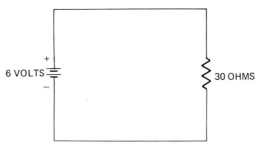

Figure 7-9. This circuit path is closed, and current is flowing.

Figure 7-10. Will the current in this circuit be greater or less than the current in the circuit of Fig. 7-9?

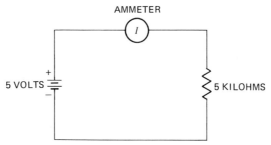

Figure 7-11. The circuit for the problem in block 22.

In the circuit of Fig. 7-11, find the current flow *in milliamperes.*
A.   1.0 milliampere (Proceed to block 8.)
B.   0.001 milliampere (Proceed to block 6.)

23.   *The correct answer to the question in block 14 is **A**. This symbol is a convenient way to remember the relationships between voltage, current, and resistance. After working a number of problems in Ohm's law, you will find that you no longer need the symbol.* Here is your next question.
In the circuit of Fig. 7-8, what is the circuit current?
A.   2.33 amperes (Proceed to block 9.)
B.   0.4 ampere (Proceed to block 2.)

24.   *Your answer to the question in block 2 is **A**. This answer is wrong. The voltage across a resistor is obtained by multiplying the resistance value by the current through the resistor. Rework the problem.* Then proceed to block 10.

25.   *Your answer to the question in block 10 is **B**. This answer is wrong. If a material has a negative temperature coefficient, its resistance decreases when its temperature increases.* Proceed to block 3.

26.   *The equation is **E** = **IR**. Sometimes the symbol* V *is used to indicate a voltage drop and* E *is used to indicate a power supply voltage. If you used this method to answer the question in block 8, the answer would be **V** = **IR**. This indicates that the answer is a voltage drop rather than a source voltage.*
You have now completed the programmed questions. The next step is to put some of these ideas to work in laboratory experiments. Proceed to the "Experiment" section that follows.

Experiment

(The experiment described in this section may be performed on the circuit board described in Appendix C or on a similar laboratory setup.)

EXPERIMENT 1

Purpose — To become familiar with circuit grounding and the use of bus lines.

Theory — For a complete circuit it is necessary for a conductor to carry the electrical energy *from* the *power source to* the *load*, and it is also necessary for a conductor to return the energy *from* the *load* back to the power source. Unless the electrical path is complete around the circuit, the circuit will not work.

The fact that there must be a conducting path both *to* and *from* the load seems to indicate that there must be two conductors connected to the load. There are two wires used for carrying power to a light or to a television set, and this supports the idea that two conductors must be present for the load.

In some cases it *appears* that there is only one conductor used for carrying electricity to the load. Take, for example, the clearance light (used for trucks) which is shown in Fig. 7-12. There is only one wire connected to this light. This wire is connected to the battery through a switch and is used for carrying

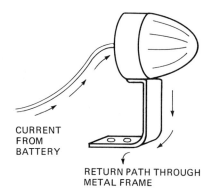

Figure 7-12. A clearance light of the type used on trucks. Note that there is only one wire for this light.

CURRENT
FROM
BATTERY

RETURN PATH THROUGH
METAL FRAME

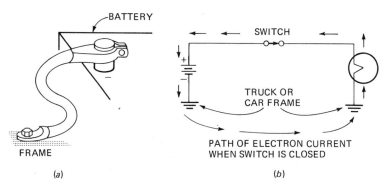

Figure 7-13. Instead of using two wires, the frame of the car or truck is used for a return path. Thus, only one wire is needed. (a) *One of the battery cables is connected to the car (or truck) frame. This is called the* **ground connection.** (b) *Schematic of light circuit with a common ground connection.*

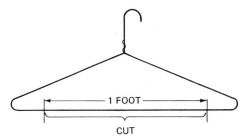

Figure 7-14. A 1-foot section of coat hanger will be used for a busbar.

energy from the battery to the light. The electrical return path is through the frame of the light, and through the metal frame of the truck, as shown by the arrows in Fig. 7-12.

The electrical connection and current path for the clearance light are shown in Fig. 7-13. Figure 7-13*a* shows the electrical connection of the battery with one terminal connected to the frame. Figure 7-13*b* shows the electric circuit for the system. The return path is referred to as the system *ground*—not to be confused with earth ground.

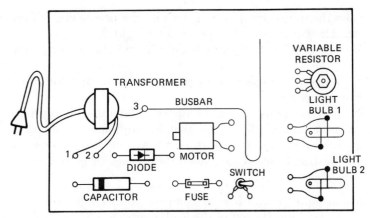

Figure 7-15. Layout plan for the busbar.

In some circuits a *bus line* (also called a *busbar*) is used for a return path. This is simply a conductor, usually a bare wire, copper bar, or copper tube, that is used for the return path to the generator. In this experiment you will lay out an electrical system using a busbar.

Test Setup — Scrape the paint or other coating off a coat-hanger wire. Cut a straight length of the wire about 1 foot long from the hanger, as shown in Fig. 7-14. You are going to use this length of wire as a busbar.

Bend and fasten the busbar onto the board, using string through the holes of the board as shown in Fig. 7-15.

Procedure —

Step 1 — Connect the busbar to pin 3 of the transformer. Also, connect one lead of the motor and each of the lights to the busbar.

Step 2 — Finish connecting the components as shown in Fig. 7-16. Notice that only one lead goes to each component. Figure 7-16 shows the circuit in schematic form.

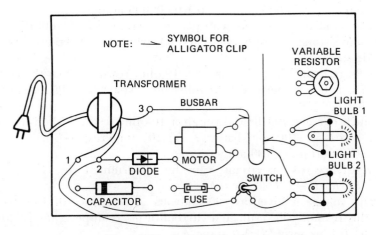

Figure 7-16. Test setup for the experiment.

Step 3 — When the transformer is plugged into the power line, all the components operate. The switch must be operated for light 2.

Step 4 — Disconnect light 1. Note that light 2 and the motor continue to operate.

Step 5 — Disconnect the busbar from pin 3 and note that the motor and light 2 both cease to operate.

This is an important point. If *both* of the components stop functioning, it must be due to some part of the circuit that is common to both parts. This could be the power source or the transformer. As you just demonstrated, it could also be the busbar, which is also called the *common* or *ground* line.

Conclusion — By using a ground line or common connection, you reduce the number of wires in a circuit. If each of the leads connected to the busbar in Fig. 7-16 was returned separately to pin 3 of the transformer, more wire would be needed. This would make the circuit more expensive, and the wiring would not be as neat.

Self-Test with Answers

(Answers with discussions are given in the next section.)

1. The current in a circuit is represented by the symbol (*a*) *I;* (*b*) *C;* (*c*) *P;* (*d*) *E.*
2. The current in a circuit can be determined by (*a*) $I = E/R;$ (*b*) $I = ER;$ (*c*) $I = R/E;$ (*d*) $I = E + R.$
3. The relationship between voltage, current, and resistance is given by (*a*) Doan's law; (*b*) Roman law; (*c*) Henry's law; (*d*) Ohm's law.
4. Which would cause the greater current flow? (*a*) A 200-ohm resistor across a 10-volt battery; (*b*) A 500-ohm resistor across a 10-volt battery.
5. A current of 5 milliamperes is the same as a current of (*a*) 5,000 amperes; (*b*) 0.5 ampere; (*c*) 0.005 ampere; (*d*) 0.000005 ampere.
6. Ohm's law will not work for calculating currents in circuits that have (*a*) linear resistors; (*b*) copper busbars; (*c*) ground connections; (*d*) nonlinear resistors.
7. What voltage drop will occur across a 10-ohm resistor when the current through it is 0.5 ampere? (*a*) 20 volts; (*b*) 5 volts; (*c*) 0.05 volt; (*d*) 50 volts.
8. How much current will flow through a 10-ohm resistor when it is connected across a 100-volt battery? (*a*) 10 amperes; (*b*) 0.1 ampere; (*c*) 1,000 amperes; (*d*) 100 amperes.
9. The resistance of a circuit can be determined by (*a*) $R = I/E;$ (*b*) $R = EI;$ (*c*) $R = E/I;$ (*d*) $R = E - I.$
10. A certain voltage is given as 100 kilovolts. A resistance of 100 kilohms across this resistor will result in a current of (*a*) 0 ampere; (*b*) 1,000 amperes; (*c*) 0.1 ampere; (*d*) 1.0 ampere.

Answers to Self-Test

1. (*a*) — You might feel that *C* would be a better symbol for current, but this would be easily confused with the *C* used for capacitance.
2. (*a*)
3. (*d*)
4. (*a*) — It is very important for you to remember that the smaller resistance will result in a larger current flow.
5. (*c*) — According to Table 7-1, the value in milliamperes is converted to the value in amperes by moving the decimal point to the left three places. For the number 5, the decimal point is assumed to be immediately after the 5. Thus, it can be written as 5.0. As you move the decimal point to the left, you must add zeroes as they are needed.
6. (*d*) — Problems involving circuits with nonlinear resistors are not included in this book.
7. (*b*) — $E = IR = 0.5 \times 10 = 5$ volts.
8. (*a*) — $I = E/R = 100/10 = 10$ amperes.
9. (*c*)
10. (*d*) — 100 kilovolts = 100,000 volts
     100 kilohms = 100,000 ohms

$$I = \frac{E}{R} = \frac{100,000}{100,000} = 1.0 \text{ ampere}$$

8.
How Is Electricity Generated?

Introduction

There are three basic factors of importance in circuits: *voltage, current,* and *resistance.* Our interest in this chapter is in the voltage. As you know, the voltage can be considered as being the pressure that forces the current through the circuit.

You may be surprised to learn that there are only six basic methods of generating a voltage. These methods will be discussed in this chapter. For each of the methods discussed, there will be some practical applications given.

You will be able to answer these questions after studying this chapter:

How is friction used for generating voltages?
How is heat used for generating voltages?
How is light used for generating voltages?
How is pressure used for generating voltages?
How is chemical action used for generating voltages?
How is mechanical motion used for generating voltages?

Instruction

HOW IS FRICTION USED FOR GENERATING VOLTAGES?

In an earlier chapter you learned that rubbing dissimilar nonconducting materials together produces a static charge. This is the method of generating voltages by *friction*. Examples are rubbing a comb through your hair and rubbing a cat's fur.

Static charges can be very troublesome in some industrial systems. Consider, for example, the problem of running large rolls of paper through a printing press. This paper moves at very high speeds. Since paper is a nonconductor, the amount of static electricity generated can be considerable and troublesome. Special provisions must be made for removing the static charge; otherwise the paper becomes extremely difficult, if not impossible, to handle when it comes out of the press.

In small presses, the charges are removed by placing metal fingers in contact with the paper as it moves out of the press. For large printing presses, more drastic methods of removing the static charges must be used.

You should not get the idea that static electricity is *always* a troublesome and undesirable thing. Special generators—called *electrostatic generators*—have been designed to produce very high voltages by the friction method. An example of such a generator is shown in Fig. 8-1. This is a *Van de Graaff generator*, and in some versions it is capable of producing voltages as high as 10 million volts (depending upon how the generator is constructed).

The simplified generator of Fig. 8-1 consists of a wide belt of nonconducting material. Sheet rubber or silk is sometimes used for making this belt. The belt moves over two pulleys. One of the pulleys (pulley *B*) is turned by a motor or by a hand crank. As the belt moves over pulley *B*, a static charge is generated due to the friction between the belt and the pulley. It is important to

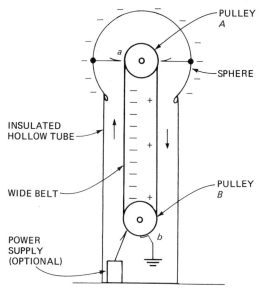

Figure 8-1. Construction of a Van de Graaff generator.

know that the two pulleys are made of different types of material. The charge generated when the belt moves over pulley *B* is negative. The material of pulley *A* is such that it causes a positive charge to be developed as the belt moves over it.

Negative charges as the belt goes over pulley *B* are carried by the belt to point *a*, which is a group of metal contacts shaped like a comb and connected to the sphere. If the Van de Graaff generator has a slow-moving belt, the contacts actually touch the belt, but for high-speed belts there is a small space between the belt and the contacts. (When the contact does not actually touch the belt, the charges jump in the form of a spark across the small air gap.) The negative charges are delivered through the contact to the sphere and are trapped there. The positive charges are carried away from the sphere and grounded through contact *b*.

Much higher voltages can be produced with this type of generator if an optional power supply is included which delivers additional negative charges to the belt.

Van de Graaff generators are used in physics laboratories for nuclear physics experiments, and they are also used to produce very high voltages for some types of machines. However, the Van de Graaff generator, or for that matter any electrostatic generator, would not be suitable for delivering power to your home. For one thing, any machine that requires friction for its operation is naturally inefficient. Friction represents losses which must be made up by the motor or hand crank that turns the generator. Furthermore, if a conductor is connected to this sphere and current is allowed to flow, the charge is almost immediately reduced to zero. Another way of saying this is that the Van de Graaff generator can produce a voltage but it cannot produce a continuous current flow for its circuits.

HOW IS HEAT USED FOR GENERATING VOLTAGES?

One of the basic rules in the study of electricity states that *whenever two dissimilar metals are joined, and the junction is heated, a voltage is generated.*

The junction of the metals is called a *thermocouple.* Figure 8-2 shows how

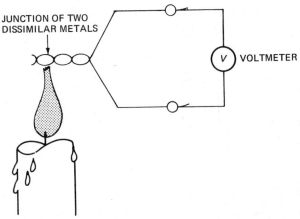

Figure 8-2. When the thermocouple is heated, a voltage is generated. This voltage can be measured with a voltmeter (V).

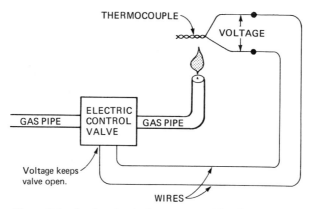

Figure 8-3. A safety circuit that operates with a thermocouple. This type of safety system is used in some types of gas hot-water heaters.

the thermocouple works. Modern science has developed combinations of metals for thermocouples that can produce an appreciable amount of electricity—enough electricity, in fact, to operate small circuits.

Figure 8-3 shows an example of a simple circuit that can be operated with a thermocouple. The junction of the thermocouple is inserted into a gas flame. As a result, a voltage is generated across the terminals of the thermocouple, and this voltage is delivered to an electric control valve. As long as the flame is present, the voltage keeps the control valve *open* so that there is a continuous flow of gas through the pipe.

Suppose the flame in Fig. 8-3 is blown out. It would be undesirable to have gas continuing to flow even though there is no flame. In fact, this would be very dangerous. In the system of Fig. 8-3, if the flame goes out, there is no longer a voltage generated by the thermocouple. Therefore, the voltage at the electric control valve drops to zero and the valve closes.

An interesting effect that is related to thermoelectricity is shown in Fig. 8-4. Here, the junction of the thermocouple is inserted in the flame, but the two metals are also twisted together at the other end. When one of these two junctions is heated as shown, *the other junction becomes cooler.* This effect (which is known as the *Peltier effect*) has been known for a long time, but up until recently it was just considered a lab curiosity. However, scientists have developed metals

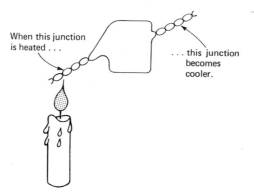

Figure 8-4. An illustration of the Peltier effect.

that can produce a sufficient amount of cooling to be used for refrigerators and air conditioners. The advantage of making a refrigerator or air conditioner by this method is that no moving parts are required. The first refrigerator made on this principle was made in Japan for the purpose of cooling drugs in tropical climates where industrial electricity is not readily available for operating ordinary refrigerators.

HOW IS LIGHT USED FOR GENERATING VOLTAGES?

There are certain types of materials which will release electrons from their surface when they are exposed to light. Examples of such materials are selenium, germanium, cadmium, and lead sulfide. The fact that electrons are released from a material by exposing it to light is called the *photoelectric effect.*

There are a number of different ways that the photoelectric effect can be put to use. Electric eyes that are used for opening doors operate on this principle. Also, the electric eye in sound-movie projectors, which converts light variations on a film into speech or music, is based on the photoelectric effect. Our interest is in how this effect can be used for generating voltages. To do this, two plates are used to make a *photocell.* One of the plates is made of a material that releases electrons when exposed to light, and the other one receives the electrons. The result is one of the plates is positive because it has lost electrons and the other is negative because it collects the electrons. Since there is a positive and a negative plate, there is a voltage that can be used for operating circuitry.

An example of the use of a photocell is shown in Fig. 8-5. This is a simple light meter of the type used by some photographers. The light shines on the photocell and establishes a voltage. This voltage is delivered across the meter. The greater the amount of light on the photocell, the greater the amount of voltage and the further the meter needle moves upscale. A photographer aims

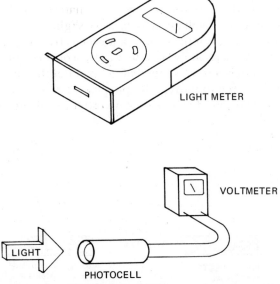

Figure 8-5. The photocell establishes a voltage across the meter. The greater the amount of light, the greater the voltage. This is how the light meter works.

the exposure meter at the point where he wishes to take the picture. The scale of the meter is calibrated in numbers which are of use to photographers. These tell him how to set the camera in order to get the best picture for the amount of light present.

SUMMARY

1. There are six basic methods used for generating a voltage.
2. The frictional method of generating voltage can be used for producing as much as 10 million volts. However, machines built on this principle—such as the Van de Graaff generator—cannot efficiently deliver a continuous current for a circuit.
3. Whenever two dissimilar metals are used to form a junction, and the junction is heated, a voltage is produced.
4. The junction of dissimilar metals used for producing a voltage when heated is called a *thermocouple*.
5. Some materials will emit electrons when they are exposed to light. These materials are used for making *photocells*.
6. So far you have studied three ways of producing a voltage: by *friction*, by *heat*, and by *light*.

HOW IS PRESSURE USED FOR GENERATING VOLTAGES?

There are certain materials which will produce a voltage when a pressure is exerted upon them. These materials are said to be *piezoelectric*. Quartz crystal is one example of a piezoelectric material. Modern technology has produced a number of *ceramic* materials which are piezoelectric.

Figure 8-6 illustrates how the piezoelectric voltage is generated. The crystal or ceramic material is anchored at one point and is free to move or vibrate at the other end. Any pressure exerted on the end that is free to move will cause a voltage to be produced across the piezoelectric material. The amount of pressure required to produce a voltage may be very slight.

Figure 8-7 shows an example of how a piezoelectric material can be used to make a *microphone*. (A microphone is a device that converts sound waves to electrical impulses.) There is a horn-shaped attachment at the end of the piezoelectric material which is free to move. The purpose of this attachment is to intercept sound waves. The sound waves are converted to voltages as a result of the

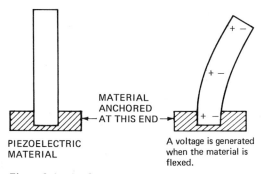

Figure 8-6. A voltage is generated whenever pressure is exerted on a piezoelectric material.

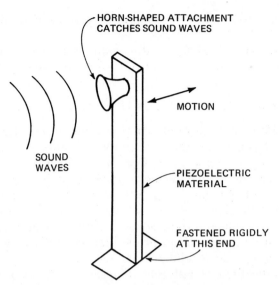

Figure 8-7. A microphone can be made from a thin slab of piezoelectric material.

vibrations they produce in the piezoelectric material. This is how a *crystal microphone* is made.

If you look closely at a phonograph record, you will note that it is comprised of wavy grooves which the needle of the record player pickup arm follows. When the needle follows the groove, it moves back and forth as shown in Fig. 8-8. The back-and-forth motion causes a voltage to be produced in the piezoelectric material. This voltage, which is relatively small, is amplified (increased) and delivered to a loudspeaker. The loudspeaker converts the voltage variations into sound.

The term "transducer" is used in electricity to mean a device which converts energy from one form to another. A microphone converts *sound* energy into *electrical* impulses, and a record player pickup arm converts needle *vibrations* into *electrical* impulses. These are both examples of transducers. A loudspeaker

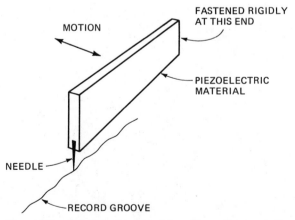

Figure 8-8. A piezoelectric material may be used as a record-player pickup.

converts *electrical* impulses into *sound* impulses, and this is another example of a transducer.

Piezoelectric transducers are used in industry to detect vibrations, and in applications where it is desired to convert a pressure into an electrical signal.

HOW IS CHEMICAL ACTION USED FOR GENERATING VOLTAGES?

Whenever two different kinds of electrical conductors are inserted into an acid or alkali, a voltage is generated. A very simple example of this is shown in Fig. 8-9. This illustration shows a coat-hanger wire, which is made of iron, and a copper wire inserted into a lemon. Lemon juice contains citric *acid,* and therefore, the simple arrangement of Fig. 8-9 will produce a voltage which can be measured by a voltmeter.

The amount of voltage produced by the simple lemon cell of Fig. 8-9 is quite small, but it illustrates the principle that dissimilar conducting materials inserted into an acid cause a voltage to be developed.

The two conducting materials that are inserted into the acid are called *poles* of the cell, and the acid or alkali is called the *electrolyte.* Figure 8-10 shows how a simple dry cell (often called a "dry battery") used for flashlights is constructed. You will note that the two conducting materials used for poles are *carbon* (for the positive terminal) and *zinc* (for the negative terminal). The electrolyte for this type of cell, which is known as a "LeClanche cell," is sal ammoniac. To be technically accurate you should make a distinction between the terms "cell" and "battery." A *cell* is a single unit comprised of a positive and a negative pole separated by an electrolyte. When you connect several cells together in order to obtain a desired voltage or current reading, then you have a *battery.* Figures 8-9 and 8-10 are actually illustrations of cells.

All cells and batteries can be placed in one of two categories. Those that can be *recharged* are called *secondary cells,* and those that cannot be recharged are called *primary cells.* Dry cells that you use in flashlights—that is, LeClanche

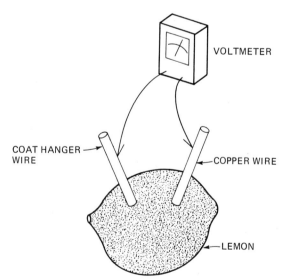

Figure 8-9. A simple cell made by inserting two different kinds of metal into an acid.

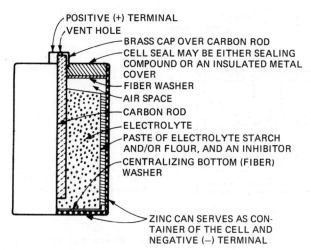

Figure 8-10. Construction of a dry cell.

cells—are primary cells because they cannot be recharged. You may have seen devices advertised which are said to recharge these cells. However, it would be more technically correct to say that they rejuvenate rather than recharge the cell. Recharging a cell or battery means that you reverse the chemical action in the battery or cell for a period of time. The process of recharging a battery involves forcing a current to flow through it in a direction opposite to its normal flow. The chemical action in the lemon cell and the dry cell cannot be reversed, and therefore, these devices cannot actually be recharged.

Figure 8-11 shows how an alkaline cell (again often called an "alkaline battery") is constructed. Alkaline cells are often made in the same type of case as LeClanche cells and can be used in many applications where dry cells are used. Examples are transistor radios and tape recorders. Since the alkaline cell is a secondary cell, it can be recharged and can provide longer service than an ordinary dry cell. It can actually be recharged, although in some cases the price of a new cell is so low that it does not pay to take the time to recharge it.

The battery that is used for starting a car is another example of a secondary battery. It is comprised of a number of secondary cells connected together. In the case of a 6-volt battery, three cells are used. In the case of a 12-volt battery, six cells are used. They are called *lead-acid cells* because the poles are made

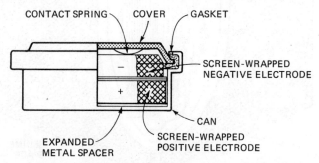

Figure 8-11. Alkaline cells are popular because they last longer than dry cells, and they can be recharged.

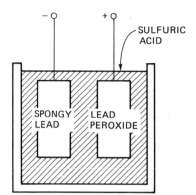

Figure 8-12. The materials used in the lead-acid storage battery.

of lead and the electrolyte is made of a solution of sulfuric acid and water. Figure 8-12 shows the materials used.

Figure 8-13 shows how the battery of an automobile is constructed. Batteries such as this are capable of delivering a tremendous amount of power for a short period of time. In fact, as far as a portable source of power is concerned, nothing has ever been found that is better than a battery. The battery makes it possible to start the engine of a car. Once the engine is going, you could really

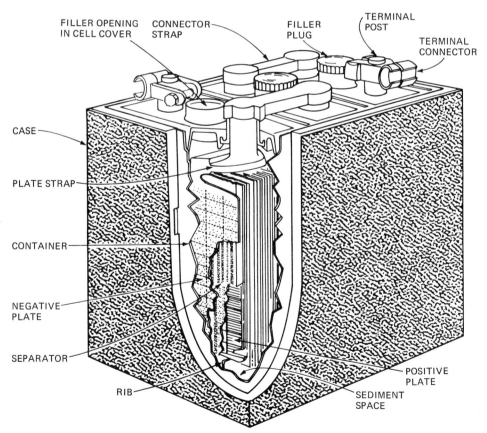

Figure 8-13. Construction details of a lead-acid battery.

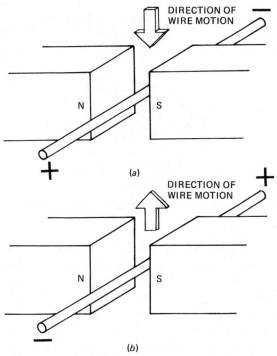

Figure 8-14. The polarity of induced voltage depends on the direction of motion through the magnetic field. (a) *When the conductor moves in one direction, the polarity is as marked.* (b) *Reversing the direction of motion reverses the voltage polarity.*

get along without the battery. The generator in the car (to be discussed in the next section) develops enough electricity to operate all the equipment in the car. Thus, the most important function of the battery in the car is for getting the engine started.

HOW IS MECHANICAL MOTION USED FOR GENERATING VOLTAGES?

A very important rule in electricity states that *whenever a conductor is moved through a magnetic field, a voltage is generated across the ends of the conductor.* This rule is illustrated in Fig. 8-14. In Fig. 8-14*a* the conductor is moved down between the north and south poles of a magnet. Notice that the conductor must move through the flux lines which are presumed to be between the north and south poles. As the conductor is moved through the flux lines, one end of the conductor becomes positive and the other becomes negative.

As shown in Fig. 8-14*b*, if you reverse the direction that the conductor is moving through the magnetic field, you also reverse the polarity of the voltage across the conductor.

A generator is a machine that makes use of the principle illustrated in Fig. 8-14. Figure 8-15 shows how a simple generator is constructed. The conductor is bent into a loop which is called the *armature*. It is caused to rotate in the magnetic field. The two ends of the armature wire are connected to round conductors which are called *slip rings. Brushes*, which are usually made of carbon, are used for taking the generated voltage from the slip rings.

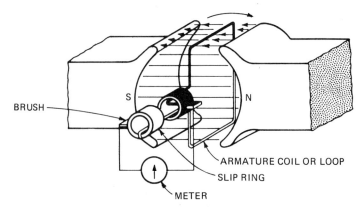

Figure 8-15. Construction of a simple generator.

To summarize the action of the generator, as the armature is rotated in the magnetic field, a voltage is generated which appears at the slip rings. This voltage is taken from the slip rings by brushes and delivered to a meter.

Figure 8-16 shows how the voltage varies as the armature is rotated in the magnetic field. In Fig. 8-16*a*, notice that the ends of the coil are moving parallel to the flux lines. Since the conductor is moving *with* the flux lines and not cutting *across* them, no voltage is generated at this time.

In Fig. 8-16*b* the armature has rotated 90° and is now moving through the flux lines. Note that the black side of the armature is moving *down* through the flux lines, and the white side is moving *up*. Therefore, the voltage generated in these two sides will be of opposite polarities, as explained in Fig. 8-14. The voltage of one polarity appears at one slip ring and the voltage of the opposite polarity appears at the other slip ring. These two voltages are opposite in polarity, and therefore, will produce a voltage difference as measured by the meter.

In Fig. 8-16*c* the wires of the armature are again moving parallel to the flux lines. Since these wires are not cutting across the flux lines, it follows that no voltage is being generated and the meter indicates 0 volt. In Fig. 8-16*d* the armature has rotated so that the black side is moving up through the flux lines and the white side is now moving down through the flux lines. Note that this causes the polarity of voltage developed on the black and white sides respectively to be opposite to that which was developed in Fig. 8-16*b*. These voltages which appear at the black and white slip rings again cause a deflection of the meter. The needle deflecting to the left indicates that the polarity of the voltage is opposite to that which was developed in Fig. 8-16*b*.

The important thing to remember about the illustration is that the armature is caused to rotate through a magnetic field, and a voltage is produced. This is the way the voltage is generated by the generator in your car. This is also the way in which the power companies generate voltage for use in homes and industry. You should remember that the simplified drawing of Fig. 8-16 does not show all the construction details of the actual generator used.

We will discuss the different types of generators and their descriptions in greater detail in a later chapter. Our point here is that voltage is generated by moving a conductor through a magnetic field.

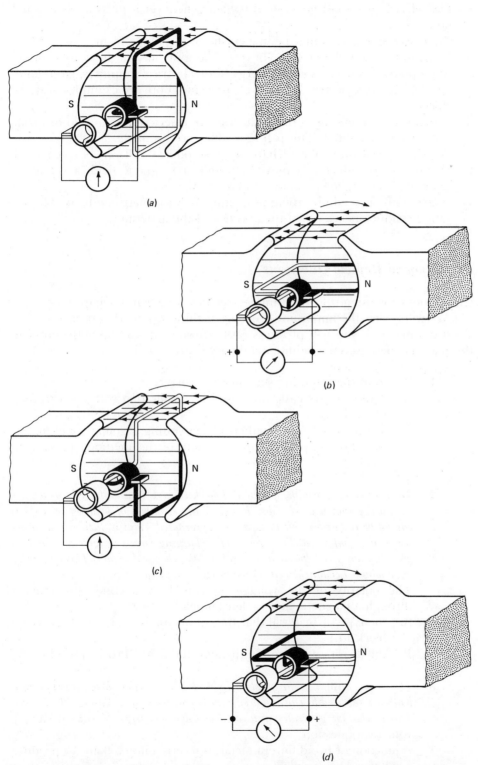

Figure 8-16. *Voltages produced by rotating the armature.* (a) *When the armature is in this position, its conductors are moving parallel to the flux lines. No voltage is generated.* (b) *In this position, the armature wires are moving at right angles to the flux lines, and maximum voltage is generated.* (c) *The armature wires are again moving parallel to the flux lines, and no voltage is generated.* (d) *In this position, the wires are cutting the flux lines in the opposite direction to that of Fig. 8-16* b. *Note that the voltage polarity is reversed.*

SUMMARY

1. Piezoelectric materials generate a voltage whenever a pressure is exerted on them.
2. Piezoelectric materials may be in a crystal or a ceramic form.
3. A transducer is a device that converts energy from one form to another. Microphones convert sound energy into electrical impulses, and loudspeakers convert electrical impulses into sound. Both are examples of transducers.
4. When two different kinds of conductors are inserted into an acid or alkali, a voltage is generated. This is how a voltage cell is made.
5. A primary cell *cannot* be recharged. A secondary cell *can* be recharged.
6. Whenever a conductor is moved through a magnetic field, a voltage is generated across the ends of a conductor.
7. A generator is made by rotating conductors in a magnetic field. The rotating conductors are called the *armature* of the generator.

Programmed Review Questions

(Instructions for using this programmed section are given in Chap. 1.)

We will review the important concepts of this chapter. If you have understood the material, you will progress easily through this section. Do not skip this material, because some additional theory is presented.

1.   Which of the following statements is correct?
   A.   There are six basic ways used for generating voltages. (Proceed to block 7.)
   B.   There is only one method of generating a voltage, and all the ways used in practice are variations of this method. (Proceed to block 17.)

2.   *The correct answer to the question in block 18 is **A**. Flashlight cells, which are technically known as "LeClanche cells," are examples of primary cells. They cannot be recharged, but they can be rejuvenated—that is, their chemical action can be improved. The process of recharging means to force a current to flow through a cell backwards to reverse the chemical action. This cannot be done with a LeClanche cell.* Here is your next question.
   The polarity of the voltage generated by a conductor moving through a magnetic field is determined by
   A.   the type of material that the conductor is made of. (Proceed to block 14.)
   B.   the direction of motion through the field. (Proceed to block 6.)

3.   *The correct answer to the question in block 11 is **A**. The chemical action of a battery is reversed by causing a current to flow backwards through the battery. This is what the generator in a car does when it charges the battery.* Here is your next question.
   A junction of two different metals is inserted into a flame to produce a voltage. This is called

A.   a thermocouple. (Proceed to block 18.)

B.   a thermopile. (Proceed to block 10.)

4.   *Your answer to the question in block 20 is* ***A***. *This answer is wrong. The electrical equipment in a car can be operated from its generator (or alternator) once the engine is running.* Proceed to block 12.

5.   *The correct answer to the question in block 21 is* ***B***. *The Van de Graaff generator is an example of a generator that operates by producing static charges due to friction. It is capable of generating a million or more volts.* Here is your next question.

A battery consists of

A.   two different kinds of metal inserted into an acid. (Proceed to block 19.)

B.   a number of cells connected together. (Proceed to block 11.)

6.   *The correct answer to the question in block 2 is* ***B***. *Reversing the direction of motion of a conductor in a magnetic field can reverse the polarity of the voltage across the conductor.* Here is your next question.

The moving conductor in a generator is called

A.   a brush. (Proceed to block 24.)

B.   an armature. (Proceed to block 16.)

7.   *The correct answer to the question in block 1 is* ***A***. *The six methods of generating a voltage are all different.* Here is your next question.

Materials that generate a voltage whenever there is a pressure on them are called

A.   piezoelectric materials. (Proceed to block 15.)

B.   pressure-sensitive materials. (Proceed to block 23.)

8.   *Your answer to the question in block 15 is* ***A***. *This answer is wrong. A conductor will transfer energy from one point to another, but it does not serve to convert energy from one form to another.* Proceed to block 21.

9.   *Your answer to the question in block 16 is* ***A***. *This answer is wrong. Piezoelectric materials generate a voltage when there is a pressure exerted on them.* Proceed to block 20.

10.   *Your answer to the question in block 3 is* ***B***. *This answer is wrong. A thermopile was not discussed in the "Instruction" section of this chapter. For a definition of "thermopile,"* proceed to block 18.

11.   *The correct answer to the question in block 5 is* ***B***. *The terms "cell" and "battery" are often used incorrectly. A battery consists of a number of cells connected together.* Here is your next question.

In order to charge a battery,

A.   the chemical action of the battery must be reversed. (Proceed to block 3.)

B.   the chemical action of the battery must be stopped. (Proceed to block 25.)

12. *The correct answer to the question in block 20 is* **B**. *A lead-acid battery—which is sometimes called a "storage battery"—can deliver a large amount of electricity, and therefore, can be used for starting a car engine. Once the engine of the car is running, it turns the generator, and the battery is not required. Even at idling speeds, alternators (which are a form of generator) can provide enough electricity to operate the electrical system in a car.* Here is your next question.
Make a list of the six methods of generating a voltage, then proceed to block 26 to check your answer.

13. *Your answer to the question in block 21 is* **A**. *This answer is wrong. As you will learn in a later chapter, it would be difficult to make a generator that can produce 1 million volts by moving a conductor through a magnetic field. The only generator that was discussed in the "Instruction" material which is capable of generating very high voltages is the Van de Graaff generator.* Proceed to block 5.

14. *Your answer to the question in block 2 is* **A**. *This answer is wrong. The* **size and type of material** *of the conductor has an influence on the* **amount** *of voltage generated, but not on its polarity.* Proceed to block 6.

15. *The correct answer to the question in block 7 is* **A**. *Examples of piezoelectric materials are quartz, Rochelle salts, and barium titanate.* Here is your next question.
A device that converts energy from one form to another is called a
A.   conductor. (Proceed to block 8.)
B.   transducer. (Proceed to block 21.)

16. *The correct answer to the question in block 6 is* **B**. *The simple generator shown in Fig. 8-15 has only a single loop of wire for an armature. In practical generators many loops are used, and they are combined to give the generator a greater output.* Here is your next question.
A material that emits electrons when a light shines on it is said to be
A.   piezoelectric. (Proceed to block 9.)
B.   photoelectric. (Proceed to block 20.)

17. *Your answer to the question in block 1 is* **B**. *This answer is wrong. There are six different ways of generating a voltage. Each of these is based on a different principle.* Proceed to block 7.

18. *The correct answer to the question in block 3 is* **A**. *A thermopile is a number of thermocouples combined to produce a larger voltage.* Here is your next question.
A cell that cannot be recharged is called a
A.   primary cell. (Proceed to block 2.)
B.   secondary cell. (Proceed to block 22.)

19. *Your answer to the question in block 5 is* **A**. *This answer is wrong. Two different kinds of metal inserted into an acid will make a cell. A battery is a number of cells connected together.* Proceed to block 11.

20. *The correct answer to the question in block 16 is **B**. Photoelectric materials are used in transducers that convert light into electricity.* Here is your next question.
    A car battery is needed for
    A. operating the lights, heater, and other electrical equipment. (Proceed to block 4.)
    B. starting the engine. (Proceed to block 12.)

21. *The correct answer to the question in block 15 is **B**. There are many types of transducers used in electricity.* Here is your next question.
    For generating a voltage of 1 million volts, you would most likely use
    A. a generator based on a conductor moving through a magnetic field. (Proceed to block 13.)
    B. a generator based on producing static charges due to friction. (Proceed to block 5.)

22. *Your answer to the question in block 18 is **B**. This answer is wrong. An example of a secondary cell is the lead-acid battery used in a car. The generator (or alternator) in a car recharges the battery by forcing a current to flow backwards through it.* Proceed to block 2.

23. *Your answer to the question in block 7 is **B**. This answer is wrong. The expression "pressure-sensitive" is not used to describe a material that produces a voltage when under pressure.* Proceed to block 15.

24. *Your answer to the question in block 6 is **A**. This answer is wrong. Brushes are used in generators to pick off the voltage from the armature and deliver it to the circuit.* Proceed to block 16.

25. *Your answer to the question in block 11 is **B**. This answer is wrong. It would be useless to stop the chemical action of a battery.* Proceed to block 3.

26. *Here are the six methods of generating a voltage. (They can be listed in any order.)*
    1. *Friction*
    2. *Heat*
    3. *Light*
    4. *Pressure*
    5. *Chemical*
    6. *Mechanical*
    You have now completed the programmed questions. The next step is to put some of these ideas to work in laboratory experiments. Proceed to the "Experiment" section that follows.

Experiment

(The experiment described in this section may be performed on the circuit board described in Appendix C or on a similar laboratory setup.)

EXPERIMENT 1

Purpose – To show how a simple cell can be constructed.

Theory – Whenever two different kinds of metals are immersed into an acid, a voltage is generated between them. This voltage is generated as a result of chemical action between the acid and the metals.

The amount of voltage generated depends upon the types of metal used and the kind of acid used. You may find that adding a little salt to the acid will improve the results of this experiment. The salt increases the chemical action.

For one of the metals you will use the copper from a wire. For the other metal you will use the metal from a "tin" can. Actually, tin cans are made mostly of steel. You can prove this by seeing if a tin can is attracted by a magnet. As you know, magnets attract ferromagnetic materials. Steel is ferromagnetic. You can readily determine that the tin can is ferromagnetic by testing it with a magnet.

The ability of a battery to deliver current to a circuit is related to the area of the plates, or poles, used as conductors. In this experiment you should try to get the largest possible area of metal for your battery.

Test Setup – Obtain a jar or glass to be used for a battery case. Fill it with lemon juice, grapefruit juice, lime juice, or the juice of any other fruit with acid. You can mix the juices if necessary. As a general rule, the higher the acid content (related to how sour the juice is), the better the battery it will make.

Bend a strip of metal from a tin can to follow the contour of the glass. This is shown in Fig. 8-17. It is important to remove paint and all coatings from the metal. One way is to suspend it over a flame to burn off the undersired coatings.

> **WARNING: DO NOT ATTEMPT TO HOLD THE METAL WHILE BURNING OFF THE COATING. INSTEAD, RIG A HOLDER OUT OF HEAVY WIRE. ALLOW PLENTY OF TIME FOR COOLING BEFORE TOUCHING THE METAL.**

Remove the insulation from about 1 foot of number 8 or number 10 solid copper wire. Bend the wire as shown in Fig. 8-17. The purpose of this bend is to get a larger area of wire into the solution.

Insert the poles into the battery as shown in Fig. 8-18. It is very important that no part of the plates touch. If they do touch, the battery will be *internally shorted,* and no external voltage will be observed.

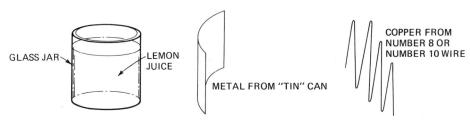

Figure 8-17. Parts for making a lemon cell.

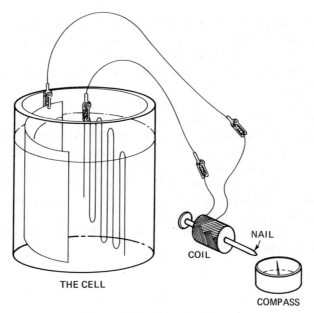

Figure 8-18. The test setup.

Procedure—

Step 1—Connect the coil to the cell with alligator clips as shown in Fig. 8-18. Insert a soft-iron nail into the coil and note that it becomes magnetized and deflects the compass needle slightly.

Step 2—Experiment by adding salt to the solution. (Use another unmagnetized nail every time you change the experiment.) Move the cell plates closer together.

Conclusion—A simple lemon cell can produce enough current, with an accompanying magnetic field, to cause a slight deflection of a compass.

Self-Test with Answers

(Answers with discussions are given in the next section.)

1. A battery that cannot be recharged by reversing the current flow through it is called (*a*) an alkaline battery; (*b*) a primary battery; (*c*) a lead-acid battery; (*d*) a secondary battery.
2. Which of the following generates voltage by friction? (*a*) A secondary cell; (*b*) A piezoelectric crystal; (*c*) A Van de Graaff generator; (*d*) A thermopile.
3. Two dissimilar metals are twisted to form a junction at both ends. One end is heated, and as a result the other becomes cooler. This is known as (*a*) the piezoelectric effect; (*b*) Hoagland's law; (*c*) Ronald's effect; (*d*) the Peltier effect.
4. Batteries are used in cars (*a*) to help the generator supply enough voltage for the lights; (*b*) that do not have generators or alternators; (*c*) as a port-

able power supply for starting the car; (*d*) to provide heat for the heater.

5. The acid or alkali fluid in a battery is known as the (*a*) electrolyte; (*b*) insulation; (*c*) semiconductor; (*d*) polar fluid.

6. The rotating conductors in a generator are called the (*a*) brushes; (*b*) slip rings; (*c*) armature; (*d*) field.

7. A device that converts energy from one form to another is called (*a*) an inverter; (*b*) an inducer; (*c*) a reducer; (*d*) a transducer.

8. The polarity of the voltage across a wire that is moving through a magnetic field depends upon (*a*) the size of the conductor; (*b*) the direction that the conductor is moving through the field; (*c*) the type of material that the conductor is made of; (*d*) the strength of the magnetic field.

9. Which of the following is a transducer? (*a*) A newspaper; (*b*) A copper wire; (*c*) An acid; (*d*) A microphone.

10. LeClanche cells are (*a*) secondary cells; (*b*) lead-acid cells; (*c*) primary cells; (*d*) alkaline cells.

Answers to Self-Test

1. (*b*) — Alkaline and lead-acid batteries can be recharged, and are examples of secondary batteries.

2. (*c*) — The Van de Graaff generator can produce very high voltages by the friction method.

3. (*d*) — Experts say that refrigerators and air conditioners of the future will employ this principle.

4. (*c*)

5. (*a*) — An important characteristic of the electrolyte in a battery is that it must be able to conduct electricity.

6. (*c*)

7. (*d*) — This is the definition of a transducer.

8. (*b*) — The strength of the magnetic field is a factor that determines the amount of voltage generated, but not the polarity of the voltage generated.

9. (*d*)

10. (*c*) — The cells used for most flashlights are LeClanche cells.

9.
What Is Meant by the Expression "Electric Power"?

Introduction

An important part of your learning in electricity is the proper use of the terminology. You know, for example, that it is not proper to say "the voltage through a circuit is so many volts." Voltage does *not* pass through a circuit. As you know, voltage exists between two points, and therefore you should say "the voltage across a circuit" or "the voltage between two points in the circuit."

Likewise, it is not proper to say "the current across a circuit is so many amperes." Current does *not* exist across two points. Instead, it flows through the wires and through the circuit. You should say "the current through the circuit" or "the current through the resistor."

Just as it is important for you to use the terms "voltage" and "current" correctly, it is also important for you to use the terms "power," "work," and "energy" according to their true technical meaning.

A person who is not educated in science or electricity uses the terms "work" and "energy" without regard to their true technical meaning. As far as science is concerned, there is a specific application for these terms which you should learn to use correctly.

Have you ever heard someone say "it takes a lot of energy to push a car," or "a good breakfast gives me power for working all morning"? You may have heard that certain types of candy bars have "energy," and you may have heard a man who works at a desk job say he has "worked hard" all day. These expressions are used with only a vague relationship to the technical meaning of the terms "work" and "energy."

In this chapter you will learn about work, energy, and power. You will study the true scientific meanings of the terms, and you will also learn how they are related to work in electricity.

You will be able to answer these questions after studying this chapter:

What is the technical meaning of the word "work"?
What is the technical meaning of the word "energy"?
What is the technical meaning of the word "power"?

Instruction

WHAT IS THE TECHNICAL MEANING OF THE WORD "WORK"?

Technically a force is defined as *a push or pull*. When a force is exerted, there must always be some form of opposition, and the purpose of the force is usually to overcome or equalize the opposition.

As far as science is concerned, you are doing work only when you exert a force through a distance. However, the force must be in the same direction as the motion that it creates in order to determine the amount of work. By definition:

$$WORK = FORCE \times DISTANCE$$

Consider the man in Fig. 9-1. He is exerting a force on the box and is pushing it through a distance. He is, in the scientific meaning of the word, doing work. However, the man in Fig. 9-2 is *not* doing work. He is exerting an upward force against the box, but the direction of motion is horizontal. Remember that work is only accomplished when the force is in the same direction as the motion. Therefore, even though the man in Fig. 9-2 may be more tired at the end of an hour than the man in Fig. 9-1, he has not done any work.

You might wonder if it is important to study about work in a book dealing with electricity. When experimenters started to work with electricity, they found it was necessary to be able to measure it if any progress was to be made. The laws for mechanical things—such as work—were already established, so the electrical experimenters simply used these laws for setting up their units of

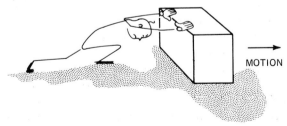

Figure 9-1. Work is done when a force is exerted through a distance.

Figure 9-2. Surprisingly, this man is not doing work in the technical meaning of the word.

measurement. For example, the volt is based on the measurement of work. Also, your electric bill is based on kilowatt-hours, and this is a unit of work.

Figure 9-3 shows how the volt was defined. There are two charged plates—one positive and one negative. A negative charge between these plates will be attracted toward the positive plate and will try to move away from the negative plate. If you move this negative charge from the positive plate to the negative plate, you must do work on it. In other words, you must exert a force to overcome the natural tendency of the charge to go in the opposite direction, and you must exert this force throughout the distance between the plates.

The voltage in a circuit is actually a measure of the amount of work done in moving a negative charge of 1 coulomb from the most positive point in the circuit to the most negative point in the circuit.

At one time, the term "electromotive force" was used to describe a voltage. The term was very misleading because voltage is actually a measure of work, and not a measure of force. However, you have been told in earlier chapters that you can consider the battery as being the force that moves electricity through a circuit. This is a convenient way of thinking of a battery or any source of voltage provided you understand that the actual measurement of voltage is a measurement of work done. Today, the term "electromotive force" is no longer in popular use.

SUMMARY

1. It is important to use technical terms correctly.
2. Voltage exists *across* a component, or *between* two points, but it is *not* proper to speak of the voltage "through" a circuit.
3. Current flows *through* a circuit, but it is *not* proper to speak of the current "across" a circuit.
4. A force is defined as being a push or a pull. A force is always exerted against some form of opposition.

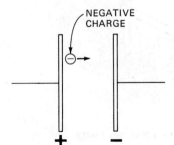

NEGATIVE
CHARGE

+   −

Figure 9-3. How the unit of 1 volt was established.

5. Work is accomplished when a force is exerted through a distance. By definition, *WORK = FORCE × DISTANCE.*
6. Voltage is a measure of the work done in moving a negative charge from a positive point to a negative point.

WHAT IS THE TECHNICAL MEANING OF THE WORD "ENERGY"?

Energy is defined as the capacity to do work. Energy has different forms, and you are no doubt familiar with most of them. We speak of "electrical energy" because electricity is capable of performing work. Of course, you need some kind of electrical component to convert electrical energy into some other form of energy to accomplish the work. For example, an electric motor is used to exert a vertical force on an elevator through a distance. It is a common practice to say that the electricity is actually doing the work.

Heat is another form of energy. (An example of the way in which heat can perform work is the steam engine.) Other examples of energy are light and atomic energy.

One of the first types of energy studied by early experimenters was mechanical energy. Mechanical energy exists in two forms: *potential energy* and *kinetic energy*. Potential energy is the energy that a body has by virtue of its position. Consider, for example, the box resting on a table shown in Fig. 9-4. It took a certain amount of work to get the box on the table. A vertical force had to be exerted through a distance to accomplish this. Once the box is resting on the table, it is capable of doing work simply because of its position. In other words, it has potential energy.

If the box is knocked off the table, it will fall and strike the ground with an impact. This is shown in Fig. 9-5. Suppose there is a nail resting under the box where it strikes. If the box is heavy enough, it can actually push the nail into a wooden floor. Since this requires exerting a force through a distance, the box is capable of doing work.

Although the box resting on the table has potential energy, it will do no work until this potential energy is converted into some other form. As the box is moving through space it has energy by virtue of its motion. This is called *kinetic energy*. Both kinetic and potential energy are forms of energy which represent the capacity to do work. When energy is converted into work, some of it may be lost in the form of heat. The energy and the amount of work done are

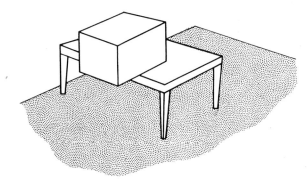

Figure 9-4. A box resting on a table has energy.

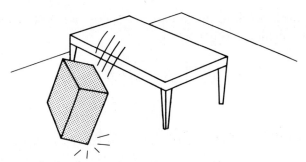

Figure 9-5. A falling box has energy. In this case it has a different kind of energy than it had in Fig. 9-4.

equivalent whenever you can neglect the heat losses. It stands to reason that they can be measured by the same units.

Let us summarize by saying that when there is energy, there is a capacity to do work, and energy can be converted into work. Energy and work are measured in the same units.

Electrical energy is found by multiplying the voltage times the current times the time:

ELECTRICAL ENERGY = VOLTS × AMPERES × TIME (IN SECONDS)

You do not need to memorize this equation, because you will not be required to find the energy in a circuit. The equation is given here because it will be used later to explain the meaning of electric power.

SUMMARY

1. Energy is defined as the capacity to do work.
2. Energy is in many forms, such as electrical, heat, light, and atomic energy.
3. Mechanical energy exists in two forms: *potential energy* and *kinetic energy*.
4. Potential energy is the energy that a body has because of its position.
5. Kinetic energy is the energy that a body has because of its motion.
6. Energy and work are closely related, and they are measured in the same units.

WHAT IS THE TECHNICAL MEANING OF THE WORD "POWER"?

Power is the rate at which work is done or the rate at which energy is being used. If the man in Fig. 9-1 moves the box through a distance of 10 feet in 1 hour, he will be doing a certain amount of work. He will be doing exactly the same amount of work if he moves the box through a distance of 10 feet in 1 minute. It is obvious there must be some difference between moving this box the distance of 10 feet in 1 hour and moving it 10 feet in 1 minute. The difference is in the amount of power that is used.

A greater amount of power is needed to move the box in a short period of time than is needed for moving it in a greater period of time. Whenever you talk about power you are always implying that there is a certain amount of *time* involved.

The amount of energy in an electric circuit is obtained by multiplying the voltage times the current times the time (ENERGY = E × I × t). The amount of electric power is equal to the amount of energy *per* unit of time. The word "per" means to divide. In other words,

$$POWER = \frac{ENERGY}{TIME}$$

$$= \frac{E \times I \times \cancel{t}}{\cancel{t}}$$

Notice that the two *t*'s cancel, and the equation for power becomes

$$P = E \times I$$

This is the basic equation for electric power. When the voltage is in volts and the current is in amperes, the power is given in *watts.*

> *Example 9-1* — How much power is radiated by a resistor when the voltage across it is 12 volts and the current through it is 5 amperes?
>
> *Solution* — The power will be radiated in the form of heat. Remember that heat is always generated when current flows through a resistor.

$$POWER = VOLTS \times AMPERES$$
$$= 12 \times 5$$
$$= 60 \text{ watts} \qquad\qquad \textit{Answer}$$

Much of the power in electric circuits is radiated in the form of heat. This is why resistors are given power ratings. The power rating of a resistor is simply a measure of how much heat the resistor can radiate safely. A 10-watt resistor can radiate more heat safely than a 5-watt resistor.

The unit of power measurement is the watt. This unit is too small for many applications, such as in power company circuits, and so the *kilowatt* is often used. One kilowatt is equal to one thousand watts.

Remember that power is a *rate* at which work is done or energy is used.

$$POWER = \frac{ENERGY}{TIME}$$

If you multiply the power times the time, you will end up with energy again.

$$\frac{ENERGY}{\cancel{TIME}} \times \cancel{TIME} = ENERGY$$

If you use a kilowatt of power in your house for 1 hour, then you have used 1 *kilowatt-hour* of electricity. The amount of kilowatt-hours is obtained simply by multiplying the amount of power times the amount of time that you used it. Since multiplying power times time gives energy, it follows that kilowatt-hours are a unit of energy. This is an important point to remember. You do not pay the power company for the power you use, but rather you pay them for the energy that you use. Since energy and work are measured in the same units, it follows that kilowatt-hours are also a measure of *work.*

SUMMARY

1. Power is the rate of doing work. Power is also the rate of using energy.
2. Power measurement always means that time is involved. Power is the work per second or the energy used per second.

3. When current flows through the resistor, power is radiated in the form of heat. The greater the amount of heat radiated per second, the greater the power.
4. The power in watts is equal to the voltage in volts times the current in amperes.
5. Resistors are rated in watts. The greater the wattage rating of a resistor, the greater the amount of heat it can safely radiate.

Programmed Review Questions

(Instructions for using this programmed section are given in Chap. 1.)

We will review the important concepts of this chapter. If you have understood the material, you will progress easily through this section. Do not skip this material, because some additional theory is presented.

1. How many kilowatt-hours of electricity are used by ten 100-watt light bulbs glowing for ½ hour?
    A. 100 kilowatt-hours (Proceed to block 7.)
    B. ½ kilowatt-hour (Proceed to block 17.)

2. *Your answer to the question in block 9 is **B**. This answer is wrong. The current in amperes is a measure of how many electrons pass a point in the circuit per second. There is no involvement with a measurement of work.* Proceed to block 24.

3. *The correct answer to the question in block 16 is **A**. The first step is to find the circuit current:*

$$I = \frac{E}{R}$$

$$= \frac{10 \ volts}{10 \ ohms}$$

$$= 1 \ ampere$$

The power can now be found:

$$P = E \times I$$
$$= 10 \ volts \times 1 \ ampere$$
$$= 10 \ watts \qquad\qquad\qquad Answer$$

Here is your next question.
Which of the following measurements is a unit of work?
    A. Watts (Proceed to block 13.)
    B. Kilowatt-hours (Proceed to block 6.)

4. *Your answer to the question in block 18 is **A**. This answer is wrong. Did you arrive at your answer by adding the voltage and current? Remember that the power in watts is equal to the number of volts times the number of amperes. Rework the problem.* Then proceed to block 23.

5. *The correct answer to the question in block 12 is **B**. The voltage across the resistor must be determined first. Then the current can be found. Here are the calculations.*

    For the Power in A:

    $$THE\ VOLTAGE\ E = IR$$

    $$= \frac{1}{2} \times 100$$

    $$= 50\ volts$$

    $$POWER = E \times I$$

    $$= 50\ volts \times \frac{1}{2}\ ampere$$

    $$= 25\ watts$$

    For the Power in B:

    $$THE\ VOLTAGE\ E = IR$$
    $$= 4 \times 15$$
    $$= 60\ volts$$
    $$POWER = E \times I$$
    $$= 60\ volts \times 4\ amperes$$
    $$= 240\ watts$$

    *The resistor in **B** is obviously radiating the most heat, since its power in watts is greater.* Here is your next question.
    How is a resistor's ability to radiate heat indicated? (Proceed to block 26.)

6. *The correct answer to the question in block 3 is **B**. Power companies charge for kilowatt-hours, which is a unit of energy or work.* Here is your next question.
    Which of these statements is correct?
    A. The voltage through a battery is measured in volts. (Proceed to block 21.)
    B. The voltage across a battery is measured in volts. (Proceed to block 14.)

7. *Your answer to the question in block 1 is **A**. This answer is wrong.* For the correct method of calculating the number of kilowatt-hours, proceed to block 17.

8. *The correct answer to the question in block 14 is **A**. Power is defined as the **rate** of doing work.* Here is your next question.
    A force of 9 pounds is exerted through a distance of 11 feet. The work done is
    A. 20 foot-pounds. (Proceed to block 19.)
    B. 99 foot-pounds. (Proceed to block 12.)

9. *The correct answer to the question in block 17 is **B**. Work and energy are closely related, and they are measured in the same units.* Here is your next question.

Which of the following electrical units is based on the amount of work done in moving a unit of charge from one point to another?
A. Volts (Proceed to block 24.)
B. Amperes (Proceed to block 2.)

10. *Your answer to the question in block 23 is **B**. This answer is wrong. **Never** say "the current **across** a circuit"! Proceed to block 16.*

11. *Your answer to the question in block 24 is **B**. This answer is wrong. Although energy and work are closely related, the **rate** at which energy is used is not the same as work. Proceed to block 18.*

12. *The correct answer to the question in block 8 is **B**.*

$$\text{WORK} = \text{FORCE} \times \text{DISTANCE}$$
$$= 9 \text{ pounds} \times 11 \text{ feet}$$
$$= 99 \text{ foot-pounds}$$

Note that the unit of work is given as "foot-pounds." This is a unit of work in the British system. Here is your next question.
Which of the following resistors will radiate the most heat? (*Hint:* Find the power in watts for each resistor. The greater the power, the greater the amount of heat.)
A. A 100-ohm resistor with ½ ampere of current flowing through it. (Proceed to block 22.)
B. A 15-ohm resistor with 4 amperes of current flowing through it. (Proceed to block 5.)

13. *Your answer to the question in block 3 is **A**. This answer is wrong. The power in watts is a measure of the **rate** at which work is done (or the rate at which energy is used). Proceed to block 6.*

14. *The correct answer to the question in block 6 is **B**. Voltage exists **across** a component, or **between** two points.* Here is your next question.
The rate of doing work is called
A. power. (Proceed to block 8.)
B. energy. (Proceed to block 25.)

15. *Your answer to the question in block 17 is **A**. This answer is wrong. Power is measured in **watts**. Energy is not measured in watts. Proceed to block 9.*

16. *The correct answer to the question in block 23 is **A**. The current **through** a circuit (not **across** a circuit) is measured in amperes.* Here is your next question.
In the circuit of Fig. 9-7, how much power is radiated by resistor *R*?
A. 10 watts (Proceed to block 3.)
B. 100 watts (Proceed to block 20.)

17. *The correct answer to the question in block 1 is **B**. Ten 100-watt light bulbs will use 1,000 watts of power (10 × 100 = 1,000). One thousand watts is 1 kilowatt.*

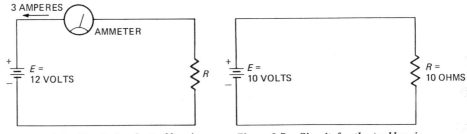

Figure 9-6. Circuit for the problem in block 18.

Figure 9-7. Circuit for the problem in block 16.

The number of kilowatt-hours is simply the number of kilowatts (in this case, 1) times the number of hours (in this case, ½):

$$KILOWATT\text{-}HOURS = 1 \times \tfrac{1}{2}$$
$$= \tfrac{1}{2} \ kilowatt\text{-}hour \qquad Answer$$

Here is your next question.
Which of the following are measured in the same units?
A.   Power and energy (Proceed to block 15.)
B.   Work and energy (Proceed to block 9.)

18. *The correct answer to the question in block 24 is **A**. Power is defined as the rate of using energy or doing work.* Here is your next question.
In the circuit of Fig. 9-6, how much power is radiated by the resistor?
A.   15 watts (Proceed to block 4.)
B.   36 watts (Proceed to block 23.)

19. *Your answer to the question in block 8 is **A**. This answer is wrong. Work is **force times distance**. Rework the problem.* Then proceed to block 12.

20. *Your answer to the question in block 16 is **B**. This answer is wrong. Did you get your answer by multiplying the voltage by the resistance? This is not the correct way to find power. Power is equal to the voltage times the current. You are not given the current in the circuit of Fig. 9-7, but you can find it by Ohm's law. Once you have found the current, you can find the power.* Proceed to block 3.

21. *Your answer to the question in block 6 is **A**. This answer is wrong. You should never say "the voltage **through** a circuit."* Proceed to block 14.

22. *Your answer to the question in block 12 is **A**. This answer is wrong. Calculate the power for each resistor.* Then proceed to block 5.

23. *The correct answer to the problem in block 18 is **B**.*

$$P = E \times I$$
$$= 12 \ volts \times 3 \ amperes$$
$$= 36 \ watts \qquad Answer$$

Here is your next question.
Which of these statements is correct?

A. The current through a circuit is measured in amperes. (Proceed to block 16.)

B. The current across a circuit is measured in amperes. (Proceed to block 10.)

24. *The correct answer to the question in block 9 is **A**. Voltage is a measure of work done in moving a negative charge from a positive point to a negative point.* Here is your next question.
The rate at which energy is used is called

A. power. (Proceed to block 18.)

B. work. (Proceed to block 11.)

25. *Your answer to the question in block 14 is **B**. This answer is wrong. Energy is the **capacity** to do work. It is not the **rate** of **doing** work.* Proceed to block 8.

26. *The wattage rating of a resistor indicates its ability to radiate heat. A 10-watt resistor can radiate more heat than a 5-watt resistor. The wattage rating of a resistor is indicated by its physical size. Naturally, the larger the physical size, the greater the amount of heat it can radiate. There is no universally accepted method of relating size to wattage rating, and it is necessary to obtain this information from manufacturers' data. You can always use a larger wattage rating to assure that the resistor will be capable of radiating heat, but you should never use a smaller wattage rating.*
You have now completed the programmed questions. The next step is to put some of these ideas to work in laboratory experiments. Proceed to the "Experiment" section that follows.

Experiment

(The experiment described in this section may be performed on the circuit board described in Appendix C or on a similar laboratory setup.)

EXPERIMENT 1

Purpose—To show how ac and dc can be identified.

Theory—There are two kinds of electric power: alternating current (ac) and direct current (dc). There are applications where ac power is needed, and there are other applications where dc power is needed. In a later chapter you will study the characteristics of alternating current, but for now the important thing to understand is that the two kinds of power exist.

When electrons in a circuit always flow in the same direction, the current is said to be dc (direct current). When the electrons flow in one direction for a moment, then flow in the opposite direction for a moment (and periodically reverse their direction of flow), the current is said to be ac (alternating current).

Circuits that get their electric power from a battery have direct-current flow. So far, there has not been a battery invented that produces alternating

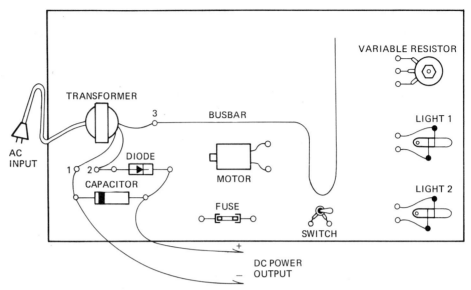

Figure 9-8. A rectifier circuit converts ac to dc.

current. Automobiles use a battery as a portable source of power. That is why the electrical system in the car is dc.

When power is transmitted over miles of wires, alternating current has advantages over direct current. The reason for this will become clearer after you have studied alternating currents and transformers.

Some types of electrical equipment operate only on dc, and others operate only on ac. A *rectifier* is used for converting ac to dc so that dc electrical equipment can be operated from ac power. The output of the rectifier is dc, as shown in Fig. 9-8.

How can you tell if an electric current is dc or ac? There are instruments that you can use, but imagine that you are an experimenter of early times who cannot afford an expensive instrument. In this experiment you will learn a simple method of distinguishing between ac and dc.

Test Setup — For this experiment you will need two flat strips of copper. If flat strips are not available, use heavy copper wire. You will also need a potato which is not so old that it has lost its moisture.

Procedure —

Step 1 — Cut the potato in half and insert the copper strips (or wires) into one-half of the potato as shown in Fig. 9-9. The strips should be close together, but they must not touch.

Step 2 — Wire the circuit as shown in Fig. 9-10. The power supply has a dc output. It is the rectifier that converts the dc to ac.

Step 3 — Allow the direct current from the power supply to flow through the potato for a few minutes. You will note that in the potato the area around *one* of the copper strips has turned black.

Step 4 — Reverse the leads from the power supply. In other words, exchange the alligator clips connected to the copper strips so that lead *a* is connected to strip *y* and lead *b* is connected to strip *x*. After the current has been

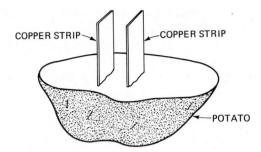

Figure 9-9. Push the copper strips into the potato as shown here.

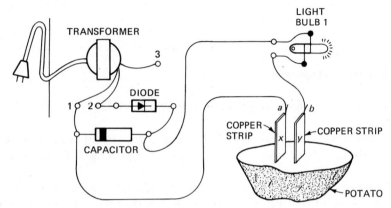

Figure 9-10. With this setup the power supply sends direct current through the potato. The purpose of the light is to limit the current.

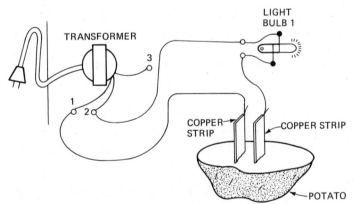

Figure 9-11. With this setup the transformer delivers alternating current to the potato. The light is used to limit current.

flowing for a few minutes, note that the area in the potato around the lead that was white has now turned black.

Step 5 — Insert the copper strips into the unused half of the potato. Wire the circuit as shown in Fig. 9-11. In this case the rectifier is no longer in the circuit, and an alternating current is flowing through the potato.

With an alternating current, one lead is positive and the other is negative for a very short period of time. Then the polarity of voltage reverses so that the lead which was positive becomes negative, and the one which was negative becomes positive. The polarity of the leads swings rapidly from positive to negative, and from negative to positive. If the alternating current from the power company is 60 hertz, each of the leads will become positive 60 times per second. (The same lead also swings to negative 60 times per second.)

Step 6—Note that both strips of copper have caused the potato to turn black. This indicates that an alternating current is flowing.

Conclusion—From this experiment you can conclude that the effects of dc and ac are different. This is due to the fact that with dc the current always flows in the same direction, and with ac the current flows back and forth.

Self-Test with Answers

(Answers with discussions are given in the next section.)

1. In which of the following two cases is the greater amount of work done? (*a*) Lifting a 10-pound weight 12 feet; (*b*) Lifting a 100-pound weight 1 foot.
2. Which of the following units of measurement is based on the amount of work done in moving a charge of electricity? (*a*) Amperes; (*b*) Volts; (*c*) Ohms; (*d*) Watts.
3. Energy exists in many different forms. What are the two kinds of mechanical energy? _____ energy and _____ energy.
4. If the voltage across a resistor is 2 volts, and the current through the resistor is ½ ampere, then how much power is radiated by the resistor in the form of heat? _____watts.
5. Two kinds of electricity are direct current and _____ current.
6. Which of the two kinds of electricity mentioned in Question 5 is supplied by a battery?
7. Which of the two kinds of electricity mentioned in Question 5 is supplied by the power company that delivers electric power to your home?
8. The rate at which work is done, or the rate at which energy is used, is called _____.
9. Your electric bill is a charge for so many *kilowatt-hours* of electricity. This is a unit of electrical (*a*) power; (*b*) energy; (*c*) current; (*d*) resistance.
10. Which of the following resistors will be able to radiate the greater amount of heat? (*a*) A 10-watt resistor; (*b*) A 1-watt resistor.

Answers to Self-Test

1. (*a*)—When a 10-pound weight is lifted 12 feet, the work done is $10 \times 12 = 120$ foot-pounds. When a 100-pound weight is lifted 1 foot, the work done is $100 \times 1 = 100$ foot-pounds. Thus, there is a greater amount of work done in lifting the 10-pound weight in this case.

2. (*b*)—A volt is the amount of work done in moving a coulomb of charge between two points. Although the volt is a unit of work, the applied voltage in a circuit can be considered to be the force that moves electricity through a circuit.

3. *Potential* energy and *kinetic* energy.

4. One watt. The power is equal to the voltage times the current.

$$P = E \times I$$

$$= 2 \times \frac{1}{2}$$

$$= 1 \text{ watt} \qquad \qquad \textit{Answer}$$

5. *Alternating current.* With *direct* current the electrons always move in the same direction through a circuit. With *alternating* current the electrons move back and forth in the circuit.

6. *Direct current.* All batteries supply direct current.

7. *Alternating current.*

8. *Power.* Remember that power always implies that *time* is a factor. The power rating of a resistor is a measure of how much heat energy it can safely radiate per second of time.

9. (*b*)

10. (*a*)—The greater the wattage rating, the greater the amount of heat that it can radiate.

10.
How Do DC Measuring Instruments Work?

Introduction

Electricity did not become a really useful source of energy until methods were devised for measuring current, voltage, and resistance. In the very early experiments, before the measurements were possible, electricity was highly unpredictable. Many of the results of electric current flow were clearly seen, but practical uses were almost nonexistent.

Try to imagine that you are an early experimenter, and no instruments for measuring electricity are available. You wish to set up a simple experiment to prove the current law, sometimes called *Kirchhoff's current law.* (The current entering a junction equals the current leaving the junction.) You cannot go to the store and buy resistors for your experiment—you must make your own. You can obtain wire, but only at a very high expense. You cannot buy a battery in a store to supply the electricity for your experiment. Even if you could obtain the components, you would have no method of measuring the amount of current entering the junction or the amount of current leaving the junction to prove that these values are equal.

How, then, would you go about setting up an experiment to prove that the sum of the currents entering the junction equals the sum of the currents leaving the junction? If you can imagine yourself in this position, you will have more respect for the early experimenters who were faced with such problems. Once you have the necessary supplies, you can demonstrate the current law in a few minutes, but this experiment took weeks in the early days.

Ohm's law tells us that the amount of voltage across the circuit is equal to the amount of current in the circuit times its resistance ($E = IR$). There are, then, three variables in an electric circuit which can be measured: *voltage, current,* and *resistance.* However, if two of the quantities are known, the third can be determined. For example, if you have instruments for measuring the current and the voltage in a circuit, you can find the resistance by Ohm's law ($R = E/I$). Thus, our problem is to be able to measure *two* of the three variables. The purpose of this chapter is to show how some of the basic instruments for measuring voltage and current are made, and to show how these instruments are actually used for making measurements in circuits.

You will be able to answer these questions after studying this chapter:

What are the characteristics of electron current flow that can be measured?
How is the magnetic field of a current used for measuring current?
How is heat used for measuring current?
How can one meter movement be used for measuring both large and small currents?
How can a current meter be used for measuring voltage?
How are ammeters, milliammeters, and microammeters used for measuring current?
How are voltmeters used for measuring voltages?

Instruction

WHAT ARE THE CHARACTERISTICS OF ELECTRON CURRENT FLOW THAT CAN BE MEASURED?

It is not possible to make a direct measurement of electron current flow. To do this, you would need to count the number of electrons that flow past a point in a circuit each second. If the current is 1 ampere, 6.24×10^{18} electrons would flow past a point each second. You would not, of course, count this many, even if you *could* see the electrons (which is *not* possible).

In order to measure current, then, you must make use of one of the *effects* of current. Let's review what they are. *First,* whenever there is a current flow, there is always a magnetic field associated with the current. We can use this magnetic field to set up an opposing force with another magnetic field, and measure the amount of opposition. This will be explained further as we go along in the chapter. *Second,* whenever there is an electric current flowing through a resistor, there is always heat generated. The amount of heat is related to the amount of current flow. A thermocouple can be used to convert the heat into a voltage which can be measured. *Third,* whenever a current flows through a resistance, there is a voltage drop across the resistor. If you can measure this voltage drop, you can determine how much current is flowing through the resistor. This, of course, presumes that we know the exact value of resistance and can measure voltage. This method is used extensively as a method of determining current when only a voltmeter is available.

SUMMARY

1. Electricity did not become a useful source of energy until methods were devised for measuring current, voltage, and resistance.
2. If two of the electrical quantities (current, voltage, and resistance) are known, the third can be determined by using the equation for Ohm's law ($I = E/R$).
3. It is not possible to make a direct measurement of current. For a current of 1 ampere, it would be necessary to count 6.24×10^{18} electrons flowing past a point in 1 second. Not only is it impossible to count this rapidly, but also the electrons are not visible.
4. Instead of measuring current by counting electrons, indirect methods are used. In other words, one of the effects of the current is measured and related to the amount of current.
5. The effects of current used for measurement are magnetic field, heat, and voltage drop.

HOW IS THE MAGNETIC FIELD OF A CURRENT USED FOR MEASURING CURRENT?

Figure 10-1 shows a simple method of indicating current flow by placing the needle of a compass near a current-carrying conductor. The magnetic field of the current reacts with the compass needle and causes it to turn. The amount of turning is proportional to the current. The *restoring force*—that is, the force that causes the needle to go back to its original position—is provided by the earth's magnetic field. Thus, the angle through which the needle turns is related to *the amount of current* and also to *the amount of restoring force*.

The disadvantage of the very crude method of measuring current shown in Fig. 10-1 is immediately obvious. The wire in which the current is flowing should be placed so that it points north and south in order to start the needle at the zero position. It would not always be possible, when making a measurement, to get the wires to run in the desired direction.

Another obvious disadvantage of the system in Fig. 10-1 is in the fact that a very large current is needed to get even a small deflection of the needle. The ability of a meter as an instrument to measure very small quantities is called its *sensitivity*. Thus, the simple instrument of Fig. 10-1 has a *low sensitivity*.

The sensitivity of the instrument in Fig. 10-1 can be increased considerably

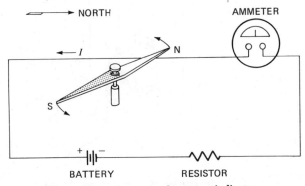

Figure 10-1. *A very simple current indicator.*

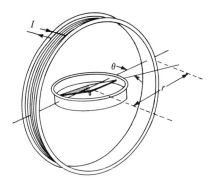

Figure 10-2. A tangent galvanometer can be used for measuring a current (I).

by looping a wire for carrying the current around the compass, as shown in Fig. 10-2. This instrument is more sensitive, and it has many uses in laboratory and classroom experiments. It is called a *tangent galvanometer*. The current to be measured is caused to flow through the loops of wire around the compass, and this sets up a magnetic field which causes the compass to deflect. The amount of deflection is measured in degrees, and then the amount of current is obtained with a mathematical equation. One disadvantage of the tangent galvanometer is that it is necessary to know the exact strength of the earth's magnetic field at the point where the measurement is being taken. This is sometimes difficult to determine, especially in buildings with steel girders, which tend to distort the earth's magnetic field and change the field strength from point to point.

The most important disadvantages in making measurements with the in-

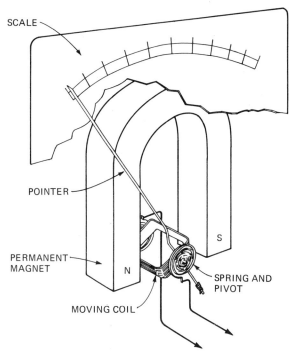

Figure 10-3. An instrument that does not depend on the earth's magnetic field.

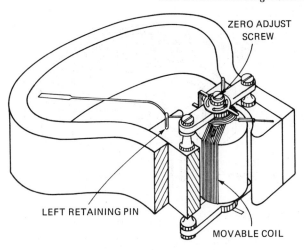

ZERO ADJUST
SCREW

LEFT RETAINING PIN

MOVABLE COIL

Figure 10-4. Construction details of a simple permanent-magnet, moving-coil mechanism.

struments of Figs. 10-1 and 10-2 are the facts that the earth's magnetic field is very weak and its exact value is unknown. To get around these problems, an instrument can be made which does not use the earth's magnetic field, but instead uses the magnetic field of a permanent magnet. Such an instrument is shown in Fig. 10-3. Here the permanent magnet provides a very strong magnetic field, one which is so strong that the earth's magnetic field can be ignored completely. The coil in the magnetic field carries the current to be measured. This current sets up a magnetic field in the coil which reacts with the magnetic field of the permanent magnet and causes the needle to turn up scale. The amount of turning is proportional to the amount of current. Refinements in this method of measurement are used extensively for measuring electric currents.

Figure 10-4 shows the basic construction of a *permanent-magnet, moving-coil meter movement.* This movement is used in modern measuring instruments. (It is often incorrectly referred to as the "D'Arsonval movement.")

Figure 10-5 shows the major parts used in the construction of a permanent-magnet, moving-coil meter movement. Figure 10-5*a* shows the permanent magnet and pole pieces used to establish the magnetic field. (The pole pieces serve to make the magnetic field lines parallel as they cross the coil [not shown]. The reason for doing it is to improve the accuracy of the meter.)

Figure 10-5*b* shows the construction of the coil and the hairsprings used for providing the restoring force to return the pointer to zero. The counterweights are used for making the action of the pointer smoother.

Figure 10-5*c* shows the pivot (called a *jewel*) upon which the movable coil rests. There are actually two pivots used—one at each end of the movable-coil form. Figure 10-6 shows an exploded view of the parts used for making a moving-coil type movement.

Instead of mounting the coil on a jeweled pivot, a more recent method is to use a *taut band.* This method is shown in Fig. 10-7. Current flowing through the coil causes a stretched band to twist, and it is this twisted band which provides the restoring force for returning the needle to zero. An advantage of the taut-band method is that there is no friction between moving parts, and this improves the accuracy of measurement.

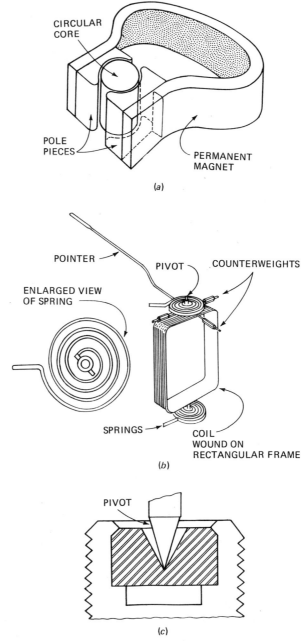

Figure 10-5. *Major parts used in the construction of a permanent-magnet, moving-coil meter movement.* (a) *The permanent magnet and pole pieces used to establish the magnetic field.* (b) *Construction of the coil and the hairsprings.* (c) *The pivot upon which the movable coil rests.*

The permanent-magnet, moving-coil type meter movement is usually made for measuring very small currents. A typical maximum current value is 50 microamperes—that is, 50-millionths of an ampere. In order to measure larger currents, or to measure a voltage, this sensitive meter movement must be placed in circuits which limit the current through it to a maximum value of 50-

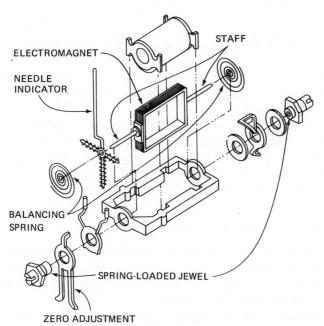

Figure 10-6.   This exploded view shows how the parts of a permanent-magnet, moving-coil mechanism are assembled.

millionths of an ampere. As was stated earlier, in order to measure voltage, current, and resistance, it is necessary to be able to measure only two of the quantities—voltage and current—and the third (resistance) can be determined mathematically. With a sensitive meter movement, an instrument can be made that will measure both voltage and current. The meter movement most often used is the permanent-magnet, moving-coil type.

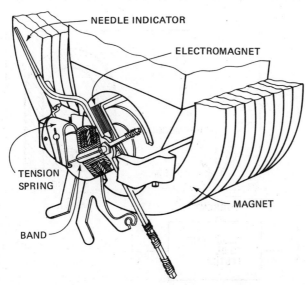

Figure 10-7.   A taut-band meter movement.

SUMMARY

1. A compass needle can be used for making a simple instrument to indicate current.
2. When the wire is looped around the compass, the magnetic field of the current is stronger and the instrument is more sensitive. This is how a tangent galvanometer is made.
3. A very sensitive meter movement can be made by using a strong permanent magnet instead of using the earth's magnetic field. This is how a permanent-magnet, moving-coil meter movement is made.
4. The restoring force for a permanent-magnet, moving-coil movement may be a very fine spring, or it may be a taut band.

HOW IS HEAT USED FOR MEASURING CURRENT?

In an earlier chapter you learned that there are devices called *thermoelectric* (TE) generators for converting heat energy into electrical energy. A TE generator is a junction of two dissimilar metals (called a *thermocouple*). When heated, it produces a voltage across its terminals.

The *thermocouple ammeter* employs a thermocouple for converting the heat which is generated by a current flowing through a resistance into a voltage value. This thermocouple voltage causes a current to flow through a sensitive meter movement.

Figure 10-8 shows the basic construction. The current to be measured is caused to flow through the wire between points *A* and *B*. This wire has a known value of resistance which is quite small. Resistance must be present so that heat will be generated when the current flows. A thermocouple is welded to the wire so that the temperature produces a voltage across terminals *C* and *D* of the thermocouple. The voltage causes a small amount of current to flow through a resistance (wire *A*) and meter movement. (The meter movement may be the jeweled or taut-band type described in the previous section.) The resistor (*R*) limits the current through the sensitive meter. The amount of current flow in

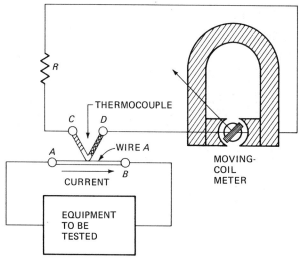

Figure 10-8. A thermocouple ammeter.

the meter circuit is directly related to the amount of current flowing between points *A* and *B* — that is, it is related to the current to be measured. The meter is calibrated to read the amount of current being measured.

Thermocouple meters are very rugged, and they are used in industrial applications.

HOW CAN ONE METER MOVEMENT BE USED FOR MEASURING BOTH LARGE AND SMALL CURRENTS?

Suppose you have a meter movement that deflects to full scale — that is, deflects so that the needle is all the way to the maximum point on the scale — when a current of 50 microamperes flows through it. If you connect this meter into a circuit that has more than 50 microamperes of current, the needle will slam against the right side of the scale and will likely be damaged. Furthermore, excessive current flow through the delicate coil mechanism is likely to cause the wire to overheat and be destroyed.

In order to use a 50-microampere meter movement in a circuit for measuring a larger current, it is necessary to provide a bypass path for the excessive current to flow around the meter movement. This is illustrated in Fig. 10-9, where it is desired to measure a maximum current of 200 microamperes. A parallel path is provided around the meter movement, so that 50 microamperes flows *through* the meter and causes full-scale deflection. The balance of the current, or 150 microamperes, flows through the parallel path (R_{SH}), which is called the *shunt resistance*.

In practice, this shunt resistance has such a low value of resistance that it is not a resistor in the same form as the ones shown on the circuit board in Appendix C. Instead, it is usually just a bar of metal.

The arrangement in Fig. 10-9 provides that 200 microamperes will produce full-scale deflection on the meter. Anything less than 200 microamperes can also be measured with this arrangement. The meter scale is numbered so that full scale indicates 200 microamperes, as shown in Fig. 10-10. At half-scale deflection 100 microamperes is flowing, and for one-fourth–scale deflection 50 microamperes is flowing.

The use of a shunt increases the range of measurements that can be made with a delicate meter movement. Some commercial meters are sold with a

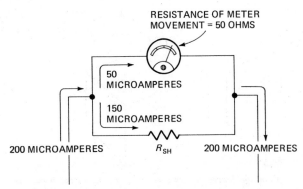

Figure 10-9. How a shunt is used to extend the range of measurements that can be made with a given meter movement.

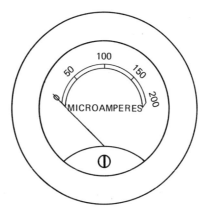

Figure 10-10. This meter movement has a scale that shows a maximum current of 200 microamperes. Such a meter is said to have a "full-scale deflection of 200 microamperes."

number of different ranges for measuring current. Such meters employ a different value of shunt resistance for each different range of current to be measured.

HOW CAN A CURRENT METER BE USED FOR MEASUSING VOLTAGE?

You will remember that a voltage is considered to be the force which pushes electric current through a circuit. A voltage may exist without a current flowing, as shown in Fig. 10-11. When you purchase a battery at a gas station, for example, there is a 6- or 12-volt potential across the terminals even when the battery is not connected into a circuit.

Suppose you wish to measure the amount of terminal voltage and all that is available is a delicate meter movement of the type that has been discussed. One way of doing this is to put a large resistor in series with the meter movement to limit the amount of current flow to a maximum value of 50 microamperes. This resistor, which is marked R_S in Fig. 10-11, is called a *multiplier* or *series multiplier*. The meter face is marked to read voltage values.

Suppose the battery in Fig. 10-11 has a terminal voltage of 12 volts, and the meter movement is such that it cannot carry more than 50 microamperes of current. The resistance value must be such that it limits the current flow from

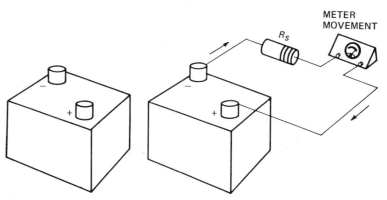

Figure 10-11. A voltage source causes a current flow whenever a current path is present. A voltage exists across the battery terminals at all times. A large resistance value across the voltage source causes a small amount of current flow.

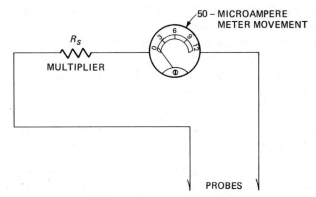

Figure 10-12. Voltmeter circuit.

the 12-volt source to 50 microamperes. Ohm's law can be used to find the value of resistance. If we neglect the resistance of the meter,

$$R = \frac{E}{I}$$

$$= \frac{12 \text{ volts}}{50 \text{ microamperes}}$$

(Remember that you cannot divide volts by microamperes. To use Ohm's law, it is necessary to change the current to amperes. 50 microamperes = 0.00005 ampere.)

Substituting 0.00005 ampere for 50 microamperes:

$$R = \frac{12}{0.00005}$$

$$= 240,000 \text{ ohms}$$

$$= 240 \text{ kilohms} \qquad \qquad \textit{Answer}$$

With the value of 240 kilohms in series with the meter movement, the needle will deflect to a full-scale reading when placed across 12 volts. Anything less than 12 volts will cause the meter to deflect to a lower value.

Figure 10-12 shows the circuit for a voltmeter with the multiplier in series with the meter movement. The meter scale has been divided to read a maximum value of 12 volts full scale for the particular example used. If this voltmeter is used to measure the voltage across a 6-volt battery, the meter will deflect to half-scale value. In other words, it will read 6 volts on the meter face.

Commercial-type meters which have more than one voltage scale have some provision for switching in different values of series-multiplier resistors in series with the meter movement.

SUMMARY

1. A thermocouple can be used to convert heat into a voltage. When the thermocouple is welded to a resistance wire, it can be used to indicate the amount of current flowing through the wire.
2. A sensitive meter movement can be placed in a parallel circuit for measuring larger current values.

3. The parallel path that permits a sensitive meter movement to be used for measuring larger currents is called a *shunt*.
4. A sensitive meter movement can be placed in series with a resistor to limit the amount of current through it.
5. A voltmeter is made by placing a sensitive meter movement in series with a resistor. The resistor is called a *multiplier*.

HOW ARE AMMETERS, MILLIAMMETERS, AND MICROAMMETERS USED FOR MEASURING CURRENT?

Figure 10-13 shows how a current-measuring device is used for measuring the amount of current flowing through a circuit. Note that in order to employ the meter for measuring the *current*, it is necessary to open the circuit and place the meter in series with the circuit. In other words, the current that is flowing through the circuit must flow through the meter if the meter is to be able to measure it.

The terminals on an ammeter (or milliammeter, or microammeter) are usually marked with a minus and a plus sign. The meter must be inserted into the circuit so that the electron current flows through the meter from its minus terminal toward its plus terminal. As shown in Fig. 10-13, the negative side of the meter must go toward the negative side of the circuit, and the positive side of the meter must go toward the positive side of the circuit. This is not difficult

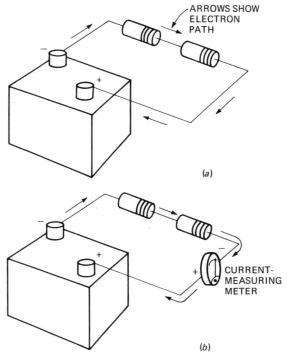

Figure 10-13. To measure current, the meter is placed in the circuit in such a way that the current being measured flows through it. (a) *To measure the **current** in this circuit,* (b) *the circuit must be opened and the meter must be placed in series with the circuit components.*

to remember if you keep in mind the fact that the current must flow *through* the meter toward the plus terminal.

Before you connect a current instrument into a circuit you should have an approximate idea of how much current is flowing. This can usually be calculated quickly by Ohm's law, and you should do this in order to determine that the amount of current does not exceed the rating of the meter. In other words, if the meter is designed to read a maximum current of 1 ampere, and your calculations show that the current through the circuit is 1½ amperes, you should *not* connect the meter into the circuit, or it will be destroyed.

For meters that have a number of different scales, the procedure is to start with the largest scale—that is, the scale that provides for the maximum current reading—and try the measurement to see if the current can be measured. If the current is too small to deflect the meter, the next step is to move to the next smallest scale, and then to the next smallest scale in turn, until you reach a scale in which the deflection is between ¼ and ¾ of full-scale deflection. By starting with the high scale and working down, you will not exceed the current rating of the meter. On the other hand, if you started on a lower scale and the current was excessive, the meter would be destroyed.

There is one very important precaution that you must remember when using an ammeter, milliammeter, or microammeter. These instruments are purposely designed so that they do not offer much opposition to the current flowing through them. If the meter offered a large opposition to current flow, it would change the amount of current in the circuit and the instrument would not be accurate. For this reason a very small amount of current will produce a large deflection on the meter movement. If you should accidentally connect a current-reading instrument across a voltage source, the instrument would act as a short circuit. Thus, it would act in such a way that there would be practically no resistance across the voltage, and a large amount of current would flow. This situation is illustrated in Fig. 10-14. Remember this very important rule:

NEVER CONNECT A CURRENT-READING INSTRUMENT IN A PLACE WHERE THERE IS A VOLTAGE PRESENT. THESE INSTRUMENTS ARE DESIGNED TO BE CONNECTED *IN SERIES WITH THE CIRCUIT, NEVER* ACROSS THE CIRCUIT!

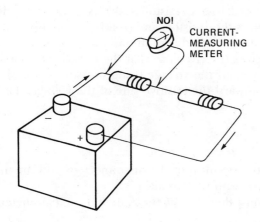

Figure 10.14. YOU MUST NEVER CONNECT AN AMMETER, MILLI-AMETER, OR MICROAMMETER ACROSS A VOLTAGE AS SHOWN IN THIS ILLUSTRATION! IF YOU DO, THE METER WILL BE DESTROYED!

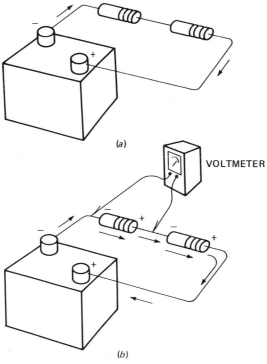

Figure 10-15.   This illustration shows how a voltmeter is used for measuring current. (a) ***To measure the voltage across a component,*** (b) ***place the voltmeter across the component.***

HOW ARE VOLTMETERS USED FOR MEASURING VOLTAGES?

A voltmeter is made by connecting a delicate meter movement in series with a multiplier resistor. Voltmeters are specifically designed for measuring differences in potential. Since a difference in potential exists *across* two points, it stands to reason that the voltmeter must be connected *across* the same two points. In contrast to an ammeter, a voltmeter is connected *across* a voltage source as shown in Fig. 10-15. Note that it is not necessary to disconnect any part of the circuit in order to make a voltage reading. The arrows in the circuit of Fig. 10-15 show the direction of electron current flow.

The polarity of the voltage drop across a resistor is always negative at the side where the electron current enters the resistor, and positive at the opposite side. The voltmeter terminals are often marked with minus and plus signs, and the voltmeter is to be connected across the circuit so that the negative terminal of the voltmeter is to the negative side of the voltage being measured, and the positive side of the voltmeter goes toward the positive side of the voltage. This is illustrated in Fig. 10-15.

SUMMARY

1. In order to measure current, the circuit must be disconnected, and the instrument must be placed in *series* with the circuit.
2. The current being measured must flow *through* the ammeter, milliammeter, or microammeter.

3. To measure a voltage, the voltmeter is connected *across* the voltage. There is no need to disconnect the circuit to make a voltage measurement.
4. A current meter is connected into a circuit so that the electron current enters the negative terminal of the meter and leaves the positive terminal.
5. A voltmeter is connected so that its negative lead goes to the negative side of the voltage being measured, and the positive lead goes to the positive side.

Programmed Review Questions

(Instructions for using this programmed section are given in Chap 1.)

We will review the important concepts of this chapter. If you have understood the material, you will progress easily through this section. Do not skip this material, because some additional theory is presented.

1. When an ammeter is connected into a circuit, the positive terminal of the ammeter goes toward
A. the positive side of the source. (Proceed to block 7.)
B. the negative side of the source. (Proceed to block 17.)

2. *Your answer to the question in block 4 is* **B**. *This answer is wrong. For the very simple type of measuring instrument called the "tangent galvanometer," the restoring force is provided by the earth's magnetic field. However, for the permanent-magnet, moving-coil meter, the restoring force is provided by a spring or by a taut band.* Proceed to block 18.

3. *Your answer to the question in block 7 is* **B**. *This answer is wrong. A multiplier is used with a sensitive current-reading instrument to make a voltmeter.* Proceed to block 19.

4. *The correct answer to the question in block 20 is* **A**. *Remember that the value of* E *must be in volts and the value of* I *must be in amperes in order to use this equation.* Here is your next question.
The restoring force in a permanent-magnet, moving-coil meter may be provided by a fine spring or by
A. a taut band. (Proceed to block 18.)
B. the earth's magnetic field. (Proceed to block 2.)

5. *The correct answer to the question in block 23 is* **A**. *As shown in Fig. 10-16, you should always* **start** *by measuring with the meter set on its highest scale. This is true for both voltmeters and current-reading meters.* Here is your next question.
A certain meter movement deflects full scale when a current of 100 microamperes flows through it. You wish to use this meter movement to measure currents up to 1,000 microamperes. Which of the following would you use?
A. A multiplier that limits the current to 100 microamperes (Proceed to block 24.)
B. A shunt that carries a current of 900 microamperes when measuring a 1,000-microampere current (Proceed to block 12.)

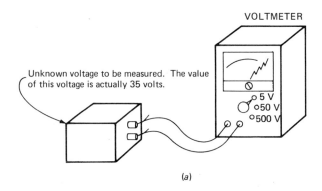

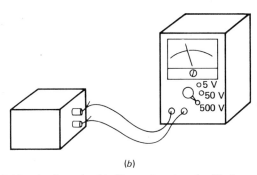

Figure 10-16. As shown in this illustration, you should always start your measurement with the highest scale on a meter. (a) *When you set the voltmeter on the 5-volt scale and read a voltage greater than 5 volts, the meter is likely to be permanently damaged. In this illustration, the "unknown" voltage actually has a value of 35 volts. Connecting the meter across this voltage when it is on the 5-volt scale will cause the meter needle to slam against the pin and may permanently damage the meter movement.* (b) *When connecting the meter across an unknown voltage value, always start with the highest scale. Your reading will tell you if it is OK to go to the next lower scale.*

6. *The correct answer to the question in block 12 is **A**. If you connect a voltmeter into a circuit "backwards"—that is, if you connect it so that its **positive lead** is at the negative side of the voltage being measured and its **negative lead** is at the positive side of the voltage being measured—you can damage the meter movement. Current flowing in the wrong direction through the meter movement causes the needle to slam against the pin at the zero point. The same thing happens if current flows through a current-reading meter in the wrong direction.* Here is your next question.

   A certain ammeter is designed to read a maximum value of current of 1 ampere. This ammeter is connected into a circuit in which there is a current of 500 milliamperes flowing. Which of the following statements is correct?

   A.   The ammeter will be damaged. (Proceed to block 13.)
   B.   The ammeter will not be damaged. (Proceed to block 22.)

7. *The correct answer to the question in block 1 is **A**. Remember that the meter must be placed in the circuit so that electrons will flow through it from its negative terminal toward its positive terminal.* Here is your next question.

In order to convert a delicate 50-microampere meter movement into a meter for measuring amperes, you would use a

A.   shunt. (Proceed to block 19.)

B.   multiplier. (Proceed to block 3.)

8.   *Your answer to the question in block 16 is **B**. This answer is wrong. In order to use an ammeter, milliammeter, or microammeter, it is connected in series with the circuit so that the circuit current must flow through the instrument.* Proceed to block 20.

9.   *Your answer to the question in block 12 is **B**. This answer is wrong. A very small amount of current flows through a voltmeter from its negative to its positive lead. In order for this current to flow, the negative test lead must be connected to the negative side of the voltage being measured, and the positive lead must be connected to the positive side of the voltage being measured.* Proceed to block 6.

10. *The correct answer to the question in block 22 is **B**. The sensitivity of a meter movement may be defined as the amount of current required for full-scale deflection.*

Another method of rating the sensitivity is by the amount of resistance that must be placed in series with the meter to cause full-scale deflection when 1 volt is applied. Thus, if a 20,000-ohm resistor must be placed in series with the meter so that full-scale deflection occurs when 1 volt is applied, the sensitivity is said to be 20,000 ohms per volt. The higher the ohms-per-volt rating of a meter movement, the greater its sensitivity. Here is your next question.

A current of 50 milliamperes is flowing through a 100-ohm resistor. Can the voltage across this resistor be measured with a voltmeter that has a full-scale reading of 1 volt? (yes or no). (Proceed to block 26.)

_____

11. *Your answer to the question in block 19 is **A**. This answer is wrong. Shunts are used in **ammeters,** not in voltmeters.* Proceed to block 16.

12. *The correct answer to the question in block 5 is **B**. All the excess current must flow through the meter shunt. In the example given, more current flows in the shunt than flows through the meter movement. It follows, then, that the shunt resistance is less than the meter-movement resistance.* Here is your next question.

When connecting a voltmeter across a voltage,

A.   the negative lead of the voltmeter goes to the negative side of the voltage being measured. (Proceed to block 6.)

B.   the positive lead of the voltmeter goes to the negative side of the voltage being measured. (Proceed to block 9.)

13. *Your answer to the question in block 6 is **A**. This answer is wrong. In order to determine whether or not 500 milliamperes will damage a 1-ampere (maximum reading) ammeter, you must convert the 500 **milli**amperes to amperes. To convert milliamperes into amperes, move the decimal point to the left three places.*

Once you find out how many amperes are flowing, you will know if the meter can safely measure the current. Work this simple conversion problem. Then proceed to block 22.

14. *Your answer to the question in block 18 is **B**. This answer is wrong. A thermoelectric generator produces a voltage when it is heated, but it is not a type of measuring instrument.* Proceed to block 23.

15. *Your answer to the question in block 23 is **B**. This answer is wrong. Figure 10-16 shows why. As shown in this illustration, you have a voltmeter that has 5-volt, 50-volts, and 500-volt scales. The **maximum** amount of voltage that can be read on the 5-volt is **5 volts**. If you connect the meter across an unknown voltage that turns out to be a 35-volt source, and the meter is set on the 5-volt scale (Fig. 10-16a), you will surely damage the meter. However, if the meter is set on the 500-volt scale (Fig. 10-16b), and you connect it across an unknown voltage that turns out to be 35 volts, you will not damage the meter movement. You **will** be able to tell that the unknown voltage is less than 50 volts, and then you can safely set the meter on the 50-volt scale.* Proceed to block 5.

16. *The correct answer to the question in block 19 is **B**. Multipliers are resistors connected in series with a meter movement. The combination (meter movement and multiplier) makes a voltmeter.* Here is your next question.
Never connect an ammeter, milliammeter, or microammeter
A. across a circuit. (Proceed to block 20.)
B. in series with a circuit. (Proceed to block 8.)

17. *Your answer to the question in block 1 is **B**. This answer is wrong. Study Fig. 10-13.* Then proceed to block 7.

18. *The correct answer to the question in block 4 is **A**. Taut-band meters are more popular in the newer meter movements.* Here is your next question.
The heat generated when current flows through a resistance may be used for measuring the value of the current. An instrument that measures current this way is the
A. thermocouple ammeter. (Proceed to block 23.)
B. thermoelectric generator. (Proceed to block 14.)

19. *The correct answer to the question in block 7 is **A**. A "shunt path" is a parallel path. When a shunt is used with a sensitive current-reading instrument, it provides a path for excessive current to flow around the meter. Typically, shunts have a very low resistance value, while multipliers have a high resistance value.* Here is your next question.
In order to convert a delicate 50-microampere meter movement into a voltmeter, you would use a
A. shunt. (Proceed to block 11.)
B. multiplier. (Proceed to block 16.)

20. *The correct answer to the question in block 16 is **A**. This is a very important point. Meters that are designed to read current values should never be con-*

nected **across** *a circuit where a voltage is present.* Here is your next question.

If you measure the voltage across a resistor and the current through that resistor, then you can determine the resistance value from the equation

A. $R = E/I$. (Proceed to block 4.)

B. $R = EI$. (Proceed to block 21.)

21. *Your answer to the question in block 20 is* **B**. *This answer is wrong. You may have been confused by the equation* $P = EI$ (*power equals voltage times current*). *Proceed to block 4.*

22. *The correct answer to the question in block 6 is* **B**. *A current of 500 milliamperes is equal to 0.5 ampere. This is only ½ of the maximum amount of current that the meter can safely carry!* Here is your next question.

The ability of a meter to measure very small quantities is called its

A. selectivity. (Proceed to block 25.)

B. sensitivity. (Proceed to block 10.)

23. *The correct answer to the question in block 18 is* **A**. *A thermocouple ammeter is useful for certain types of current measurements.* Here is your next question.

Which of the following rules should be followed when using a meter?

A. When measuring with a meter that has more than one scale, always start by using the highest scale. (Proceed to block 5.)

B. When measuring with a meter that has more than one scale, always start by using the lowest scale. (Proceed to block 15.)

24. *Your answer to the question in block 5 is* **A**. *This answer is wrong. Remember that multipliers are used with* **voltmeters,** *not with current-reading meters. Proceed to block 12.*

25. *Your answer to the question in block 22 is* **A**. *This answer is wrong. The word "selectivity" does not describe a meter's* **sensitivity** *to small currents. Proceed to block 10.*

26. *To answer the question in block 10, it is necessary to find the voltage across the resistor by using Ohm's law. The current value of 50 milliamperes is converted to amperes so the Ohm's law equation can be used.*

$$50 \text{ milliamperes} = 0.05 \text{ ampere}$$

According to Ohm's law, voltage equals current times resistance.

$$E = IR$$

The current (I) is 0.05 ampere, and the resistance (R) is 100 ohms. Substituting these values into the equation:

$$E = IR$$
$$= 0.05 \times 100$$
$$= 5 \text{ volts}$$

Since the voltage across the resistor is 5 volts, the meter cannot be used for measuring this voltage, because its maximum reading is 1 volt.

You have now completed the programmed questions. The next step is to put some of these ideas to work in laboratory experiments. Proceed to the "Experiments" section that follows.

Experiments

(The experiments described in this section may be performed on the circuit board described in Appendix C or on a similar laboratory setup.)

EXPERIMENT 1

Purpose — The purpose of this experiment is to determine the resistance of a circuit component by making voltage and current measurements.

Theory — It is not always possible to determine the approximate value of current flowing in a circuit before you connect the meter into the circuit. Thus, you will have to start by using the highest scale on the meter and working down until you obtain the reading.

In this experiment you will determine the resistance of a filament in a light bulb. Two 6-volt light bulbs will be used, as shown in Fig. 10-17. If you know the current through the light bulb (I) and the voltage across the light bulb (E), then you can calculate the resistance of the light bulb using the equation $R = E/I$.

Procedure —

Step 1 — Connect the circuit as shown in Fig. 10-17. Be sure to connect the milliammeter with the highest possible scale of reading.

Step 2 — With the meter on the highest current scale, close the switch and determine if there is any current measurement with the meter at this setting. If not, open the switch and then set the meter to the next lower scale. Continue doing this until the meter needle moves up scale when you close the switch.

Your reading on the meter will tell you if it is safe to go to the next lower scale. For example, if the meter needle shows a reading of 15 millamperes when it is on the 100-milliampere scale, then it would *not* be safe to go to a 10-milliampere scale.

When you have determined the value of current flow (I), record the value here.

$$I = \underline{\hspace{3cm}} \text{ milliamperes}$$

Step 3 — Measure the voltage across light 1 (E_1) and across light 2 (E_2). Record these voltage values here.

$$E_1 = \underline{\hspace{2cm}} \text{ volts}$$
$$E_2 = \underline{\hspace{2cm}} \text{ volts}$$

Step 4 — Using Ohm's law ($R = E/I$) find the resistance of light 1 (R_1) and of light 2 (R_2). These resistance values should be approximately equal. Record your calculated values here.

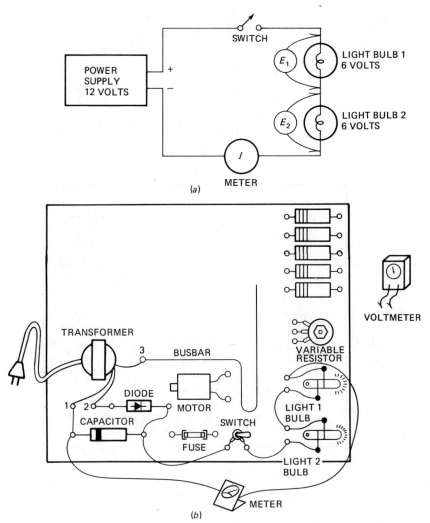

Figure 10-17. *Test setup for Experiment 1.* (a) *Test setup for determining the resistance of light filaments.* (b) *This is the way your circuit board will look if you have wired the circuit correctly.*

$R_1 =$ _____ ohms

$R_2 =$ _____ ohms

Conclusion — The resistance of a component can be determined by measuring the voltage across it and the current through it.

EXPERIMENT 2

Purpose — The purpose of this experiment is to obtain practice in measuring voltage.

Theory — Voltage is measured by placing the voltmeter across the voltage. In this experiment you will measure the power supply voltage and the voltage across a 100-ohm resistor. The voltage value that you measure across the resistor will be used for part of the next experiment.

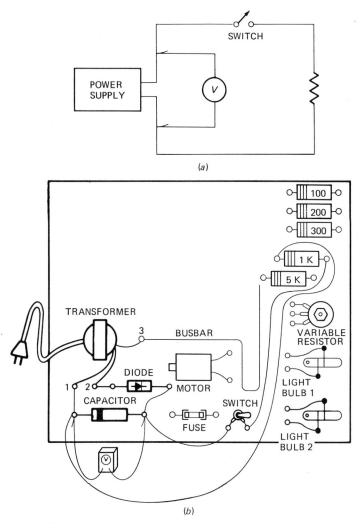

Figure 10-18. Test setup for the first part of Experiment 2. (a) *Use of the voltmeter to measure the power-supply voltage.* (b) *This is the way your circuit board will look if you have wired the circuit correctly.*

Procedure—

Step 1—With the meter connected in the circuit as shown in Fig. 10-18, close the switch and measure the amount of voltage across the power supply terminal. Remember to start your voltage measurement by starting on the highest scale and working down. Record the value here.

$$E = \underline{\hspace{2cm}} \text{ volts}$$

Step 2—Connect the circuit as shown in Fig. 10-19 and record the voltage value as obtained in this position.

$$E = \underline{\hspace{2cm}} \text{ volts}$$

You should get exactly the same value as you did in step 1. This is easy to understand, because the wires in the circuit are considered to have no resistance and essentially you are connecting the voltmeter to the same point.

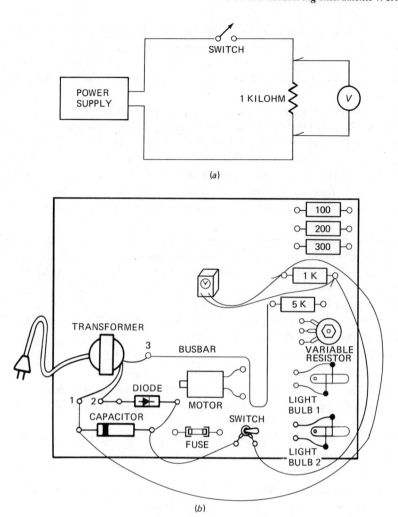

Figure 10-19. *Test setup for the second part of Experiment 2.* (a) *Use of the voltmeter to measure the voltage across a resistor.* (b) *This is the way your circuit board will look if you have wired the circuit correctly.*

Conclusion— Voltage is measured by placing the voltmeter across the voltage being measured.

EXPERIMENT 3

Purpose— The purpose of this experiment is to get experience in measuring current, and to show that the current has the same value in different parts of a simple circuit.

Theory— In this experiment you will first calculate the amount of current that is flowing in a circuit by Ohm's law. Then you will verify your answer by connecting a meter into the circuit and actually measuring the current. One of the reasons for calculating the current value before measuring it is that you get practice in working Ohm's law problems. Another important reason for doing this is that it gives you an idea of how much current is flowing in a circuit, and therefore allows you to choose the proper scale when making the current measurements. Remember that if you have no idea of how much current is flowing in a circuit, the procedure is always to start with the largest scale on the meter and work to the lowest value.

Answers are not given to the experiment problems in this section, because you will check each Ohm's law answer by actual measurement. You should not expect that your answers will be 100 percent accurate. For one thing, you will calculate the amount of current flowing by using the *rated* amount of resistance. However, resistance values of resistors are seldom exactly equal to their rated value, and therefore some measure of error will occur here. Also, your meter is not a 100 percent accurate instrument, and a slight error may exist in the meter reading.

It is not unusual to have a slight error between the calculated and measured values. Technicians working on the most complicated computers and color-television receivers use instruments which are no more accurate than the ones that you are using in these experiments. Of course, very precise instruments could be used to determine current and voltage values more accurately, but for most purposes it is not necessary to know the current flow or voltage value exactly to three decimal places.

Procedure—

Step 1—In the circuit of Fig. 10-20, a 1,000-ohm resistor is placed across a 12-volt power supply. In order to use the equation for Ohm's law, use the value of *E* that you obtained in Experiment 2.

Using Ohms's law, determine the amount of current (*I*) in this circuit.

$$I = E/R = \underline{\hspace{3cm}} \text{milliamperes}$$

Step 2—Determine which scale on your meter you will be able to use to measure the current in the circuit of Fig. 10-20.

Meter scale _____

Step 3—Connect the circuit as shown in Fig. 10-21. Note that the current meter must be used on the scale that you determined to be appropriate—that is, your answer to Step 2. Note also that the current meter is connected between the positive terminal of the power supply and the switch.

After you have determined that your circuit is connected properly, and that you have used the proper scale on the meter, close the switch and record your current reading at this point.

Meter reading _____ milliamperes

Step 4—Wire the circuit as shown in Fig. 10-22. Note that this circuit is the same as the one in Fig. 10-21 except that the meter is connected between the negative side of the power supply and the resistor instead of between the positive side of the power supply and the switch. You should measure exactly the same amount of current (*I*) with the meter in this position. Therefore, your meter should be connected on the same scale as in the previous step.

Once you have determined that the meter is properly connected and that it

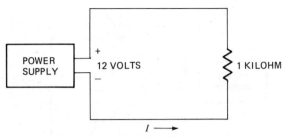

Figure 10-20. A simple resistive circuit.

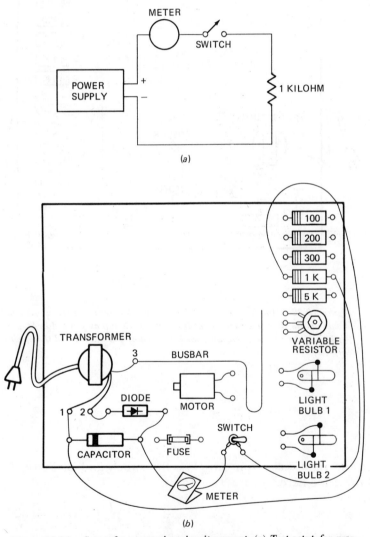

Figure 10-21. Setup for measuring circuit current. (a) *Test setup for measuring the current through the resistor.* (b) *This is the way your circuit board will look if you have wired the circuit correctly.*

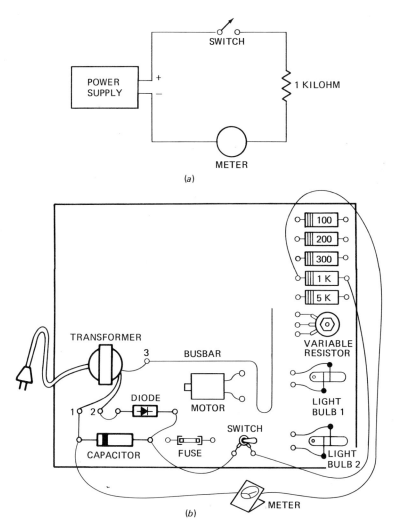

Figure 10-22. Another setup for measuring the circuit current. (a) You should obtain the same current reading with this setup as you did in Step 3. (b) This is the way your circuit board will look if you have wired the circuit correctly.

is set on the right scale, you can close the switch and measure the current. Record the value here.

$$I = \underline{\hspace{3cm}} \text{ milliamperes}$$

Conclusions—

1. The current measurements verify the current value as calculated by Ohm's law.
2. The current value in different parts of a simple circuit is the same.

Self-Test with Answers

(Answers with discussions are given in the next section.)

1. It is not possible to measure current directly. Instead, current is measured by the effects that it produces. Two of these effects are the magnetic field that it produces and (*a*) the color code of the resistor in the circuit; (*b*) the heat produced by the current flow.
2. There are two methods of providing the restoring force for the needle in a moving-coil meter. One method is to use coil springs, and the other is to use (*a*) a taut band; (*b*) a heavy weight.
3. The jewel of a meter movement is (*a*) a diamond or ruby used because of its hardness; (*b*) the pivot that the movable coil turns on.
4. In order to increase the range of current values that a sensitive current meter can measure, you would use (*a*) a shunt; (*b*) a multiplier.
5. In order to use a current-measuring meter for measuring a voltage, you would use (*a*) a shunt; (*b*) a multiplier.
6. An ammeter is connected (*a*) so that the circuit current flows through it; (*b*) so that it is across the circuit voltage.
7. When measuring a voltage in a circuit in which you are unsure of the approximate voltage value, always start with (*a*) the highest scale on the meter; (*b*) the lowest scale on the meter.
8. If you measure the current (I) through a component, and the voltage (E) across that component, then you can find the resistance of that component by using the equation (*a*) $R = E/I$; (*b*) $R = E \times I$.
9. Which of the following is *not* an effect that is always produced when current flows through a resistor? (*a*) Light is produced; (*b*) A voltage drop occurs.
10. The ability of an instrument to measure small currents is called its (*a*) sensitivity; (*b*) drive.

Answers to Self-Test

1. (*b*)—The color code of the resistor tells the resistance value, and the resistance value determines the amount of current flow. However, the current flow does not *produce* the color code.
2. (*a*)—Taut-band meters are now preferred because they are very rugged and very accurate.
3. (*b*)—Figure 10-5*c* shows a closeup view of this pivot.
4. (*a*)—Figure 10-9 shows how the shunt is used.
5. (*b*)—Figure 10-10 shows how the multiplier is connected.
6. (*a*)—*Never* connect an ammeter (or any other current-measuring instrument) *across* a voltage!
7. (*a*)
8. (*a*)—It is very important to know all the forms of Ohm's law.
9. (*a*)—It is true that light is produced in the case of a light bulb. However, there are many circuits in which no light is produced. The three effects always produced are *heat, a magnetic field,* and *a voltage drop.*
10. (*a*)

11.
What Is a Series Circuit?

Introduction

Sometimes a single power source is not capable of performing a job by itself. In such cases, more than one power source may be used. Figure 11-1 shows a simple example. One boy could not pull the wagon up the hill, but the two boys working together can pull it.

As another example, on a long uphill grade, two train engines may be used to pull the loaded cars, but one engine working alone could not pull the load.

Can you think of any other examples where more than one power source is used to accomplish a job?

In electric circuits more than one power supply may be needed. When two or more electric power sources are used in a circuit, they may be connected together in a number of different ways. One way is to connect them in *series*. This is a method that allows the sources to work together to produce a greater voltage. In this chapter you will learn how power sources may be connected in series.

There are different ways of connecting resistors in circuits. One way is to connect them in series. You will study series resistor circuits in this chapter, and what you learn here will apply to other types of loads connected in series. In this chapter you will learn how resistors are connected in series.

Figure 11-1.   A simple example of two power sources used together.

The amount of current flowing in a series circuit can be found by using Ohm's law, but it has to be changed slightly. The power used in a series circuit is important to know if you have to choose a power supply. In this chapter you will learn about current and power in a series circuit.

You will be able to answer these questions after studying this chapter:

How are power sources in series circuits connected?
How are resistors in series circuits connected?
What are voltage drops in series circuits?
How is power used in series circuits?

Instruction

HOW ARE POWER SOURCES IN SERIES CIRCUITS CONNECTED?

There are many examples in electricity where power sources are connected in series. Figure 11-2 shows a few. Electric power sources are connected in series in order to get a higher voltage. Many electrical components, such as light bulbs and motors, are rated by the amount of voltage that is required for their operation. You may need to combine power sources in series to obtain the required operating voltages for these components.

Not all power sources that are connected together are in series. Here is a simple but important rule for determining if electric power sources are con-

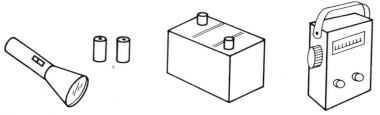

Figure 11-2.   Examples of cells connected in series. This flashlight uses two dry cells in series. A car battery is made of a number of cells in series. A transistor radio or portable tape player may use cells connected in series.

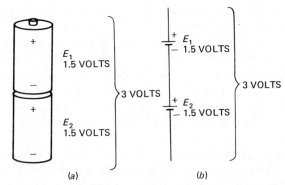

Figure 11-3. Connection of two electric power sources in series. (a) *Two dry cells connected in series.* (b) *Schematic representation of two cells in series.*

nected in series: *Whenever two or more power sources are connected in series and applied to a circuit, the same current flows through all the power sources.* This same rule will apply to any electric parts connected in series.

Here is another important rule for determining if electric power sources are connected in series: *Whenever two or more power sources are connected in series, their voltages are combined.* So, when you are studying a circuit in order to see if the power sources are connected in series, there are two questions that you should ask:

1. Does the same current flow through all the power sources?
2. Do the voltages of the power sources combine?

If the answers to both of these questions are *yes*, then the sources are in series.

Figure 11-3 shows two dry cells connected in series. Note that the *positive* terminal of one source (E_1) is connected to the *negative* terminal of the other source. This is the way electric power sources are usually connected in series. The object is to obtain a higher voltage. The *total voltage* of the combination is 3 volts, but each cell is only 1.5 volts.

The dry cells in Fig. 11-3 are said to be *series-aiding* because their voltages *add*. Figure 11-4 shows two cells connected *series-opposing*. The name comes

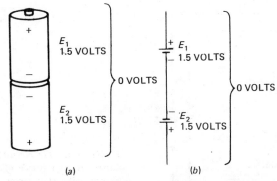

Figure 11-4. Series-opposing connection of two electric power sources. (a) *Two dry cells connected series-opposing.* (b) *Schematic of two cells connected series-opposing.*

from the fact that the cells provide voltages in opposite directions. The combined voltages of the cells shown in Fig. 11-4 is *0 volt.*

You might think of the two cells in Fig. 11-4 as being like two boys pulling a sled in opposite directions. Both boys pull, but the sled does not move.

What would happen if you put the cells in a flashlight so that they are series-opposing (as shown in Fig. 11-4)? The positive terminals of both cells would try to pull electrons from the light. Since they are pulling in opposite directions, no electron will flow. The flashlight will not work. Try it some time and see for yourself.

The voltage of the two cells in Fig. 11-4 is obtained by subtracting the cell voltages because they are connected in such a way that they try to make currents to flow in opposite directions. (When voltages are connected series-aiding, as in Fig. 11-3, they force the current to flow in the same direction.) In Fig. 11-4, 1.5 volts is subtracted from 1.5 volts to obtain 0 volt.

You should remember that there are two ways to connect electrical sources in series: *series-aiding* and *series-opposing.*

The reason for connecting sources in series-aiding is to obtain a higher voltage for operating some component. Why would anyone want to connect batteries so that they are series-opposing? There are a number of circuits that are connected this way, but we will consider only one at this time.

In order to charge an automobile battery, an electric current is made to flow through it backwards. This is shown in Fig. 11-5, which shows the circuit for a generator charging a battery. The arrows show the direction of electron current flow.

Normally electrons *leave* the negative terminal of a battery, flow through the circuit, and *return* to the positive terminal. However, in the circuit of Fig. 11-5 the electrons are flowing into the negative terminal of the battery and *away from* the positive terminals. In other words, the current is flowing backwards through the battery in order to charge it. The resistor in the circuit limits the amount of charging current.

An important thing about the circuit of Fig. 11-5 is that it shows an example of power sources connected in series-opposing. The generator and the battery try to force the electrons to flow in opposite directions. The generator has a higher voltage, so it forces the electrons to flow backward through the battery. To understand this, think about the boys pulling a sled in opposite

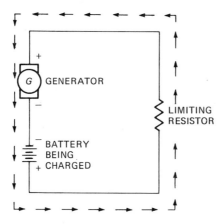

Figure 11-5. *The circuit for charging a battery is an example of electric power sources connected* **series-opposing.** *Arrows show the path of charging current.*

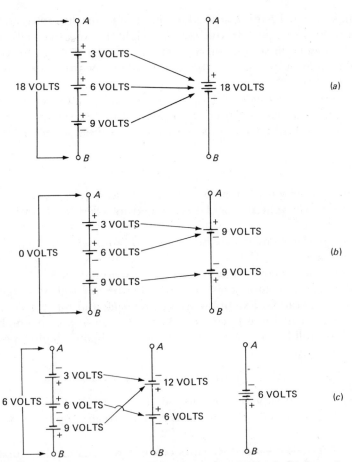

Figure 11-6. Here are some additional examples showing how voltages in series can be combined. (a) *Three batteries connected series-aiding. The terminal voltage is obtained by adding all the voltages. This group of voltage sources is equal to a single 18-volt battery.* (b) *The 6-volt and 3-volt batteries are connected series-aiding. Their combined voltage is 9 volts, which is opposite in polarity to the 9-volt battery. The overall voltage is zero. This group of voltages is equal to two 9-volt batteries connected in series-opposition.* (c) *The 3-volt and 9-volt batteries are connected series-aiding. They combine to give 12 volts, which is opposite in polarity to the 6-volt battery. The overall combination is 6 volts. Note that the negative terminal of the equivalent battery is at the top. The 9-volt and 3-volt sources can be combined into a single 12-volt source, and these sources can be combined into a single 6-volt source.*

directions. If one boy is stronger than the other, the sled will go in his direction. Likewise, the generator produces a larger force than the battery, and the current flows the way the generator requires it to go.

Figure 11-6 shows three voltage sources connected in different circuits. In each case, the overall voltage—called the *terminal voltage*—is obtained by adding all the voltages connected in one direction, and adding all the voltages connected in the opposite direction. These values are then subtracted to obtain the terminal voltage.

You cannot solve all the problems of power by simply connecting power sources in series. A 12-volt lead-acid battery can be used to start a car that has a high-compression engine. If you connect eight 1.5-volt dry cells (of the type used in flashlights) in series, you will get 12 volts. However, you could not use this combination of dry cells to start a car engine, because they cannot deliver the required *power.*

Remember this: *Connecting electric power sources in series-aiding provides a higher voltage.* You may also need more *power.* In a later chapter we will discuss methods of connecting electric power sources together to get a higher power.

SUMMARY

1.   Electric power sources may be combined in different ways.
2.   Electric power sources may be combined in series to obtain a higher voltage.
3.   Electric power sources may be connected *series-aiding* so that all their voltages are added together.
4.   Electric power sources may be connected *series-opposing.*
5.   Some electrical components are rated by the amount of voltage required for their operation. Examples are *12-volt lights* and *6-volt motors.*
6.   You have learned how to connect power sources to obtain a higher voltage. Later you will learn how to connect them to obtain a higher power.

HOW ARE RESISTORS IN SERIES CIRCUITS CONNECTED?

A resistor is a component that opposes the flow of electric current. When resistors are connected in *series,* there is a greater opposition to current flow.

Figure 11-7 shows three resistors connected in series. The total resistance of the series combination is obtained by adding the resistance values. This is always true for series resistors. Thus, we have an important rule for series resistors: *You get the total resistance of two or more resistors in series by adding resistance values.*

Earlier you learned that when power sources are connected in a series circuit, the same current flows through them. This rule also holds true for resistors in series. Figure 11-8 shows this. Remember this important fact: *The current is the same through all resistors in series.*

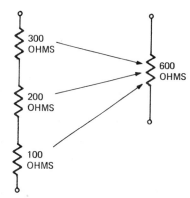

Figure 11-7. When resistors are connected in series, the total resistance is the sum of the resistance values. These three resistors connected in series have a total resistance of 600 ohms.

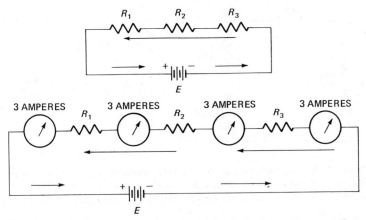

Figure 11-8. *When resistors are connected in series, the same current flows through them. When resistors are connected in series across a voltage source, the same current flows through all resistors. In this example, 3 amperes is flowing in all parts of the series circuit.*

If you are studying a circuit to see if the resistors are connected in series, there are two questions that you should ask:

1. Is the total resistance equal to the sum of the resistance values?
2. Does the same current flow through all the resistors in the combination?

If the answers to both of these questions are *yes*, then the resistors are connected in series.

One purpose of connecting resistors in series is to obtain a desired resistance value. Suppose you need a 1,000-ohm resistance, but you do not have a resistor with that value. If you have two 500-ohm resistors, you can connect them in series to get the desired value. Figure 11-9 shows this problem.

In order to find the current flowing in a series circuit, the voltage across the circuit is divided by the total resistance. This is illustrated in Fig. 11-10.

Figure 11-10*a* shows a series circuit. We wish to know the current through this circuit. The first step is to find the total resistance.

$$\text{TOTAL RESISTANCE} = 35 \text{ ohms} + 65 \text{ ohms}$$
$$= 100 \text{ ohms}$$

As far as the generator current is concerned, the series-resistance combination

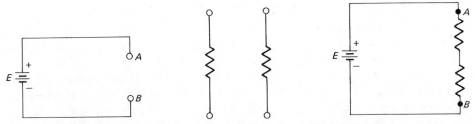

Figure 11-9. *Resistors can be combined to obtain a desired value of resistance. If you need 1,000 ohms between terminals A and B but you only have two 500-ohm resistors, they can be connected in series to obtain 1,000 ohms of resistance.*

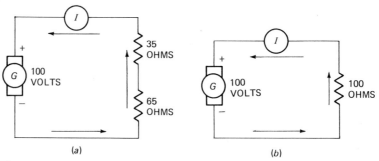

Figure 11-10.   To find the amount of current flowing in a series circuit, first find the total resistance, then use Ohm's law. (a) The amount of current flowing in this current . . . (b) . . . is the same as the amount of current flowing in this circuit.

of Fig. 11-10*a* could be replaced with a single resistor having a resistance value of 100 ohms. This is shown in Fig. 11-10*b*.

The current in the circuit of Fig. 11-10*b* is found by Ohm's law:

$$CURRENT = \frac{GENERATOR\ VOLTAGE}{RESISTANCE}$$

$$I = \frac{E}{R}$$

$$I = \frac{100\ \text{volts}}{100\ \text{ohms}}$$

$$I = 1\ \text{ampere} \qquad\qquad Answer$$

This means that 1 ampere of current is flowing through the resistor in the circuit of Fig. 11-10*b*. Therefore, there must be 1 ampere of current flowing through the circuit of Fig. 11-10*a*.

SUMMARY

1. When resistors are connected in series, their resistances add.
2. When resistors are connected in series, the same current flows through each resistor.
3. Resistors can be combined in series to obtain a desired resistance value.
4. The current in a series circuit equals the voltage divided by the total resistance.

WHAT ARE VOLTAGE DROPS IN SERIES CIRCUITS?

Whenever there is current flowing through a resistor, there is *always* a voltage drop across it. This is one of the basic rules of electricity. You have learned in an earlier chapter how to calculate this voltage drop using Ohm's law.

When a current flows through resistors in series, there is a voltage drop across each resistor. The amount of voltage drop depends upon the amount of current, and it is also dependent upon the resistance of the resistor.

The total voltage drop across resistors in series is found by adding the voltage drops across each resistor. This is illustrated in Fig. 11-11.

If you add all the voltage drops in a series circuit, the sum will always be

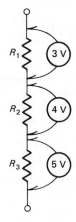

Figure 11-11. The total voltage drop in this circuit is 12 volts.

equal to the applied voltage. This can be written as an important rule: *The sum of the voltage drops around a closed circuit is equal to the applied voltage.* This rule is so important that it is a law of electric circuits. It is known as *Kirchhoff's voltage law.* We will refer to it simply as the *voltage law.*

Figure 11-12 shows how the voltage law applies to a series circuit. The applied voltage in this circuit is 100 volts. The voltage drops are 20 volts, 30 volts, and 50 volts, making a total of 100 volts. If the generator voltage is increased, the voltage drop across each resistor will also increase. The new values of voltage drops will add to equal the new generator voltage. Likewise, if the generator voltage is decreased, the voltage drops will decrease so that their sum is still equal to the generator voltage.

If you change the value of one of the resistors in the circuit of Fig 11-12, the current will change. However, the change will *always* be such that voltages will still add to equal the applied voltage.

SUMMARY

1. There is *always* a voltage drop across a resistor when there is current flowing through it.
2. The sum of the voltage drops around a closed circuit is *always* equal to the voltage applied to that circuit.
3. Changing resistance values in a series resistor circuit will alter the voltage drops, but their sum will still be equal to the applied voltage.

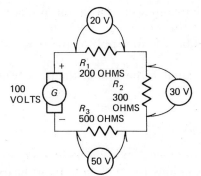

Figure 11-12. The voltage drops in a circuit will always add to equal the applied voltage.

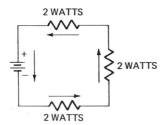

Figure 11-13. The total power used in a series circuit is equal to the sum of the powers used in each part of that circuit. This circuit is using 6 watts of power.

4. Changing the applied voltage in a series resistor circuit will alter the voltage drops, but their sum will still be equal to the applied voltage.

HOW IS POWER USED IN SERIES CIRCUITS?

Whenever current flows through a resistor, power is always present in the form of heat. Sometimes we make use of this heat. An electric stove, for example, works on the principle of current flowing through resistance elements. The resulting heat is used for cooking.

Can you give another example where heat generated by current flowing through a resistance is used for some purpose?

Although there are cases where heat generated by current flowing through a resistance is useful, it is usually considered to be a power loss. Sometimes the heat generated in a circuit is so great that special provisions must be made for cooling. It is important to be able to calculate power in circuits. You must know how much power is to be used in a circuit in order to be able to choose the power supply needed.

The total power in a series circuit is equal to the sum of the powers in each part of the circuit. This is illustrated in Fig. 11-13. Each resistor in the circuit is using 2 watts, and the total power being used is 6 watts. It follows that the battery is supplying 6 watts to the circuit.

We have shown resistors in the circuits of this chapter, but in actual circuits, lights, motors, and other loads may be used. The theories that we have discussed apply to all types of loads.

SUMMARY

1. Heat is always generated when current flows through a resistor.
2. In some cases the heat caused by current flowing through a resistor is useful.
3. It is important to be able to find the total power in a circuit so that you can choose the proper power supply for that circuit.
4. The power supply for a circuit must be able to supply all the power needed by the circuit components.
5. The total power used in a circuit is equal to the sum of the powers used by each part of the circuit.

EQUATIONS RELATED TO THIS CHAPTER

We have discussed voltages, currents, resistances, and powers in a series circuit. Up to now we have not used math equations to show how these factors are related. However, equations are convenient ways of remembering how to calculate voltage, current, resistance, and power in series circuits. The equations are given here for your convenience.

Total Voltage Drop in a Series Circuit

$$\text{TOTAL VOLTAGE } V_T = V_1 + V_2 + V_3 \ldots + V_n$$

This equation shows that the total voltage, V_T, is the sum of each voltage drop. The reason for including V_n is that any *number* of voltages may be added in series.

Total Resistance in a Series Circuit

$$\text{TOTAL RESISTANCE } R_T = R_1 + R_2 + R_3 \ldots + R_n$$

This equation shows that the total resistance in a series circuit is found by adding all the resistance values.

Current in a Series Circuit

$$I = \frac{E}{R_T}$$

This equation shows that the current in a series circuit is found by dividing the applied voltage by the total resistance, R_T, of the circuit.

Power in a Series Circuit

$$\text{TOTAL POWER } P_T = P_1 + P_2 + P_3 \ldots + P_n$$

This equation shows that the total power, P_T, dissipated in a series circuit is equal to the sum of the powers used by each of the circuit components.

Programmed Review Questions

(Instructions for using this programmed section are given in Chap. 1.)

We will review the important concepts of this chapter. If you have understood the material, you will progress easily through this section. Do not skip this material, because some additional theory is presented.

1. How many 5-watt lights can be operated in series from a power supply that is capable of delivering 20 watts of power?
   A.   4 lights (Proceed to block 7.)
   B.   5 lights (Proceed to block 17.)

2. *Your answer to the question in block 12 is* **A**. *This answer is wrong. The total resistance in a series circuit is equal to the* **sum** *of the resistors.* Proceed to block 9.

3. *The correct answer to the question in block 9 is* **B**. *With the switch open, no current can flow anywhere in the circuit. The light is not lit.* Here is your next question.
   Figure 11-18 is a circuit with a battery, a switch, and two lights. Draw lines to connect these components so that the lights are in series, and so the switch can turn them both on or both off. After you have connected the components, proceed to block 25.

4. *Your answer to the question in block 19 is **A**. This answer is wrong. The total voltage drop must be 50 volts according to the voltage law. Ten volts is dropped across* R_1, *and the rest of the voltage will be dropped across* R_2. Proceed to block 13.

5. *Your answer to the question in block 25 is **A**. This answer is wrong. You have calculated the current incorrectly. Remember that the current is equal to the voltage divided by the* **total resistance.** Proceed to block 11.

6. *Your answer to the question in block 11 is **A**. This answer is wrong. The amount of current flow does not depend upon the number of resistors in the circuit. It does depend upon the total resistance in the circuit. Proceed to block 19.*

7. *The correct answer to the question in block 1 is **A**. Only **four** of the 5-watt lights connected in series can be supplied by the power supply. Here is your next question.*
    Study the circuit of Fig. 11-14.
    A. Would you say that the resistors are connected in series? (If so, proceed to block 12.)
    B. Would you say that the resistors are not connected in series? (If so, proceed to block 14.)

8. *The correct answer to the question in block 13 is **A**. Both lights will glow with equal brightness. The same current is flowing through both of them because they are connected in series.* Here is your next question.
    Which of the following is a statement of the voltage law?
    A. The sum of the voltage around a circuit is equal to the applied power. (Proceed to block 18.)
    B. The total voltage across two resistors in series is equal to the sum of the voltages across each resistor. (Proceed to block 24.)
    C. The sum of the voltage drops around a circuit equals the total voltage applied to the circuit. (Proceed to block 23.)

9. *The correct answer to the question in block 12 is **B**. Sometimes it is helpful to draw a circuit for solving such problems. If you drew such a circuit, it probably looked like the one in Fig. 11-16.*

    $$R_T = R_1 + R_2 + R_3$$
    $$= 50 + 50 + 50$$
    $$= 150 \ ohms \qquad\qquad Answer$$

    Here is your next question.
    A switch is connected in series with a circuit component to control the current flowing through it. Figure 11-17 illustrates a circuit with the switch shown in the open position.
    A. The light is lit. (Proceed to block 10.)
    B. The light is not lit. (Proceed to block 3.)

10. *Your answer to the question in block 9 is **A**. This answer is wrong. No current can flow through the light unless there is a complete path from the nega-*

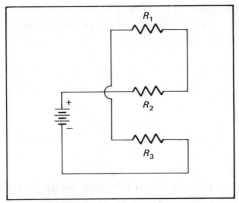

Figure 11-14. Circuit for problem in block 7.

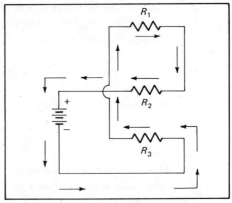

Figure 11-15. Circuit which shows current path for the circuit in Fig. 11-14.

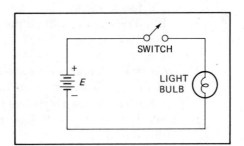

Figure 11-17. Circuit for problem in block 9.

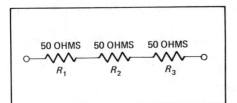

Figure 11-16. Circuit for solving the problem in block 12.

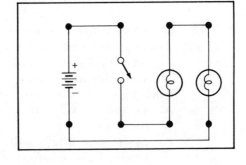

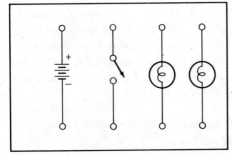

Figure 11-18. Circuit for solving the problem in block 3.

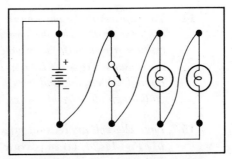

Figure 11-19. Two ways to connect the circuit in block 3.

tive terminal of the battery to the positive terminal of the battery. With the switch open, no such complete path exists. Since there is no current flowing through the light, it is not lit. Proceed to block 3.

11. *The correct answer to the question in block 25 is* **B**. *The total resistance in the circuit is 30 ohms. The current is equal to the voltage (90 volts) divided by the total resistance (30 ohms).*

$$I = \frac{90 \; volts}{30 \; ohms} = 3 \; amperes \qquad\qquad Answer$$

Here is your next question.

The two series circuits in Fig. 11-21 have equal voltages applied. Which of the following statements is correct?

A. The current in the circuit with only two resistors will be greater than the current in the circuit with three resistors. (Proceed to block 6.)

B. The current will be the same in both circuits. (Proceed to block 19.)

12. *The correct answer to the question in block 7 is* **A**. *The resistors* **are** *in series. This follows because the same current flows through all of them. The arrows in Fig. 11-15 show the current path. Note that there is only one path for current flow, and that path takes you through all the resistors.* Here is your next question.

If three 50-ohm resistors are connected in series, the total resistance is

A. 50 ohms. (Proceed to block 2.)

B. 150 ohms. (Proceed to block 9.)

13. *The correct answer to the question in block 19 is* **B**. *Since 10 volts is dropped across* R_1, *and the total drop must be 50 volts, it follows that 40 volts will be dropped across* R_2. Here is your next question.

The two lights in the circuit of Fig. 11-23 are identical. Which of the following statements is correct?

A. L_1 and L_2 will both glow with equal brightness. (Proceed to block 8.)

B. L_2 will glow brighter than L_1 because it is closer to the negative terminal of the battery. (Proceed to block 16.)

14. *Your answer to the question in block 7 is* **B**. *This answer is wrong. Do not be confused by the way the circuit is drawn. The test for determining if the resistors are in series is to note whether the same current flows through all the resistors. Trace the current path from the negative terminal of the battery, through the circuit, and back to the positive terminal of the battery. Did you go through all three of the resistors?* Proceed to block 12.

15. *Four 1½-volt dry cells in series will produce 6 volts. Your circuit should look like Fig. 11-24.* Here is your next question.

Figure 11-25 is a simple series circuit in which you are to determine the resistance of the light. Which of the following statements is true?

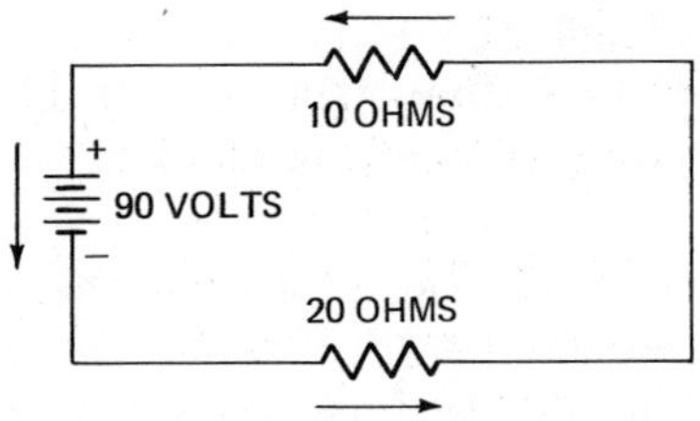

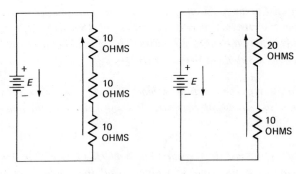

Figure 11-20. Circuit for problem in block 25.

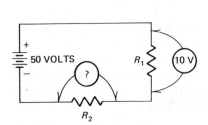

Figure 11-21. Circuit for problem in block 11.

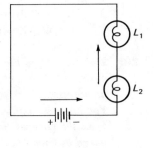

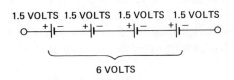

Figure 11-22. Circuit for problem in. block 19.

Figure 11-23. Circuit for problem in block 13.

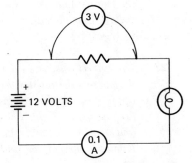

Figure 11-24. Circuit for solving the problem in block 23.

Figure 11-25. Circuit for problem in block 15.

A. There is not enough information given in the problem to find the resistance. (Proceed to block 21.)
B. The resistance of the light is 4 ohms. (Proceed to block 20.)
C. The light resistance is 90 ohms. (Proceed to block 22.)

16. *Your answer to the question in block 13 is **B**. This answer is wrong. The same current flows through all parts of a series circuit. Since the lights are identical, with the same current flowing through both, they will glow equally brightly.* Proceed to block 8.

17. *Your answer to the question in block 1 is **B**. This answer is wrong. Five lights that use 5 watts each would use a total of **25 watts**. But the power supply can only supply a total of **20 watts**. Therefore, only four lights, connected in series, can be supplied.* Proceed to block 7.

18. *Your answer to the question in block 8 is **A**. This answer is wrong. You did not read the statement carefully. Voltage cannot equal power.* Proceed to block 23.

19. *The correct answer to the question in block 11 is **B**. The total resistance in both circuits is the same (30 ohms), and the applied voltage is the same. Therefore, the current in the two circuits is the same.* Here is your next question.
What is the voltage drop across R_2 in the circuit of Fig. 11-22?
A. 50 volts (Proceed to block 4.)
B. 40 volts (Proceed to block 13.)

20. *Your answer to the question in block 15 is **B**. You may have obtained the answer of 4 ohms by dividing 12 volts by 3 volts, but this is not a correct procedure.*
*You know the **voltage** across the light if you apply the voltage law. Also, you know the **current** through the light. The resistance of the light can be determined by Ohm's law.* Proceed to block 22.

21. *Your answer to the question in block 15 is **A**. This answer is wrong. The problem **can** be solved, but you must combine several things that you have learned. For example, you know the **voltage** across the light if you remember the voltage law. Also, you know the **current** through the light. Knowing the voltage and the current, you can find the resistance by Ohm's law.* Proceed to block 22.

22. *The correct answer to the question in block 15 is **C**. Since there is a 3-volt drop across the resistor, and since the total voltage drop of the circuit must be equal to the applied voltage (12 volts), it follows that there must be a 9-volt drop across the light. The current through all parts of this series circuit is 0.1 ampere as indicated by the ammeter.*
Knowing the voltage and the current, we can find the resistance by Ohm's law:

$$R = \frac{\text{VOLTAGE}}{\text{CURRENT}}$$

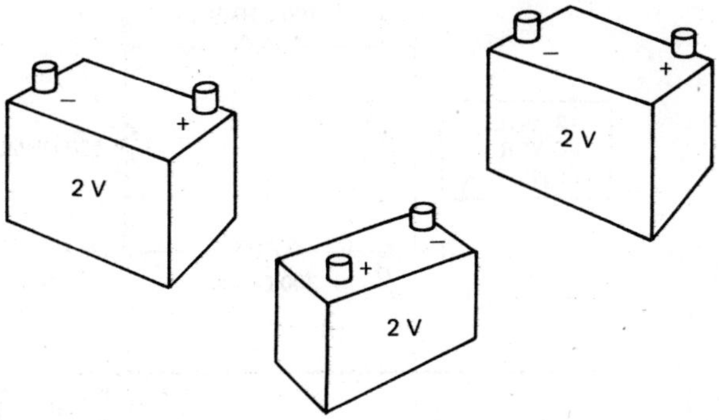

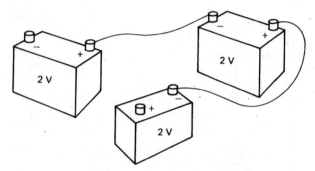

Figure 11-26. Circuit for problem in block 22.

Figure 11-27. Circuit for solving the problem in block 22.

$$= \frac{9 \text{ volts}}{0.1 \text{ ampere}}$$
$$= 90 \text{ ohms} \qquad\qquad Answer$$

Here is your next question.

The illustration of Fig. 11-26 shows three batteries. Draw lines to show how you would connect them in series. How much voltage will the three batteries in series produce? After you have completed this drawing and you have answered the question, proceed to block 26.

23. *The correct answer to the question in block 8 is* **C**. *The voltage law states that the sum of the voltage drops around a circuit equals the total voltage applied to the circuit.* Here is your next question.

How would you draw a circuit showing how a number of 1½-volt cells can be connected to make a 6-volt power source? Use the fewest number of cells possible.

After you have drawn the circuit, proceed to block 15.

24. *Your answer to the question in block 8 is* **B**. *This answer is wrong. The statement is correct, but it is not a statement of the voltage law.* Proceed to block 23.

25. *Your circuit may look like either of the circuits in Fig. 11-19.*

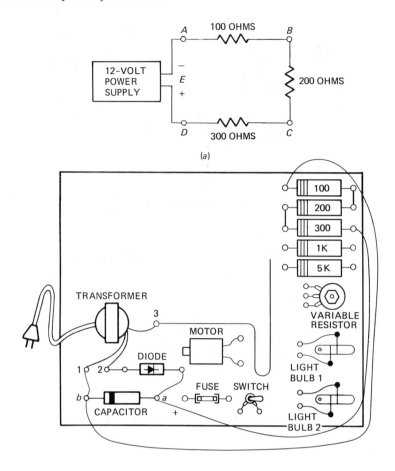

(a)

Figure 11-28.   *Test setup for Experiments 1 and 2.* (a) **Schematic of the circuit for Experiments 1 and 2.** (b) **This is the way your circuit board will look if you have wired the circuit correctly.**

*Check **your** circuit by tracing the path of electron flow from the negative terminal of the battery through the circuit and back to the positive terminal. Practice with the circuits given here. (You will have to trace through the switch even though it is in the open position.) Note that current will flow through **both** lights, so they will both light when the switch is closed.* Here is your next question.

Calculate the current flowing in Fig. 11-20. The current is

A.   9 amperes. (Proceed to block 5.)

B.   3 amperes. (Proceed to block 11.)

26.   *Three 2-volt batteries in series will produce a total of 6 volts. You must connect them with the positive terminal of one battery going to the negative terminal of the next. There are several ways to connect the three batteries in series, depending on which one you put in the middle. Figure 11-27 shows a typical answer.*

Check your answer to make sure that you have not connected two positive terminals together. Also, make sure that you have not connected two negative terminals together.

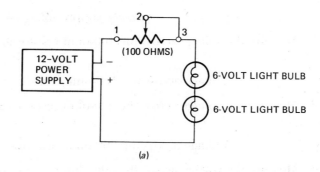

(a)

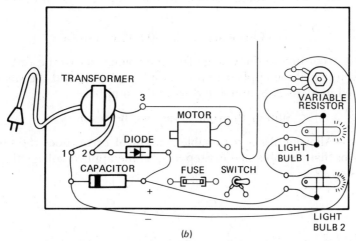

(b)

Figure 11-29. Test setup for Experiment 3. (a) **The circuit for Experiment 3.** (b) **This is the way your circuit board will look if you have wired the circuit correctly.**

You have now completed the programmed questions. The next step is to put some of these ideas to work in laboratory experiments. Proceed to the "Experiments" section that follows.

Experiments

(The experiments described in this section may be performed on the circuit board described in Appendix C or on a similar laboratory setup.)

EXPERIMENT 1

Purpose — To demonstrate the voltage law.

Test Setup — Wire the circuit board as shown in Fig. 11-28.

Procedure —
    Step 1 — With the circuit connected as shown in Fig. 11-28, measure the power supply voltage — that is, the voltage across the power supply output terminals (points *a* and *b*). Write the voltage reading here.

Power supply voltage = _____ volts

Step 2 — Measure the voltage across the 100-ohm resistor and write the voltage here.

Voltage across the 100-ohm resistor = _____ volts

Step 3 — Measure the voltage across the 200-ohm resistor and write the voltage here.

Voltage across the 200-ohm resistor = _____ volts

Step 4 — Measure the voltage across the 300-ohm resistor and write the voltage here.

Voltage across the 300-ohm resistor = _____ volts

Step 5 — According to the voltage law, the voltage across the three resistors should add to equal the applied (power supply) voltage. Add the voltages that you obtained in Steps 2, 3, and 4 and write the sum here.

Sum of the voltages across the resistors = _____ volts

Conclusion — Allowing for a reasonable amount of error, does the sum of the voltage drops that you obtained in Step 5 equal the applied voltage that you obtained in Step 1?

_____

EXPERIMENT 2

Purpose — To show that the current is the same in all parts of a series circuit.

Test Setup — Wire the circuit board as shown in Fig. 11-28.

Procedure —

Step 1 — *Calculate* the total resistance of the circuit. (Remember that the total resistance of a series circuit is obtained by adding the resistance values.) Write the total resistance here.

Total resistance = _____ ohms

Step 2 — Ohm's law tells us that the current in the circuit is equal to the applied voltage divided by the total resistance. In other words:

$$CURRENT = \frac{POWER\ SUPPLY\ VOLTAGE}{TOTAL\ RESISTANCE\ IN\ THE\ CIRCUIT}$$

Calculate the current and write the value here.

_____ amperes

How many milliamperes is this?

_____ milliamperes

Step 3 — If you worked the problem carefully, you found that the current in the circuit of Fig. 11-28 is about 0.02 ampere, or 20 milliamperes. Which scale will you use on your meter to measure this current?

_____

Show this answer to your teacher before you proceed with the experiment!
    Step 4 — Measure the current at point *A* and write the value here.

        Current measured at point *A* = _____ milliamperes

    Step 5 — Measure the current at point *B* and write the value here.

        Current measured at point B = _____ milliamperes

    Step 6 — Measure the current at point *C* and write the value here.

        Current measured at point *C* = _____ milliamperes

    Step 7 — Measure the current at point *D* and write the value here.

        Current measured at point *D* = _____ milliamperes

    Step 8 — One of the basic rules of series circuits is that the current is the same at all points. This means that the values you got for Steps 4, 5, 6, and 7 should all be the same. Are they the same?

_____

    Step 9 — Do the measured values that you got in Steps 4, 5, 6, and 7 match with the value you calculated for Step 2?

_____

Conclusion — Allowing for a reasonable amount of error, do your results show that the current is the same in all parts of a series circuit?

_____

EXPERIMENT 3

Purpose — To show how a variable resistor can be connected in a series circuit to control the current through a load.

Test Setup — Connect the circuit as shown in Fig. 11-29. The variable resistor is *in series* with the lights. When a variable resistor is connected like this, it is called a *rheostat*. When the variable resistor is adjusted, the amount of current flowing through both lights is varied. This changes the light brightness.

Procedure —
    Step 1 — Adjust the variable resistor so that the lights are *brightest*.
    Step 2 — Disconnect all the terminals of the resistor. Without changing the setting of the variable resistor, measure the resistance between terminals 1 and 2. Write the value.

        Resistance when the light is brightest = _____ ohms

Step 3 — Connect the circuit of Fig. 11-29 again. Adjust the variable resistor so that the lights are *dimmest*.

Step 4 — Disconnect all the terminals of the resistor. Without changing the setting of the variable resistor, measure the resistance between terminals 1 and 2.

Resistance when the light is dimmest = _____ ohms

Step 5 — Based on your results of Steps 2 and 4, answer this question: Is the light brightest or dimmest when the maximum resistance is in series with the light?

_____

Step 6 — Rewire the circuit of Fig. 11-29. Adjust the resistor for maximum brightness. Remove one of the lights from its socket. Does the other light stay lit, or does it go out?

_____

Conclusion — If you performed the experiment correctly, you found that the light is *dimmest* when the resistance is *maximum*. This is because the minimum current is flowing when the circuit resistance is maximum.

The series circuit of Fig. 11-29 is used in practical applications. For example, the dimming control for the dash lights of an automobile works with a rheostat.

When you removed one of the lights in the series circuit, the other one stopped glowing. This is to be expected, because the current stops flowing when the light is removed.

The two headlights of an automobile are *not* connected in series. If one headlight burns out, the other one still glows. (Actually, they are in *parallel*. You will study about parallel circuits in the next chapter.)

Self-Test with Answers

(Answers with discussions are given in the next section.)

1. Electric power sources may be connected in series to (*a*) increase the cost; (*b*) get a higher voltage.
2. Two 1½-volt cells are connected together to obtain 3 volts. The cells are connected (*a*) series-aiding; (*b*) series-opposing.
3. A battery is charged by (*a*) forcing current to flow backwards through it; (*b*) reducing the current flow through it.
4. Suppose the battery in your car is dead. You have eight 1.5-volt dry cells that are brand new. Which of these choices is correct? (*a*) You can connect them in series-aiding to get the 12 volts that you need to start the car; (*b*) The cells won't help because they cannot deliver enough power.
5. You have three 100-ohm resistors. You need a resistance of 200 ohms. How will you connect the resistors? (*a*) Connect two in series-opposing

and then put them in series with the third one; (*b*) Connect two in series and do not use the third one.

6. You want to connect three resistors in series to get 1,000 ohms. You have already connected a 300-ohm resistor in series with a 400-ohm resistor. Which of these is correct? (*a*) You need another 300-ohm resistor in order to get 1,000 ohms; (*b*) It cannot be done.

7. The sum of the voltage drops around a closed circuit is equal to the applied voltage. This is (*a*) Ohm's law; (*b*) the voltage law.

8. Two lights are connected in series. One of the lights burns out. Which of the following statements is correct? (*a*) The other light will still be lit; (*b*) If one light burns out, the other light will not be lit.

9. The taillights of an automobile are connected (*a*) in series; (*b*) not in series.

10. One reason that the headlights of a car are not connected in series is that (*a*) if one burned out, they would *both* be out; (*b*) it would cost too much.

Answers to Self-Test

1. (*b*) — The cost of power supplies in series is greater than the cost of a single supply, but that is not the *reason* for connecting them in series.
2. (*a*)
3. (*a*)
4. (*b*) — It is true that the dry cells will produce 12 volts when they are connected in series, but they cannot supply the *power* needed for starting a car.
5. (*b*)
6. (*a*) — With another 300-ohm resistor the total resistance will be 300 ohms + 400 ohms + 300 ohms = 1,000 ohms.
7. (*b*) — Remember that the voltage drops will *always* equal the applied voltage.
8. (*b*) — This was demonstrated in Experiment 3.
9. (*b*) — When one taillight burns out, the other one is still lit. This proves that they are not in series.
10. (*a*) — The cost would be the same. You would not want both lights to go out all of a sudden if you were driving at night.

12.
What Is a Parallel Circuit?

Introduction

You have learned how power sources may be connected in series to get more voltage. Electric parts need both voltage *and* current to operate. Since voltage times current equals power, we say that power must be delivered to such devices. In this chapter you will learn how power sources may be connected in parallel to get more current.

When you connect resistors or lights in series, the same current flows through them. In this chapter you will learn how to connect resistors in parallel so that they have the same voltage drop across them.

The amount of current flowing in a parallel resistor circuit can be found by Ohm's law. It is necessary to know how to find the resistance of the parallel circuit in order to find the current. In this chapter you will learn about current and resistance in parallel circuits.

Almost all the characteristics of parallel circuits are different from those of series circuits. One important thing remains the same: the method of finding the total power used in a circuit. In this chapter you will learn how to find the amount of power used in a parallel circuit.

You will be able to answer these questions after studying this chapter:

How are power sources connected in parallel?
How are resistors connected in parallel?
How are voltage drops related in parallel?
How are currents related in parallel circuits?
How is power in a parallel circuit determined?

Instruction

HOW ARE POWER SOURCES CONNECTED IN PARALLEL?

In the last chapter you learned how electric power sources may be connected in *series* to get a *higher voltage*. Electric power sources may also be connected in *parallel* in order to get a *higher current*.

Figure 12-1 shows two batteries connected in series across a load. Two batteries connected in parallel across a load are shown in the same illustration. Study these drawings carefully. Can you see the difference between the circuits?

An important rule for seeing if electric power sources are in series is *that the same current must flow through them when they are connected in a circuit.* If you follow the current paths in the parallel circuit of Fig. 12-1, you will see that this rule does *not* apply. Notice how the current divides into two *branches* as it returns to the batteries from the load. Branches are parts of a parallel circuit.

Here is an important rule for electric power sources connected in parallel: *Whenever two or more power sources are connected in parallel, and applied to a circuit, the same current does **not** flow through all the supplies.* This same rule will apply to any electric parts connected in parallel.

Here is another important rule for determining if electric power sources are connected in parallel: *Whenever two or more power sources are connected in parallel, their currents are added.*

In the parallel circuit of Fig. 12-1 you can see that the currents flowing from the negative battery terminals combine into a single current. This is like two rivers of current coming together to make one larger river.

When you are studying a circuit to determine if the power supplies are in parallel, there are two questions that you should ask:

1. Do the currents entering the positive terminal separate so that the current through one power supply is not the same current as flows through the other power supply?
2. Do the currents leaving the power supply combine into a single current?

If the answers to these questions are *yes*, then the power supplies are in parallel.

An important difference between the series and parallel circuits of Fig. 12-1 is in the way the terminals are connected. In the series circuit the positive terminal of one battery is connected to the negative terminal of the other. In the *parallel* circuit the two positive terminals are connected together, and the two negative terminals are connected together.

You connect electric power sources in series to get a higher *voltage*, and

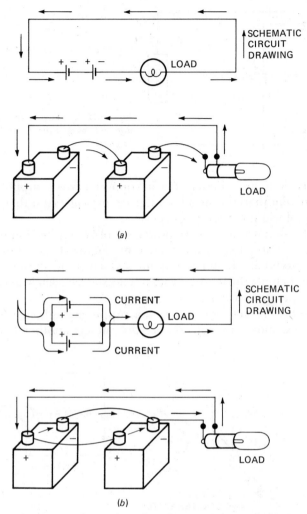

Figure 12-1. Comparison of power sources connected in series and in parallel. Arrows show direction of current flow. (a) *In this circuit the batteries are connected in series across the load. In this case the load is a light bulb.* (b) *In this circuit the batteries are connected in parallel across the load.*

you connect them in parallel if you want the sources to be able to deliver a higher *current.* However, you must *never* connect power sources in parallel *unless they have the same voltage!* Figure 12-2 shows why.

The 12-volt battery in the circuit of Fig. 12-2 is forcing current to flow backwards through the 6-volt battery. This will discharge the larger battery and overcharge the smaller one. Forcing a current to flow backwards through a storage battery charges it. If you overcharge a battery by causing too much current to flow backwards through it, *the battery will be ruined.*

Remember this very important rule: ***Never*** *connect electric power sources in parallel unless they have the same voltage.*

Batteries are shown in Figs. 12-1 and 12-2, but the same rules apply for all types of power supplies.

The amount of current flowing from any power supply is determined by

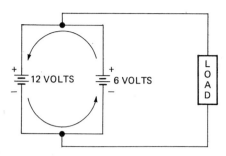

Figure 12-2. Why power sources with different voltages should not be connected in parallel.

the load connected to its terminals. If you connect a power supply to a load that needs more current than the supply can give, you *might* ruin the power supply. At least, the load will not operate properly.

Figure 12-3 shows a case where power supplies may be connected in parallel in order to be able to get more current. Figure 12-3a shows the starter system in an automobile. It shows a motor—which is the starter—connected in series with the battery through a switch. The starter switch is in the ignition system and is operated with a key.

When the battery gets old, it may not supply enough current to turn the starter. You need more current, so you can connect another battery—*with the*

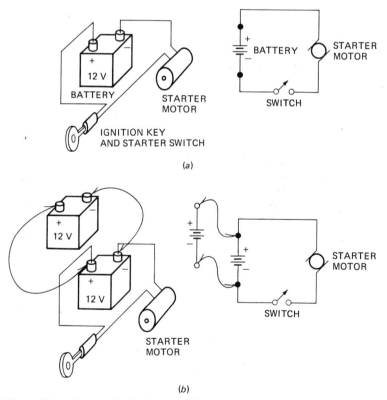

Figure 12-3. An example of using parallel power supplies. (a) Here is the starter system of an automobile. It is a simple series circuit. The battery supplies the current for operating the starter motor. If the battery cannot supply enough current, (b) another battery may be connected in parallel to increase the current flow.

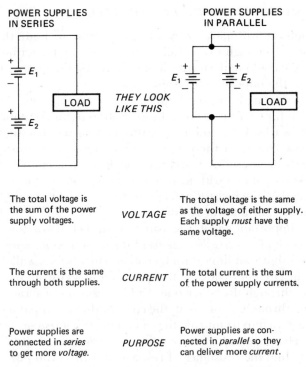

Figure 12-4. *Important features of series and parallel power supplies.*

same voltage—in parallel with the car battery. This is shown in Fig. 12-3*b*. Note that one jumper cable connects the two positive terminals, and the other jumper cable connects the two negative terminals.

When power supplies are connected in parallel, the voltage of the parallel combination is the same as for one power supply. In the circuit of Fig. 12-3, the voltage applied to the starter will be 12 volts with the two batteries connected in parallel.

Figure 12-4 shows some important things to remember about series and parallel power supplies.

SUMMARY

1. Electric power sources may be connected in parallel to obtain a higher current capability.
2. When power supplies are connected in parallel, they *must* have the same voltage.
3. The amount of current flowing in a circuit depends upon the size of the load.
4. When power sources are connected in parallel, the same current does not flow through them. Instead, the current divides so that only part of it flows through each supply.
5. If you connect a power supply to a load that needs more current than the power supply can give, it may be ruined.
6. When power supplies are connected in parallel, their combined output voltage is the same as the voltage of one of the supplies.

HOW ARE RESISTORS CONNECTED IN PARALLEL?

A two-lane highway will allow more cars to pass in an hour than a one-lane highway. A three-lane highway will allow more cars to pass per hour than a two-lane highway. We could make a rule about highways: the greater the number of lanes, the lower the opposition to traffic flow.

There is a similar thing that happens in electric circuits. The greater the number of paths for current to flow, the lower the circuit resistance. Figure 12-5 shows how this works. When the two resistors are in parallel, there is less resistance to current flow than there is with only one of the resistors connected.

No matter what value of resistance you have in a circuit, if you connect another resistor in parallel with it, you will *always* reduce the circuit resistance. This is true whether the resistor you connect in parallel has a higher resistance, an equal resistance, or a lower resistance.

Here is an important rule for resistors in parallel: *When resistors are in parallel, their resistance is always less than the smallest resistance in the parallel combination.*

Figure 12-5 shows an important fact about current in parallel resistors. You will note that the current divides so that part of it flows through one resistor and part of it flows through the other one. Unlike resistors in series, with the same current flowing through all of them, the current divides in parallel resistor connections. Here is an important rule to remember: *When resistors are in parallel, the current divides. Only a part of the current flows through each parallel resistor.*

It is easy to find the resistance of resistors in *series*—you just add their resistance values. It is not that easy to find the resistance of resistors in parallel. However, if you know a few simple rules, you can work any problem.

The first rule to remember is that whenever two *equal* resistances are in parallel, the resistance of the combination is *always* ½ of the value of either resistance.

Look at the parallel resistance circuit in Fig. 12-5. The two resistors have

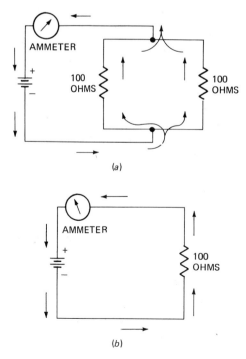

(a)

(b)

Figure 12-5. Two 100-ohm resistors in parallel have less resistance than a single resistor. (a) These resistors are connected in parallel. Since there are two paths for current to flow, there is less opposition than there is in this circuit (b), which has only one path.

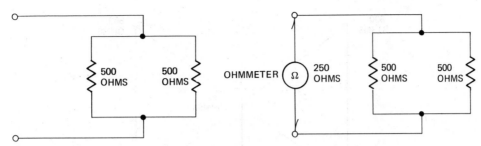

Figure 12-6. The rule for equal resistances in parallel. Whenever two equal resistance values are connected in parallel, the combined resistance is always ½ of the value of either resistor. The ohmmeter shows that two 500-ohm resistors in parallel have a combined resistance of 250 ohms.

equal resistance values of 100 ohms. The resistance of the parallel combination is ½ of 100, or 50 ohms.

Figure 12-6 illustrates the rule for finding the resistance of equal resistors connected in parallel. The ohmmeter shows that the resistance of the two is ½ of the resistance of either resistor.

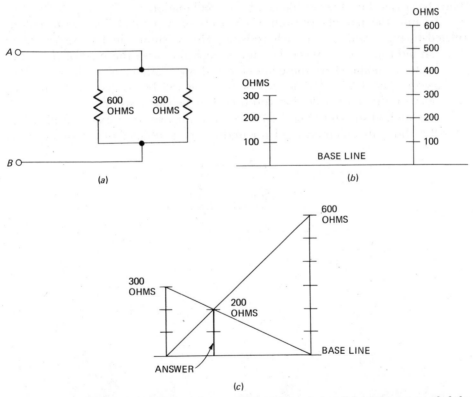

Figure 12-7. A simple way to find the resistance of two resistors in parallel. (a) *You want to find the resistance of this parallel circuit. In other words, what will an ohmmeter connected to terminals A and B read?* (b) *On a sheet of graph paper draw two parallel lines. It does not matter how far apart they are, but they must start on the same line. Make the lengths of the lines be related to the resistance values. In this case, each unit equals 100 ohms. Then 3 units equals 300 ohms and 6 units equals 600 ohms.* (c) *Connect the top of each line to the bottom of the other as shown here. At the point where the two lines cross, draw a line to the base. The length of this line represents the parallel resistance value. In this case, it is 2 units long, so the parallel resistance is 200 ohms. This means that an ohmmeter should read 200 ohms between point A and point B.*

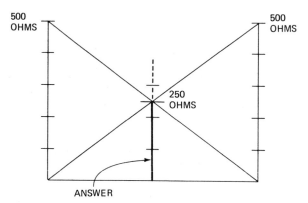

Figure 12-8. In the circuit of Fig. 12-6 there are two 500-ohm resistors in parallel. This solution shows that their resistance is 250 ohms.

If the parallel resistors do *not* have equal values, you can find their resistance with the aid of a sheet of graph paper. Figure 12-7 shows how this is done. Two unequal resistance values are connected in parallel in Fig. 12-7*a*. The object is to find the resistance of the combination.

To find the parallel resistance you draw two parallel lines with lengths related to the resistance of each resistor. This is shown in Fig. 12-7*b*. Then draw straight lines to connect the end of each line with the base of the other one. From the point where these lines cross, draw another line to the base. This is shown in Fig. 12-7*c*. The length of this line will be related to the parallel resistance on the same scale that you used for the two resistors.

The method shown in Fig. 12-7 works for any two resistors in parallel. Figure 12-8 shows this method used for finding the resistance of the two parallel

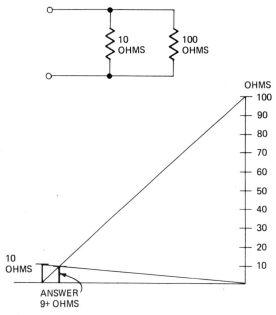

Figure 12-9. When one resistance is 10 or more times the other, you can use the resistance of the smaller one and your answer will be close enough for most work.

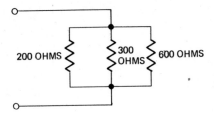

Figure 12-10. Here are three resistors in parallel. The resistance of this parallel combination must be less than 200 ohms. The resistance of any parallel combination is always less than the resistance of the smallest resistance in the circuit.

resistors shown in Fig. 12-6. Each resistor has a value of 500 ohms. Two parallel lines, each 5 units long, are drawn to the same base line. Each of these lines represents 500 ohms. The end of each line is connected to the base of the other. A line is drawn from the place where these lines cross to the base. This line is 2½ (that is, 2.5) units long. The 2.5-unit line shows that the resistance of the parallel combination is 250 ohms.

If you worked a lot of problems, you would soon find that whenever one resistor has 10 or more times the resistance of the other, you do not need to solve it. For all practical purposes, you can just take the value of the smallest one for the combination. This is shown in Fig. 12-9.

When there are more than three resistors connected in parallel, you can find the resistance of the circuit by taking them two at a time. Figure 12-10 shows a circuit with three resistors in parallel. To find the resistance of this parallel circuit, first find the resistance of any two in parallel. This is shown in Fig.

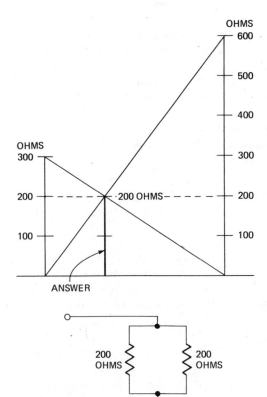

Figure 12-11. The first step in finding the resistance of the circuit in Fig. 12-10 is to replace the 300- and 600-ohm resistors with a 200-ohm resistor. The parallel combination of 300 ohms and 600 ohms has a value of 200 ohms. This means that you can replace the 300-ohm and 600-ohm resistors with a single 200-ohm resistor as shown here.

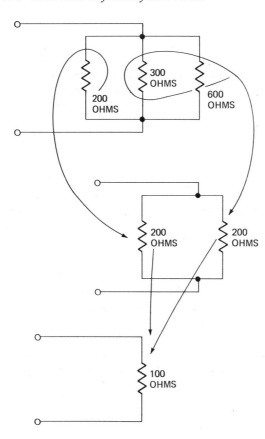

Figure 12-12. The resistance of the three resistors in parallel is 100 ohms. To find the resistance of this circuit, the 300-ohm and 600-ohm resistors are replaced with a 200-ohm resistor. Now there are two equal resistance values in parallel. Their combined resistance is 100 ohms.

12-11. In this solution, the resistance of the 300-ohm and 600-ohm resistors in parallel is found to be 200 ohms. Thus, the two resistors can be replaced with a single 200-ohm resistor.

Now you have two equal resistors in parallel. You have a rule that says that the resistance of this combination is ½ of 200 ohms, or 100 ohms. This is the resistance of the circuit.

Figure 12-12 summarizes the solution.

When you are studying a circuit to see if the resistors are connected in parallel, here are two important questions to ask:

1. Does the current divide so that only part of it flows through each resistor?
2. Is the resistance of the circuit less than the smallest resistance in the parallel resistors?

If the answers to these questions are *yes,* the resistors are in parallel.

One way to identify a parallel circuit is to start at the negative terminal of the power supply and trace through the circuit back to the positive terminal of the supply. If there is more than one path for you to take to get to the positive terminal, you have a parallel circuit.

SUMMARY

1. When two or more resistors are connected in parallel, the resistance of the circuit is always less than the smallest value of resistance in the circuit.

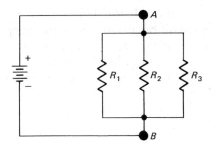

Figure 12-13. The voltage across the three resistors is the same.

2. When two resistors with equal resistances are connected in parallel, their resistance is ½ of the value of either resistor.
3. In a parallel resistor circuit, the same current does *not* flow through each resistor.
4. To find the resistance of three or more resistors in parallel, first find the resistance of any two. Once that value is found, combine it with any other resistance in the circuit. Continue to combine the resistors—two at a time—until they are all used in the process.

HOW ARE VOLTAGE DROPS RELATED IN PARALLEL?

Figure 12-13 shows a parallel resistor circuit connected to a battery. The battery voltage appears across points A and B. Since all three resistors are connected between these points, the voltage across the three resistors must be the same.

It does not matter how many resistors are connected in parallel—the voltage across every resistor in a parallel circuit will be the same.

This is a very important rule to remember: *The voltage across all parts of a parallel circuit is the same.*

Light bulbs are usually rated by the voltage needed to make them glow at full brightness, and by the power they use. Figure 12-14 shows three 6-volt light bulbs connected in parallel across a 6-volt battery. All the lights will burn at full brightness in this circuit, because there is 6 volts across each light.

If you open one of the switches in the parallel light circuit, the light in series with that switch will stop glowing. The other lights will still glow, even though one light is out.

You will remember that lights connected in series have the same current

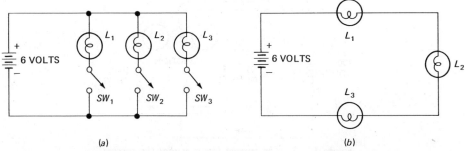

Figure 12-14. A comparison of series and parallel light circuits. (a) All three of the lights in this parallel circuit are glowing at full brightness when the switches are closed, (b) but in this circuit, there is only 2 volts across each light and the lights are not glowing at full brightness.

flowing through them. If one light burns out, all the lights in series stop glowing. If one of the lights in a parallel circuit burns out, the other lights will continue to glow. The reason is that they all have the same voltage across them. This is an important difference between series and parallel circuits.

Can you think of an example of a light circuit which is wired so that when one light burns out the other lights still glow?

SUMMARY

1. The voltage across all resistors connected in parallel is the same.
2. When light bulbs are placed in parallel, the same voltage appears across each one.
3. If one light in a parallel circuit burns out, the others will still glow.

HOW ARE CURRENTS RELATED IN PARALLEL CIRCUITS?

If you stood at the crossroad of two highways and counted cars all day, you would come to at least one conclusion: the number of cars that enter the crossroad equals the number of cars that leave the crossroad. That would not surprise you very much. After all, where would they go if they did not leave?

If you measured the amount of water entering the junction of two rivers you would find that it equals the amount of water leaving the junction. That would not surprise you very much either.

There is a basic rule in electricity that is like the rules of highways and rivers. It is known as *Kirchhoff's current law,* and it states that *all the current that flows into the junction of two wires equals all the current that flows away from the junction.* We will refer to this law as the *current law.*

Figure 12-15 shows a parallel circuit with the currents in the branches and the total current indicated. There is a current of 3 amperes entering the junction. There are two currents leaving the junction. One branch current is 2 amperes and the other is 1 ampere. These branch currents add to equal the main current of 3 amperes.

The currents in parallel branches always relate to the resistance of that branch. Note that in the circuit of Fig. 12-15 the larger current flows through the smaller resistor.

The voltage across the 50-ohm resistor can be found by Ohm's law. You

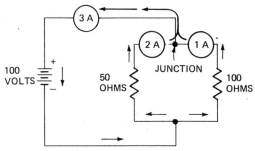

Figure 12-15. The current law says that the currents entering a junction must equal the currents leaving the junction.

know the value of current, and you know the value of resistance, so

$$VOLTAGE = CURRENT \times RESISTANCE$$
$$= 2 \text{ amperes} \times 50 \text{ ohms}$$
$$= 100 \text{ volts}$$

If you multiply the current and the resistance of the other branch, you will also get 100 volts. This shows that the voltage is the same across the parts of the parallel circuit. The currents in the branches are always the right values for making this true.

The circuit of Fig. 12-15 shows another important point regarding parallel circuits. The current does *not* have to be the same value in all branches of a parallel circuit. In this way parallel circuits are different from series circuits, in which the current is always the same value in all parts.

In the circuits that you have studied so far you can always tell which are series and which are parallel by the way the drawing is made. You will not always be able to tell so easily. The rules you are learning about series and parallel circuits will be useful later, when there are more components in the circuit.

SUMMARY

1. The current law says that the total current entering a junction equals the total current leaving that junction.
2. The largest current always flows through the smallest resistance in a parallel circuit.
3. The amount of current may not be the same value in all parts of a parallel circuit.

HOW IS POWER IN A PARALLEL CIRCUIT DETERMINED?

Figure 12-16 shows three lights connected in parallel across a voltage source. Each light is rated at 6 volts, 3 watts. The power for all the lights must come from the power supply.

The total power used in the circuit is the sum of the powers used by each light. Since there are three lights, and each uses a power of 3 watts, the total power used is

$$3 + 3 + 3 = 9 \; watts$$

This is a very important rule for you to remember: *The total power used in a parallel circuit is found by adding the powers used by all the components.*

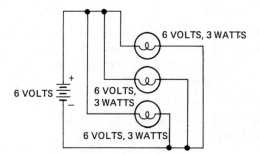

Figure 12-16. The three lights are in parallel across the power supply. What is the power used by the circuit?

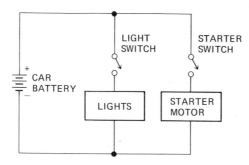

Figure 12-17. The starter motor is in parallel with the headlights.

You will remember that the total power used in a series circuit is also found by adding all the powers used by all the components. The total power in both series and parallel circuits is found the same way.

If you add another light in parallel with the other three in the circuit of Fig. 12-16, the total power used by the circuit will be increased to 12 watts. The battery must supply the increased power. If you keep adding more and more lights, you will reach a point where the battery cannot supply all the power required for the circuit.

The headlight circuit of a car is in parallel with the starter motor circuit. This is obvious because either the headlights or the starter motor may be on alone. *If they were in series, the same current would flow through both.* This would mean that the lights would always be on when the starter motor is running, and every time you turned the lights on, the starter motor would run.

Figure 12-17 shows a block diagram of the starter motor and light systems. If both switches are closed, the battery must supply power to both circuits.

The starter motor requires a lot of power. A battery can only supply so much power. If the battery is weak, it will take every bit of its power to operate the starter motor. The lights should be *off* when starting the car so that the starter motor will get all the power the battery can supply.

SUMMARY

1. The total power used in a parallel circuit is found by adding the powers used in each branch.
2. The total power used in parallel circuits is found the same way as the total power used in series circuits—that is, by *adding* them.
3. The number of parallel circuits that can be connected across a battery depends upon the ability of the battery to deliver power.

EQUATIONS RELATED TO THIS CHAPTER

We have discussed voltages, currents, resistances, and powers in a parallel circuit. Up to now we have not used math equations to show how these factors are related. However, equations are convenient ways of remembering how to calculate voltage, current, resistance, and power in parallel circuits. The equations are given here for your convenience.

Total Current in a Parallel Circuit

$$I_T = I_1 + I_2 + I_3 \ldots + I_n$$

This equation shows that the total current, I_T, flowing into a parallel branch equals the sum of the currents in each branch.

Resistance of Two Resistors in Parallel

$$R_T = \frac{R_1 \times R_2}{R_1 + R_2}$$

This equation shows that the terminal resistance, R_T, of two resistors, R_1 and R_2, connected in parallel can be found by multiplying their resistances and then dividing by their sum. If there are more than two resistors in parallel, the terminal resistance is found by pairing them off two at a time until all have been eliminated.

Power in a Parallel Circuit
TOTAL POWER IN A
PARALLEL CIRCUIT $P_T = P_1 + P_2 + P_3 \dots + P_n$

This equation shows that the total power used by a parallel circuit is equal to the sum of the powers used by each circuit component.

Programmed Review Questions

(Instructions for using this programmed section are given in Chap. 1.)

We will review the important concepts of this chapter. If you have understood the material, you will progress easily through this section. Do not skip this material, because some additional theory is presented.

1. Figure 12-18 is a circuit with three resistors. Which of the choices is correct?
   A. The three resistors are in series. (Proceed to block 7.)
   B. The three resistors are in parallel. (Proceed to block 17.)

2. *Your answer to the question in block 21 is **A**. This answer is wrong. The resistance of two resistors in parallel is **always** less than either resistance value. Your answer of 25 ohms is greater than the value of R_1 or R_2, so it cannot be right.* Proceed to block 20.

3. *Your answer to the question in block 7 is **B**. This answer is wrong. The current through all resistors connected in parallel is the same value only **if** all resistors have the same value of resistance. If the resistance values are different, the currents will not be the same.* Proceed to block 14.

4. *The correct answer to the question in block 20 is **B**. The total power is found by adding the three 6-watt power values. 6 watts + 6 watts + 6 watts = 18 watts.* Here is your next question.
   How would you draw lines to connect the two lights in parallel across the battery in Fig. 12-24? When you have completed the drawing, proceed to block 23.

5. *The correct answer to the question in block 24 is* **B**. *With the horn and lights connected in parallel, one circuit can be operated without the need for current flowing through the other circuit.* Here is your next question.
   How many 6-volt lights can be connected in parallel across a 6-volt battery?
   A. It depends upon the ability of the battery to deliver power. (Proceed to block 25.)
   B. Any number as long as the battery voltage is equal to the voltage rating of the lights. (Proceed to block 6.)

6. *Your answer to the question in block 5 is* **B**. *This answer is wrong. Every time you add a light to the circuit, the battery must supply more power. There is a limit to the amount of power that any power supply can deliver.* Proceed to block 25.

7. *The correct answer to the question in block 1 is* **A**. *The circuit is redrawn in Fig. 12-19 to show the current path through* **all** *the resistors.* Here is your next question.
   Which of the following statements is correct?
   A. The voltage across all resistors connected in parallel is always the same value. (Proceed to block 14.)
   B. The current through all resistors connected in parallel is always the same value. (Proceed to block 3.)

8. *Your answer to the question in block 14 is* **A**. *This answer is wrong. If you trace the current flow from the negative terminal of the battery back to the positive terminal, you will see that part of the current flows through resistor* R_1 *and part of the current flows through resistor* R_2. *Proceed to block 22.*

9. *Your answer to the question in block 14 is* **C**. *This answer is wrong. You have not studied a circuit that is drawn this way, but the procedure for tracing a circuit is always the same. You start at the negative terminal of the power supply and trace through the circuit back to the positive terminal of the power supply. If you will do this for the circuit in Fig. 12-20, you will see that the current divides so that part of it flows through* R_1 *and part of it flows through* R_2. *Proceed to block 22.*

10. *Your answer to the question in block 23 is* **B**. *This answer is wrong. The current through* R_1 *must add to the current through* R_2 *to make 4 amperes. Here is the way the currents are added:*

    CURRENT THROUGH R_1
    *   + CURRENT THROUGH* R_2 *= TOTAL CURRENT*
    CURRENT THROUGH R_1 *+ 3 amperes = 4 amperes*

    Proceed to block 24.

11. *Your answer to the question in block 18 is* **A**. *This answer is wrong. Rework the problem carefully.* Then proceed to block 26.

12. *Your answer to the question in block 23 is **A**. This answer is wrong. The currents through* R_1 *and* R_2 *must add to equal the main current of 4 amperes. With 3 amperes flowing through* R_2, *how much more current is needed to make 4 amperes?*

Here is the way the currents are added:

CURRENT THROUGH R_1
     + *CURRENT THROUGH* R_2 = *TOTAL CURRENT*
CURRENT THROUGH R_1 + *3 amperes* = *4 amperes*

Proceed to block 24.

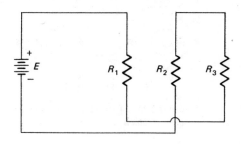

Figure 12-18. Circuit for problem in block 1.

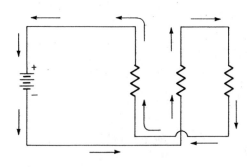

Figure 12-19. Circuit for solution of problem in block 1.

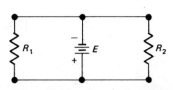

Figure 12-20. Circuit for problem in block 14.

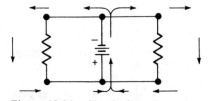

Figure 12-21. Circuit for solution of problem in block 14.

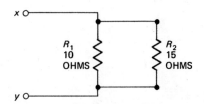

Figure 12-22. Circuit for problem in block 21.

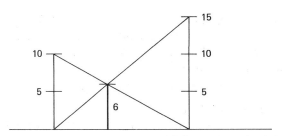

Figure 12-23. Graphical solution of problem in block 21.

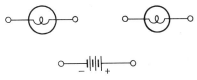

Figure 12-24. Circuit for problem in block 4.

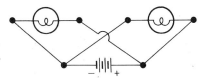

Figure 12-25. Circuit for solution of problem in block 4.

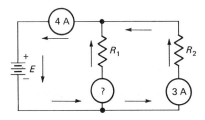

Figure 12-26. Circuit for problem in block 23.

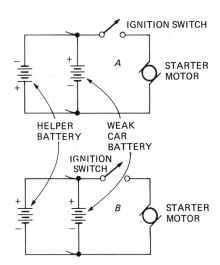

Figure 12-27. Circuit for problem in block 25.

13. *Your answer to the question in block 25 is* **A**. *This answer is wrong.* **Never** *connect batteries this way. The circuit is redrawn in Fig. 12-28. A very high current would flow around the path shown by the arrows. There is no resistance in this path. The resistance of the wires and terminals is usually ignored in problems of this type. The current would be so high that both batteries would likely be destroyed.* Proceed to block 18.

14. *The correct answer to the question in block 7 is* **A**. *The voltage across all resistors connected in parallel is always the same value.* Here is your next question.
     Study the circuit of Fig. 12-20 carefully to determine if the resistors are in series or parallel. Which of the following choices is correct?
     A. The resistors are in series. (Proceed to block 8.)
     B. The resistors are in parallel. (Proceed to block 22.)
     C. The resistors are not in series or in parallel. (Proceed to block 9.)

15. *Your answer to the question in block 22 is* **A**. *This answer is wrong. The statement is correct, but it is* **not** *a statement of the current law.* Proceed to block 21.

16. *Your answer to the question in block 20 is* **A**. *This answer is wrong. The power supplied by a battery must be equal to the sum of all powers used in the circuit.* Proceed to block 4.

17. *Your answer to the question in block 1 is* **B**. *This answer is wrong. Trace the current flow from the negative terminal of the battery, through the circuit, and back to the positive terminal. You will see that you have to go through* **all** *the resistors in order to get back to the battery.* Proceed to block 7.

18. *The correct answer to the question in block 25 is* **B**. *The positive terminal of the helper battery goes to the positive terminal of the weak battery. The negative terminal of the helper goes to the negative terminal of the weak battery. With the batteries in parallel, their currents add. If the battery is very weak, so that its terminal voltage is below normal, the helper battery will force a charging current through it and run the starter motor at the same time.* Here is your next question.
     What is the resistance of the circuit in Fig. 12-29?
     A. 2 ohms (Proceed to block 11.)
     B. 4 ohms (Proceed to block 26.)

19. *Your answer to the question in block 24 is* **A**. *This answer is wrong. If the horn and headlights were in series, the lights would come on every time the horn blew. Also, the horn would blow every time the lights were turned on. This is true because the same current flows through all parts of a series circuit.* Proceed to block 5.

20. *The correct answer to the question in block 21 is* **B**. *The resistance* **must** *be less than the smaller resistance of 10 ohms. If you solved for the answer graphically, your solution looked like Fig. 12-23.* Here is your next question.

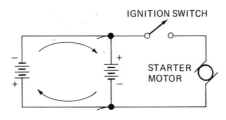

Figure 12-28. Circuit of Fig. 12-27 redrawn.

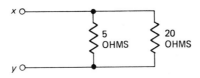

Figure 12-29. Circuit for problem in block 18.

Three 6-watt lights are connected in paralled across a battery. How much power must the battery supply?
A. 2 watts (Proceed to block 16.)
B. 18 watts (Proceed to block 4.)

21. *The correct answer to the question in block 22 is* **B**. *This is a statement of the current law.* Here is your next question.
The resistors in the circuit of Fig. 12-22 are connected in parallel. What is the resistance between X and Y?
A. 25 ohms (Proceed to block 2.)
B. 6 ohms (Proceed to block 20.)

22. *The correct answer to the question in block 14 is* **B**. *The current paths are shown in Fig. 12-21. Note how the current divides so that part flows through R_2. This means that the resistors are in parallel.* Here is your next question.
Which of the following is a statement of the current law?
A. When two resistors are connected in a parallel circuit, the larger current will flow through the smaller resistor. (Proceed to block 15.)
B. The total current entering a junction equals the total current leaving that junction. (Proceed to block 21.)

23. *Your drawing for the components in block 4 should look something like the one in Fig. 12-25. There are many ways to draw the lines to connect the parts, but there is always a way to check your answer. Start at the negative terminal of the battery and trace through the circuit to get back to the positive terminal. Can you go through* **either** *light, but not both, to get back to the positive terminal? If you can, you have drawn the lights in parallel.* Here is your next question.
What is the current through R_1 in the circuit of Fig. 12-26?
A. There is not enough information given to find the current through R_1. (Proceed to block 12.)

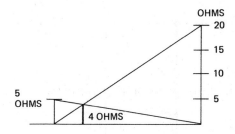

Figure 12-30. Graphical solution for problem in block 18.

B. The current through R_1 is 7 amperes. (Proceed to block 10.)
C. The current through R_1 is 1 ampere. (Proceed to block 24.)

24. *The correct answer to the question in block 23 is **C**. One ampere through R_1 plus three amperes through R_2 equals a total of four amperes.* Here is your next question.
Are the horn and headlight circuits of a car connected in series or in parallel?
A. In series (Proceed to block 19.)
B. In parallel (Proceed to block 5.)

25. *The correct answer for the question in block 5 is **A**. You cannot connect an unlimited number of loads across a power supply. There is **always** a limit to the power that can be supplied by a load.* Here is your next question.
Which of the drawings in Fig. 12-27 shows how a battery should be used with jumper cables to help start a car?
A. If you believe *A* is the correct way, proceed to block 13.
B. If you believe *B* is the correct way, proceed to block 18.

26. *The correct answer to the problem in block 18 is **B**. Figure 12-30 shows the solution to this problem.*
You have now completed the programmed questions. The next step is to put some of these ideas to work in laboratory experiments. Proceed to the "Experiments" section that follows.

Experiments

(The experiments described in this section may be performed on the circuit board described in Appendix C or on a similar laboratory setup.)

EXPERIMENT 1

Purpose—To demonstrate the current law.

Theory—The current law states that the total amount of current entering a junction equals the total amount of current leaving that junction. This very important law will be demonstrated in this experiment.

Test Setup—Wire the circuit board as shown in Fig. 12-31.

$$\frac{1}{\frac{1}{R_1}+\frac{1}{R_2}}$$

$$\frac{1}{\frac{1}{5}+\frac{1}{20}}$$

$$\frac{1}{\frac{4}{5}}$$

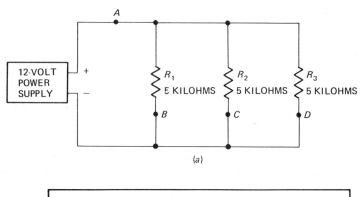

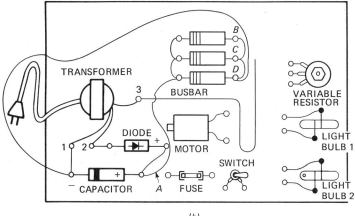

Figure 12-31. Test setup for Experiments 1 and 2. (a) The circuit for Experiments 1 and 2. (b) This is the way your circuit board will look if you have wired the circuit correctly.

Procedure—

Step 1—With the circuit board wired as shown in Fig. 12-31, measure the main current at point *A*. Use the 10-milliampere scale on the meter. This is called the *power supply current* or *total current*. Write the current value here.

Total current = _____ milliamperes

Step 2—Measure the current through R_1. This current is measured at point *B*. Write the current value here.

Current through R_1 = _____ milliamperes

Step 3—Measure the current through R_2. This current is measured at point *C*. Write the current value here.

Current through R_2 = _____ milliamperes

Step 4—Measure the current through R_3. This current is measured at point *D*. Write the current value here.

Current through R_3 = _____ milliamperes

Step 5—According to the current law, the currents through the three resistors should add to equal the total current. Add the currents that you obtained in Steps 2, 3, and 4, and write the sum here.

Sum of the currents through the resistors = _____ milliamperes

Conclusion—Allowing for a reasonable amount of error, does the sum of the currents that you obtained in Step 5 equal the power supply current that you obtained in Step 1?

_____

EXPERIMENT 2

Purpose—To show that the voltage is the same across all parts of a parallel circuit.

Theory—The fact that the voltage is the same across all resistors in a parallel circuit may be obvious from a circuit diagram. However, it is always a good idea to demonstrate such theories. This will give you practice in making measurements, and it will also make the rules easier to remember.

Test Setup—Wire the circuit board as shown in Fig. 12-31.

Procedure—

Step 1—*Calculate* the resistance of the three resistors in parallel. You could do this by the graphical method. However, note that the three resistors are identical. You know that when two identical resistors are connected in parallel, their resistance if $\frac{1}{2}$ of either value. This is because the two paths make it half as difficult for current to flow. With three identical resistors in parallel, their resistance is only $\frac{1}{3}$ as great as a single resistance value; with four identical resistors in parallel, their resistance is $\frac{1}{4}$ as great; and so on. To find the resistance of the circuit in Fig. 12-31, divide 5,000 ohms by 3. Write the resistance value here.

Resistance of circuit = _____

Step 2—Ohm's law tells us that the battery current is equal to the battery voltage divided by the circuit resistance. In other words:

$$CIRCUIT\ CURRENT = \frac{BATTERY\ VOLTAGE}{RESISTANCE\ OF\ CIRCUIT}$$

Calculate the current and record the value here.

Circuit current = _____ amperes

How many milliamperes is this?

_____ milliamperes

Step 3—The value of current that you calculated in Step 2 should be the same as the value of current that you measured in Step 1 of Experiment 1.

Allowing for some error in measurement, are the values the same?

_____

Step 4—Measure the voltage across R_1 and write the value here.

Voltage drop across R_1 = _____ volts

Step 5 — Measure the voltage across R_2 and write the value here.

Voltage drop across $R_2 = $ _____ volts

Step 6 — Measure the voltage across R_3 and write the value here.

Voltage drop across $R_3 = $ _____ volts

Step 7 — One of the basic rules of parallel circuits is that the voltage is the same across all branches. This means that the values you got for Steps 4, 5, and 6 should all be the same. Are they the same?

_____

Step 8 — Do the values you got for Steps 4, 5, and 6 equal the power supply voltage?

_____

Step 9 — Disconnect the circuit from the power supply and measure the resistance of the parallel circuit with an ohmmeter. Write the value here.

Measured value of resistance = _____ ohms

Does the measured value of resistance of Step 9 equal the calculated value of Step 1?

_____

Conclusion — Allowing for a reasonable amount of error, do your results show that the voltage is the same across all branches of a parallel circuit?

_____

EXPERIMENT 3

Purpose — To show that circuits connected in parallel work independently.

Theory — When components are connected in *series*, the same current flows through all of them. If one of the *series* components fails, the operation of the others will be affected.

When components are connected in *parallel*, each operates independently of the others. If one component fails in such a way that current no longer can flow through it — such as occurs when the filament in a light bulb is open — the operation of the other components will not be affected.

In this experiment switches are used to open parallel branches.

Test Setup — Connect the circuit as shown in Fig. 12-32.

Procedure —
Step 1 — Close both switches and adjust the variable resistor so that L_1 is giving off the maximum light.
Step 2 — Open switch 1 and note that all lights are controlled by the main switch.

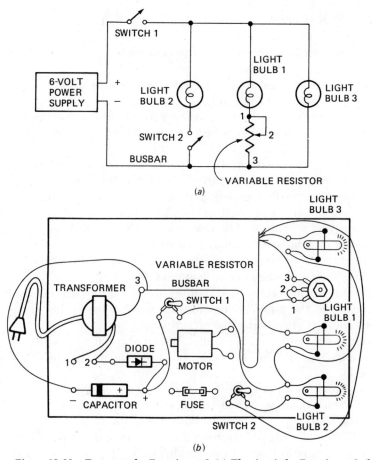

Figure 12-32. Test setup for Experiment 3. (a) *The circuit for Experiment 3.* (b) *This is the way your circuit board will look if you have wired the circuit correctly.*

The main switch in this experiment does the same job as a main fuse or circuit breaker in a house. If the fuse is burned out, or if the circuit breaker is tripped, all the lights go out.

Step 3 — Close switch 1 again and open switch 2. Note that light L_2 stops glowing but lights L_1 and L_3 are not affected. This shows that one light in a parallel branch can burn out and the others will still be lit.

Step 4 — Close switch 2 so that all lights are lit. Adjust the variable resistor and note that light L_1 changes brightness. This demonstration shows that parallel circuits work separately.

Step 5 — With all lights glowing at maximum brightness, remove light L_1 from its socket. Note that the other lights still glow. This, again, shows that parallel circuits work separately.

Conclusion — Do the results show that parallel circuits operate independently, but can all be controlled by a main switch or fuse?

Self-Test with Answers

(Answers with discussions are given in the next section.)

1. Electric power sources are connected in parallel to (a) get more current capability; (b) increase the voltage.
2. Suppose the battery in a car is too weak to turn the starter motor. To get the car started, you would connect another battery (a) in series; (b) in parallel.
3. In order to connect two batteries in parallel (a) connect the positive terminal of one battery to the negative terminal of the other; (b) connect the two positive terminals together and the two negative terminals together.
4. Three resistors are connected in parallel. They have values of 2 kilohms, 4 kilohms, and 8 kilohms. The resistance of this parallel combination is (a) 5 kilohms; (b) a little more than 1 kilohm.
5. The total current entering a junction equals the total current leaving the junction. This is (a) the current law; (b) Ohm's law.
6. Two resistors with values of 4 kilohms and 8 kilohms are in parallel. The voltage across the 8-kilohm resistor is 8 volts. What is the voltage across the 4-kilohm resistor? (a) 4 volts; (b) 8 volts.
7. Two lights are connected in parallel. Which of these statements is right? (a) If one burns out, neither will be lit; (b) If one burns out, the other one will still be lit.
8. Two resistors are in parallel in a circuit. Each resistor is using 3 watts of power. Which of these statements is right? (a) The power used by both is 1.5 watts; (b) The power used by both is 6 watts.
9. The taillights of a car are connected (a) in series; (b) in parallel.
10. Never connect power sources in parallel unless (a) they are the same size; (b) they have the same voltage rating.

Answers to Self-Test

1. (a) — They are connected in series to get more voltage.
2. (b) — The weak battery cannot supply enough *current* to run the starter motor, so another battery is connected in parallel with it.
3. (b) — Figure 12-1 shows the way to connect batteries in parallel.
4. (b) — You know it cannot be 5 kilohms because the resistance must be smaller than the smallest resistance in the parallel group.
5. (a)
6. (b) — The voltage across all parts of a parallel circuit is the same.
7. (b) — This is an important thing to remember about parallel circuits.
8. (b) — The total power used in a parallel circuit is found by adding the powers in each branch.
9. (b) — When one burns out, they do not all burn out. This shows that they are in parallel.
10. (b) — This is a very important rule to remember.

13.
What Is a Series-Parallel Circuit?

Introduction

A series-parallel circuit is simply a circuit that has some components connected in series and others connected in parallel. There are no new rules required for working with these circuits. All that is necessary is to combine the rules for series circuits and parallel circuits that you learned in Chaps. 11 and 12. Part of this chapter is devoted to a review of the principles that you learned in those chapters, and the balance of the chapter is devoted to new applications of the rules.

Let us start by reviewing some of the important characteristics of the series circuits. Components are said to be in series when the same current flows through them. The components may be batteries or resistors, or some other electrical or electronic devices.

When power sources, such as batteries, are connected in series, their voltages add. For example, a 12-volt automobile battery is obtained by connecting two 6-volt cells in series.

When resistors are connected in series, their resistance values are added together. For example, three 50-ohm resistors in series provide a total circuit resistance of 150 ohms.

When current flows through a resistor, there is always a voltage drop. The total amount of voltage drop in a series circuit must always equal the amount of voltage applied to that circuit.

The total amount of power dissipated in a series circuit is calculated by adding the amounts of power dissipated in each of the components in the circuit. The power dissipated in a resistor, whether it is connected in series or parallel, can be found by multiplying the voltage drop across the resistor by the current flowing through the resistor.

Circuit current is measured by an ammeter, milliammeter or microammeter. In order to measure circuit current with an ordinary current meter, the meter must be connected in series with the circuit. In other words, the meter must be connected in such a way that the circuit current flows through it.

Having reviewed some of the more important characteristics of series circuits, we will now review the important characteristics of parallel circuits.

When power sources, such as batteries, are connected in parallel, their ability to deliver current is increased. If a single battery can deliver 1 ampere safely to a load, two such batteries connected in parallel could deliver 2 amperes. Of course, the amount of current that is actually delivered to a circuit is dependent upon the resistance of that circuit.

When two resistors are connected in parallel, the circuit resistance is less than the smallest amount of resistance of the combination. For example, if a 100-ohm resistor is connected in parallel with a 10-ohm resistor, the combined circuit resistance must be less than 10 ohms.

The voltage is the same across all parts of a parallel circuit, but the current divides and flows into the different branches. The amount of current in each branch depends upon the amount of resistance in that branch. The largest amount of current flows through the smallest amount of resistance in a parallel circuit.

To find the total power in a parallel circuit, it is necessary only to add the powers dissipated by each of the resistors in that branch. Thus, the total powers in both series and parallel circuits are found in the same way.

To measure the voltage drop across a component, it is necessary to connect a voltmeter in parallel with that component. Remember that the voltage is across the component, and therefore, the meter must be connected across the component also.

In this chapter you will apply these basic rules of series and parallel circuits to the study of series-parallel combinations. You will learn how to calculate the resistance of a series-parallel circuit, and how to *trace* a circuit.

You will be able to answer these questions after studying this chapter:

What is a series-parallel circuit?
How do you trace a circuit?
How do you find the resistance of a series-parallel circuit?
What are some practical applications of series-parallel circuits?

Instruction

WHAT IS A SERIES-PARALLEL CIRCUIT?

Figure 13-1 shows a typical series resistor circuit. Meters in the circuit are connected for measuring the current flowing through the resistors. Since the cur-

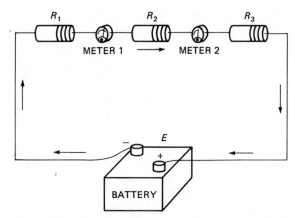

Figure 13-1. A series resistor circuit. Arrows show path of electron current flow.

rent is the same in all parts of a series circuit, we can expect that the reading on meter 1 is identical to the reading on meter 2.

If you would measure the voltage across resistor R_1, across resistor R_2, and across resistor R_3, and add these voltage drops, you would get a value of voltage which is equal to the battery voltage. This is always true, and it is expressed in Kirchhoff's voltage law: *The sum of the voltage drops around a circuit is always equal to the applied voltage.*

Figure 13-2 shows a parallel circuit. In this case, the meters are being used to measure the voltage across each part of the parallel circuit. Since the voltage is the same across all parts of the parallel branch, it follows that each meter reading must be identical.

The total current in the circuit is considered to be a battery current. It is marked I_T in Fig. 13-2. This current flows to junction A, and at that point it divides into three individual values, I_1, I_2, and I_3 in the illustration. If you would measure these three currents in the branches and add them together, you would find that they are always equal to the total current I_T. This is stated in Kirchhoff's current law: *The sum of the currents entering a junction in a circuit equals the sum of the currents leaving that junction.* A junction is any place in a circuit where currents divide or combine.

The current divides at point A so that part of the total current flows

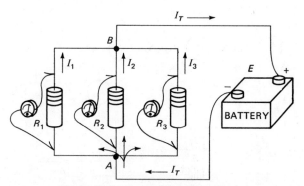

Figure 13-2. A parallel resistor circuit.

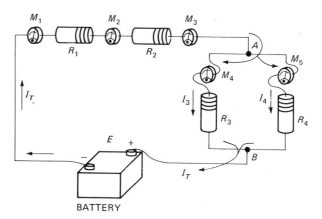

Figure 13-3. A series-parallel resistor circuit. The meters are measuring current.

through each branch. At point *B* the current recombines into a single current, I_T.

Figure 13-3 shows an example of a series-parallel circuit. Here we have two resistors (R_1 and R_2) connected in series. Since these resistive components are in series, the current readings on meters M_1, M_2, and M_3 are equal. This current is I_T, and it is considered to be the *total current* of the circuit. The total current of the circuit is the amount of current that the battery must supply.

When the total current arrives at junction *A*, it divides so that part of it is flowing through R_3 and part of it is flowing through R_4. It is a good practice to assign the same subscripts (numbers) to the currents as used for the resistors. Thus, I_3 is the current flowing through R_3, and I_4 is the current flowing through R_4.

If you add the values of I_3 and I_4, you will obtain the value of total current (I_T). At point *B* the currents recombine to obtain the total circuit current.

If you were to measure the voltage drop across R_3 and the voltage drop across R_4, you would find that these voltage drops are identical. This must be true because R_3 and R_4 are in parallel.

Figure 13-4 shows the same circuit as shown pictorially in Fig. 13-3, but in this case the drawing is made with schematic symbols. Compare the two illustrations carefully to make sure that you understand they are the same.

Figure 13-5 is a schematic drawing that shows how voltage measurements

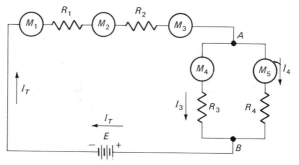

Figure 13-4. Schematic drawing of the circuit shown in Fig. 13-3.

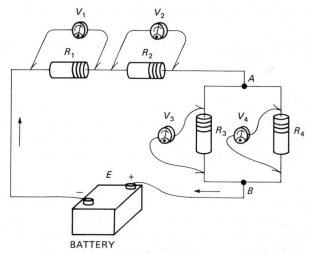

Figure 13-5. Voltage measurements in a series-parallel circuit.

are made in the circuit of Fig. 13-3. Note that the voltmeter is across the components in each case. Note also that the same subscripts are used for the voltage measurements as are used for the resistors. Thus, V_3 is the voltage across resistor R_3. As mentioned earlier, since R_3 and R_4 are in parallel, the voltage across these two resistors must be the same. In other words, V_3 is equal to V_4.

The battery voltage (E) must be equal to V_1 plus V_2 plus the voltage across the parallel branch. Since V_3 and V_4 are the same, we could say that the total voltage drop in the circuit is $V_1 + V_2 + V_3$. Remember that you cannot add both V_3 and V_4 since these voltage drops are identical and represent the voltage from point A to point B in this circuit. Figure 13-6 shows the schematic drawing which represents the voltage measurements illustrated in Fig. 13-5.

SUMMARY

1. A series-parallel circuit is a circuit in which some of the components are connected in series and the rest of the components are connected in parallel.

2. The basic rules of series circuits and of parallel circuits apply also to series-parallel circuits.

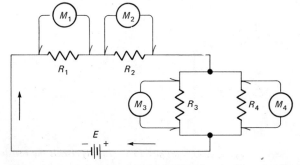

Figure 13-6. Schematic drawing of the circuit shown in Fig. 13-5.

3. In a series-parallel circuit the total current is the current supplied by the battery.
4. When adding the voltages around a closed circuit, the voltage across parallel branches is added only once.

HOW DO YOU TRACE A CIRCUIT?

You cannot tell whether resistors are in series or in parallel simply by the way they are drawn in a circuit. For example, in the circuit of Fig. 13-7 the resistors appear to be in parallel because they are side by side. However, they are actually in series.

To determine if resistors are in series or parallel, it is often necessary to *trace* the circuit. To trace a circuit, start at the negative terminal of the source and follow the electron path around the circuit and back to the positive terminal of the source. (If there is no source, such as in Fig. 13-7, one may be imagined.) If you *must* go through two resistors to get back to the positive terminal, then the resistors are in series. If you have a choice of paths, and can take any of them to get back to the positive terminal, then the resistors are in parallel.

In the circuit of Fig. 13-7 there is no power source shown. In cases like this you should *assume* a power source. Figure 13-8 shows the circuit with an imaginary battery (*E*). It does not matter which direction you put the battery into the circuit. In other words, the negative terminal of the battery can be connected to either *x* or *y*.

The arrows in Fig. 13-8 show the path of electron flow. Note that the current must go through all the resistors in order to get from the negative terminal of the battery to the positive terminal. If you reverse the battery, the current will flow in the opposite direction, but it will still have to flow through *all* the resistors to get back to the positive terminal.

In series-parallel circuits the current leaving the power source is called the *total circuit current*. If the total circuit current flows through two or more components in a circuit, they must be in series. If only part of the circuit current flows through a component, then it must be in parallel with one or more components.

It is possible to have series-connected resistors in a parallel circuit. Con-

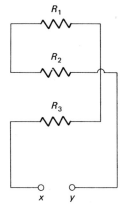

Figure 13-7.   In this circuit the resistors are in **series**.

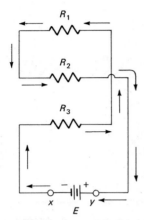

Figure 13-8. To show that the resistors are in series, a voltage is assumed and the circuit is traced.

sider, for example, the circuit of Fig. 13-9. The arrows are used to show the path of electron flow for tracing the circuit. Note that the total battery current divides at point a. This indicates that there is a parallel circuit. Part of the divided current flows through R_2, and the rest flows through both R_3 and R_4. Since the branch current flows through both R_3 and R_4, it follows that these resistors are in series, and this series combination is in parallel with resistor R_2.

All the current flows through R_1, and it is said to be in series with the circuit comprised of R_2, R_3, and R_4.

HOW DO YOU FIND THE RESISTANCE OF A SERIES-PARALLEL CIRCUIT?

Figure 13-10 shows a resistive network comprising eight resistors connected in series-parallel. This is a typical series-parallel circuit, and we are going to find the resistance of the complete combination. The symbols beside each number in some illustrations are Greek letters omega (Ω). This symbol is used to designate ohms. In other words, 1.6 Ω means a resistance of 1.6 ohms.

The procedure for finding the resistance of any series-parallel circuit, like the one shown in Fig. 13-10, is to simplify it by combining the series and the parallel resistors into *equivalent resistors*. The equivalent resistance of a series or parallel resistor combination is a single value of resistance that can be used to replace the resistors. For example, when two 10-ohm resistors are connected in parallel, they have an equivalent resistance of 5 ohms.

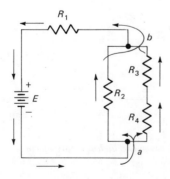

Figure 13-9. Circuit tracing shows that resistors R_3 and R_4 are in series.

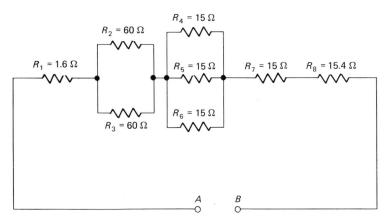

Figure 13-10.   The resistance of this series-parallel circuit is to be determined.

Until you gain a little experience working with circuits, it may be difficult to determine when resistors are connected in series and when they are connected in parallel. As we mentioned earlier in this chapter, one of the best techniques is to assume a voltage across the circuit terminals and then trace the electron current through the circuit. This is done in Fig. 13-11, and the electron current path is illustrated by arrows.

Note that the total current flows through R_1, R_7, and R_8; therefore, these three resistors must be presumed to be in series. However, the total current must divide in order to flow through R_2 and R_3, and it must also divide in order to flow through R_4, R_5, and R_6. Therefore, these are parallel resistance combinations.

Since R_1, R_7, are R_8 are in series, their resistance values can be combined by simple addition. As far as the circuit is concerned, the three resistors can be replaced with a single resistor that has a value equal to the three series resistors in combination.

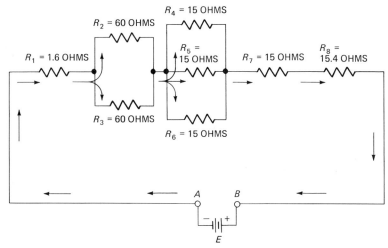

*Figure 13-11.   You can assume battery **E** to determine which resistors are in series and which are in parallel.*

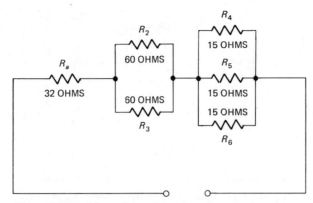

*Figure 13-12. The series resistors have been combined into a single equivalent resistor (**R_a**).*

We will call this single resistance R_a. Mathematically

$$R_a = R_1 + R_7 + R_8$$

This is true because resistors in series are combined by simple addition.

$$R_a = 1.6 + 15 + 15.4$$
$$= 32 \text{ ohms}$$

Figure 13-12 shows the circuit redrawn with the three series resistors R_1, R_7, and R_8 replaced by the single equivalent resistance R_a. It is important for you to understand that as far as the battery is concerned, the amount of current supplied will not be affected by replacing the three resistors with the single equivalent resistor.

You might wonder, then, why not use a single resistor in the first place. There are a number of reasons. For one thing it might not be possible to obtain a single resistor that has exactly the desired amount of resistance that you are looking for. Also, there are a number of circuits in which it is necessary to connect the resistors in series in order to produce a desired voltage drop. Other reasons for series connections will be apparent as you extend your study of electricity.

At this time we are trying to make a specific point. *Whenever two or more resistors are connected in **series**, you can always replace the resistors with a single resistor which has a resistance value equal to the sum of the series resistance value.* This is what was done in Fig. 13-12.

In Fig. 13-12 you have a resistor (R_a) in series with one parallel combination comprising R_2 and R_3, and another parallel combination comprising R_4, R_5, and R_6.

Let's look first at the combination R_4, R_5, and R_6. Note from the illustration that each of these resistors has a resistance value of 15 ohms. When all the resistors are equal in value, the three resistors in parallel have ⅓ of the value of a single resistor. In other words, with three resistors in parallel, each having a value of 15 ohms, the equivalent resistance is 5 ohms. If you do not remember the rule for parallel resistors that allows us to replace the three resistors with a single resistor of 5 ohms, you can solve the parallel resistance circuit using the graphical solution or the equation method. In any event, you will find that the three resistors in combination have a parallel resistance value of 5 ohms. Figure

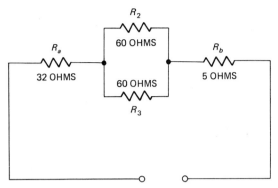

Figure 13-13.   The three parallel resistors (R_4, R_5, and R_6) have been combined into a single equivalent resistor (R_b).

13-13 shows the equivalent circuit of Fig. 13-12. Equivalent resistor R_b has replaced the parallel combination of R_4, R_5, and R_6. Note also that R_a and R_b are now in series.

Resistor R_b, having a value of 5 ohms, can be added to the resistance of R_a, having a value of 32 ohms. This is done in Fig. 13-14. In this illustration R_c is equal to the combination of R_a and R_b in series.

$$R_c = R_a + R_b$$
$$= 32 + 5$$
$$= 37 \text{ ohms}$$

Resistor R_c is an equivalent resistance. It will offer the same amount of resistance in the circuit as the two resistors offered in the circuit of Fig. 13-13.

The equivalent circuit in Fig. 13-14 shows R_c in series with the parallel combination of R_2 and R_3. Resistors R_2 and R_3 have identical values of 60 ohms each. When the two resistors are in parallel, and they have identical values, then the resistance of the parallel circuit is equal to $\frac{1}{2}$ of either of the resistors in the parallel circuit. This means that the parallel resistance of R_2 and R_3 must be $\frac{1}{2}$ of 60 ohms, or 30 ohms.

R_d is what we are calling an equivalent resistance of R_2 and R_3, and we are showing it in series with R_c in Fig. 13-15. Now we have two resistors (R_c and R_d) in series, and the resistances combine to produce the equivalent circuit resistance (R_{EQ}) value of 67 ohms.

Figure 13-16 shows R_{EQ}, which is the equivalent resistance of the complete

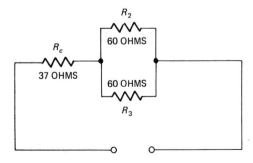

Figure 13-14.   A simple signal-tracing procedure for the circuit of Fig. 13-13 shows that R_a and R_b are in series. In this circuit they have been combined into a single equivalent value (R_c).

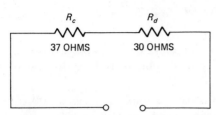

Figure 13-15. Parallel resistors R_2 and R_3 have been combined into a single equivalent resistor (R_d).

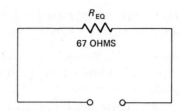

Figure 13-16. Resistors R_c and R_d are combined to produce R_{EQ}. This is the equivalent resistance value of the complete circuit shown in Fig. 13-10.

network shown in Fig. 13-10. If you connect a battery between terminals *A* and *B* of Fig. 13-10, it will supply a certain amount of current to the resistive circuit. If you connect the same battery to the circuit terminals of Fig. 13-16 in which the equivalent resistance is shown, you will find that exactly the same amount of current flows. This is the reason for the term "equivalent resistance." As far as the battery is concerned, the resistance of one circuit is equivalent to the resistance of the other.

We have taken you step by step through the solution of a typical series-parallel circuit. Remember that whenever series resistances are combined, an equivalent resistance can be used to replace the series resistors. Likewise, when parallel resistors are combined, they can be replaced with the equivalent resistance. In the step-by-step procedure, the series resistances are added. Parallel resistances are combined, and the equivalent resistances are then added to the series resistances. This procedure can take place as many times as necessary in order to reduce the series-parallel network to a simple series resistance value.

It would be advisable if you would review this problem a number of times before you have completed this book so that you understand this procedure thoroughly. However, it will not be necessary for you to work a large number of problems in order to learn the procedure better. Most of the resistive branches that you work with will be mostly simple series or simple parallel. When series-parallel branches are encountered, they are usually much less complicated than the one shown in Fig. 13-10.

SUMMARY

1. In order to trace a circuit, follow the path that the electron current takes through the circuit.
2. When the same current flows through two or more components, they are in series.
3. When the current divides so that part of it flows through each component, they are in parallel.
4. When components are in series-parallel, some of them are in series and some of them are in parallel.
5. To find the resistance of a series-parallel circuit, combine the series resistors by simple addition and combine the parallel resistances by the graphical solution (or by rules and equations). Do this as many times as required to reduce the circuit to a single equivalent resistance value.

*WHAT ARE SOME PRACTICAL APPLICATIONS OF
SERIES-PARALLEL CIRCUITS?*

Figure 13-17 shows how a series-parallel circuit makes it possible to use resistors with a lower power rating, thus making it possible to reduce the cost of making the circuit.

In the circuit of Fig. 13-17a a 100-ohm resistor is placed across a 100-volt source. The resistor in this circuit must be capable of dissipating 100 watts of power. In most electric circuits that is a very large amount of power to be dissipated by one resistor. The cost of a resistor is related to its power rating—that is, it is related to the number of watts of power (in the form of heat) that the resistor can dissipate safely. While a 100-watt resistor would be very large and expensive, 25-watt resistors are cheaper and readily available.

In the circuit of Fig. 13-17b, four 100-ohm, 25-watt resistors are used in a series-parallel combination across the 100-volt source. The resistance of this series-parallel circuit is 100 ohms. In other words, the resistance of the circuit in Fig. 13-17b is exactly the same as for the circuit in Fig. 13-17a. However, each resistor in the series-parallel circuit must dissipate only 25 watts. This means that the resistors used are cheaper and more readily available, and yet the circuit offers the same amount of resistance to the battery circuit.

To summarize, one purpose for using series-parallel resistance networks is that it allows you to use resistors with a lower power rating.

Figure 13-18 shows another example of a series-parallel circuit. Light 1 is rated at 6 volts, 2 amperes; light 2 is rated at 6 volts, 1 ampere. The rating of a light tells how much voltage must be across it and how much current must flow through it for the light to glow at full brightness. Hence, 2 amperes *must* flow through light 1, and this must be the total circuit current. However, if 2 amperes is allowed to flow through light 2, it will burn out quickly. Therefore, a resistor (R_{SH}) is placed in parallel with light 2. Its resistance value is chosen so that 1 ampere of current will flow around light 2 and 1 ampere will flow through it.

The circuit is similar to any series-parallel resistive circuit. You will remember the filament in a light bulb is actually a small resistance.

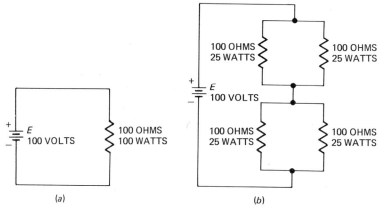

Figure 13-17. One reason for connecting resistors in series-parallel. (a) *In this circuit the resistor must dissipate 100 watts.* (b) *With this arrangement, each resistor dissipates only 25 watts.*

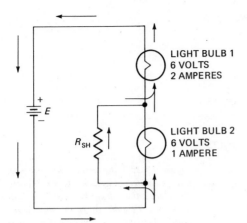

Figure 13-18. A practical series-parallel circuit.

The shunt resistor in Fig. 13-18 is very similar to the shunt resistor used in current-measuring meters. It permits a small current meter to be used for measuring large current values.

Figure 13-19 shows another way of connecting lights in a series-parallel circuit. In this case the lights are all placed in parallel, so they can all be presumed to have the same voltage rating.

In the circuit of Fig. 13-19, the voltage rating of the lights is lower than the applied voltage (E). The resistor (R_s) is placed in series with the parallel combination to provide a voltage drop so that only the rated voltage will appear across the lights. Remember that this is one of the purposes of resistors in circuits: to provide a voltage drop.

When lights are connected in parallel, as shown in Fig. 13-19, it is necessary only that they have the same *voltage* rating. Each light can draw a different amount of current, and it will not affect the circuit operation. The total amount of current through R_S is equal to the total of the current values through each light. (This is a statement of Kirchhoff's current law.)

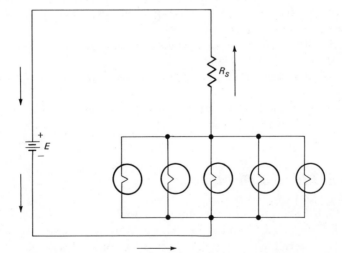

Figure 13-19. Resistor R_s is sometimes called a "dropping resistor."

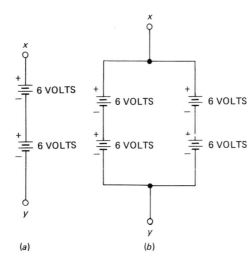

Figure 13-20.   Batteries, like resistors, may be connected in series-parallel. (a) *In this circuit the voltage between terminals* **x** *and* **y** *is 12 volts.* (b) *In this circuit the voltage between terminals* **x** *and* **y** *is also 12 volts, but the circuit can deliver a greater amount of current.*

Figure 13-20 shows why voltage sources are connected in series-parallel. Figure 13-20*a* shows two voltage sources in series. The voltage between *x* and *y* is the sum of the voltages of each source, or 12 volts. (This arrangement was discussed in an earlier chapter.)

A disadvantage that may occur with two voltage sources in series, as shown in Fig. 13-20*a*, is that it may not be able to supply a desired amount of current. By arranging the voltage sources in a series-parallel combination, as shown in Fig. 13-20*b*, a larger amount of current can be delivered to the external circuit. In this case, there are two parallel branches, each containing 12 volts (two 6-volt sources in series). Each branch can deliver current to the external circuit, and the total current delivered to the external branch is larger than can be delivered by either branch independently.

SUMMARY

1. Resistors may be connected in series-parallel to reduce the cost of the circuit. For example, four 25-watt inexpensive resistors can be used in place of a more expensive 100-watt resistor.
2. Series-parallel circuits are used when it is desired to connect components with different current ratings in series.
3. Series-parallel resistor circuits are used to obtain desired voltage relationships.
4. Voltages are connected in series-parallel in order to get a larger voltage and current-delivering capability.

Programmed Review Questions

(Instructions for using this programmed section are given in Chap. 1.)

We will now review the important concepts of this chapter. If you have understood the material, you will progress easily through this section. Do not skip this material, because some additional theory is presented.

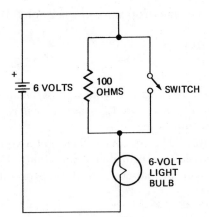

Figure 13-21. How will operation of the switch affect the brightness of the light?

1. Regarding the circuit of Fig. 13-21, which of the following statements is correct?
    A. The light will glow more brightly when the switch is closed. (Proceed to block 7.)
    B. The light will glow more brightly when the switch is open. (Proceed to block 17.)

2. *The correct answer to the question in block 15 is **A**. If you imagine a voltage source across terminals* x *and* y, *and trace the path of electron current through the circuit, you will see that the three resistors are in parallel. When three resistors, all having the same resistance value, are placed in parallel, the circuit resistance is* ⅓ *of the resistance of one resistor.* Here is your next question.
    In order to get a larger output voltage, batteries are connected
    A.   in series. (Proceed to block 12.)
    B.   in parallel. (Proceed to block 18.)

3. *The correct answer to the question in block 16 is **B**. Three amperes **must** flow through light 2 in order for it to glow at full brightness, but if three amperes is allowed to flow through light 1, it will burn out. Thus, a shunt is placed across light 1 to allow 1½ amperes of current to flow **around** light 1.* Here is your next question.
    In order to trace a circuit, always start at
    A. the negative terminal of the source and follow the electron path around the circuit and back to the positive terminal. (Proceed to block 15.)
    B. the positive terminal of the source and determine which is the shortest path back to the negative terminal. (Proceed to block 21.)

4. *The correct answer to the question in block 20 is **A**. It is important to remember that the total power dissipated in any circuit is always found the same way. It does not matter whether it is a series circuit, parallel circuit, or series-parallel circuit. The procedure is to add all the powers dissipated in each resistor in the circuit.* Here is your next question.

In the circuit of Fig. 13-25 the voltage drop across R_2 is 2 volts. What is the voltage drop across R_1?

A. 8 volts (Proceed to block 10.)

B. 4 volts (Proceed to block 16.)

5. *The correct answer to the question in block 12 is **B**. One-half ampere of current in* L_a *plus one-half ampere of current in* L_b *makes a total current of one ampere.* Here is your next question.
Show how you would connect four 1½-volt dry cells to get a voltage of 3 volts and the maximum possible current-delivering ability. Draw your circuit on a separate sheet of paper, and then proceed to block 11.

6. *Your answer to the question in block 15 is **B**. This answer is wrong. If the resistors were in series, their combined resistance value would be 450 ohms. However, the resistors in Fig. 13-27 are not in series.*
Assume a voltage across terminals x *and* y *and determine whether the resistors are in parallel or series parallel. Determine the resistance value.* Then proceed to block 2.

7. *The correct answer to the question in block 1 is **A**. With the switch closed, all the electron current will flow through the switch circuit and light. Remember that electron current takes the easiest path. Thus, all the current flows through the closed switch rather than through the 100-ohm resistor. With no resistance in the circuit with the light, it will glow more brightly than it would with 100 ohms in series with it.* Here is your next question.
In the circuit of Fig. 13-22,

A. light 1 will not glow as brightly as light 2 or light 3. (Proceed to block 8.)

B. light 1 will glow more brightly than light 2 or light 3. (Proceed to block 14.)

8. *Your answer to the question in block 7 is **A**. This answer is wrong. If you will trace the electron flow in this series-parallel circuit, you will note that **all** the battery current **must** flow through light 1.* Proceed to block 14.

9. *Your answer to the question in block 16 is **C**. This answer is wrong. The rating of light 1 is 1½ amperes, and the rating of light 2 is 3 amperes. You **must** have a current of 3 amperes through light 2 in order for it to glow at full brightness. That is too much current for light 1, so you will need a method of bypassing the extra current around light 1.* Proceed to block 3.

10. *Your answer to the question in block 4 is **A**. This answer is wrong. The sum of the voltage drops across* R_1 *and* R_2 *must equal the applied voltage* (E). *In other words, the drops across* R_1 *and* R_2 *must add to equal 6 volts. Since 2 volts is dropped across* R_2, *the rest must be dropped across* R_1. *Recalculate the voltage drop across* R_2. *Then proceed to block 16.*

11. *The correct circuit for connecting the four dry cells is shown in Fig. 13-29. Note that there are two branches, with two cells in series in each branch.*

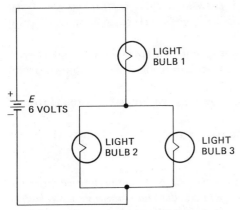

Figure 13-22. A series-parallel light circuit.

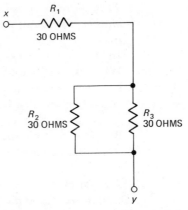

Figure 13-23. The problem is to find the resistance between terminals x and y.

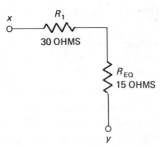

Figure 13-24. This is the circuit of Fig. 13-23 with the parallel branch replaced with an equivalent resistance (R_{EQ}).

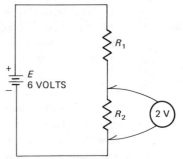

Figure 13-25. The problem is to find the voltage drop across R_1.

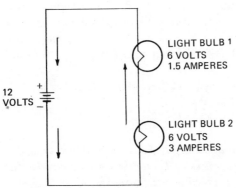

Figure 13-26. This circuit will not work properly.

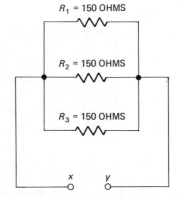

Figure 13-27. A parallel circuit with three resistive branches.

Thus, there is 3 volts for each parallel branch. Connecting the 3-volt branches increases the current-delivering ability. Here is your next question.

In the circuit of Fig. 13-30 each resistor radiates 10 watts of power. What is the amount of power radiated by the complete circuit between points *x* and *y*?

A. The power radiated by the complete circuit is 15 watts. (Proceed to block 23.)

B. The power radiated by the complete circuit is 30 watts. (Proceed to block 19.)

12. *The correct answer to the question in block 2 is **A**. Connecting batteries (or cells) in series gives an output voltage that is the sum of the individual battery (or cell) voltages. Remember that a 12-volt car battery is made by connecting six 2-volt cells in series.* Here is your next question.

In the circuit of Fig. 13-28, what is the amount of current flowing through light *b* (that is, through L_b)?

A. The amount of current flowing through L_b is 1.5 amperes. (Proceed to block 24.)

B. The amount of current flowing through L_b is 0.5 ampere. (Proceed to block 5.)

13. *Your answer to the question in block 14 is **B**. This answer is wrong. In order to solve this problem, assume a voltage across terminals* x *and* y. *Follow the electron current path and note that resistors* R_2 *and* R_3 *are in parallel. Since their resistance values are the same, the parallel combination has an equivalent resistance equal to* 1/2 *of the value of either* R_2 *or* R_3.

Replace R_2 *and* R_3 *with a single resistor that has a resistance equivalent to* R_2 *and* R_3. *Then combine this equivalent resistance value with* R_1. Proceed to block 20.

14. *The correct answer to the question in block 7 is **B**. The amount of light depends upon the amount of current flowing through the filament. Since all the battery current is flowing through light 1, but only part of the current flows through light 2 and light 3, it follows that light 1 is glowing brightest.* Here is your next question.

In the circuit of Fig. 13-23 the resistance between terminals *x* and *y* is

A. 45 ohms. (Proceed to block 20.)

B. 90 ohms. (Proceed to block 13.)

15. *The correct answer to the question in block 3 is **A**. It is a good idea to start at the negative terminal of the power source and trace the electron current path through **all** branches of the circuit.* Here is your next question.

In the circuit of Fig. 13-27, what is the resistance between terminals *x* and *y*?

A. The resistance between terminals *x* and *y* is 50 ohms. (Proceed to block 2.)

B. The resistance between terminals *x* and *y* is 450 ohms. (Proceed to block 6.)

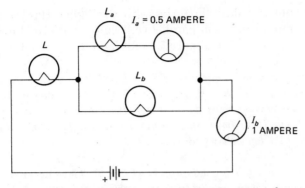

Figure 13-28. The problem is to determine the amount of current through light L_b.

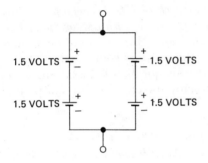

Figure 13-29. A series-parallel battery circuit.

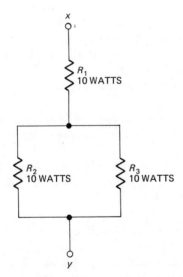

Figure 13-30. The problem is to find how much power is dissipated in the complete circuit.

16. *The correct answer to the question in block 4 is* **B**. *The voltage across* R_1 *(4 volts) plus the voltage across* R_2 *(2 volts) must add to equal the applied voltage (6 volts). This is an application of Kirchhoff's voltage law.* Here is your next question.

The circuit of Fig. 13-26 will not work properly because the current ratings of the lights are different. In order to make this circuit work properly,

A.   put a resistor in series with the lights. (Proceed to block 22.)

B.   put a resistor in parallel with light 1. (Proceed to block 3.)

C.   put a resistor in parallel with light 2. (Proceed to block 9.)

17.   *Your answer to the question in block 1 is **B**. This answer is wrong. To under-stand why, trace the circuit of Fig. 13-21 with the switch open. Note that the electron current must flow through the 100-ohm resistor and through the light. The 100-ohm resistor reduces the current flow, thus reducing the bright-ness of the light.* Proceed to block 7.

18.   *Your answer to the question in block 2 is **B**. This answer is wrong. Batteries are connected in parallel in order to obtain a greater current-delivering abil-ity.* Proceed to block 12.

19.   *The correct answer to the question in block 11 is **B**. The total power dissipated in any circuit is found by adding the powers of each resistor. In the circuit of Fig. 13-30, the total power is 30 watts.* Here is your next question. In the circuit of Fig. 13-31 are the resistors connected in series, par-allel, or series-parallel? Answer this question by assuming a voltage across terminals x and y and tracing the electron current path. Then proceed to block 26.

20.   *The correct answer to the question in block 14 is **A**. Figure 13-24 shows the circuit after R_2 and R_3 have been replaced by the equivalent resistor (R_{EQ}). Note that we now have a simple series circuit with R_1 and R_{EQ} in series. Their combined series resistance is obtained by simple addition. In this case the combined resistance value is 45 ohms.* Here is your next question. Which of the following explains how to find the total amount of power dissipated in a series-parallel circuit?

A.   Add all the powers dissipated by each resistor in the circuit. (Proceed to block 4.)

B.   Subtract the powers dissipated in the parallel branches from the powers dissipated in the series branches. (Proceed to block 25.)

21.   *Your answer to the question in block 3 is **B**. This answer is wrong. The short-est path may not be the only path. It is important to follow **all** circuit paths, not only the shortest path.* Proceed to block 15.

22.   *Your answer to the question in block 16 is **A**. This answer is wrong. Note that the lights will each drop 6 volts, and they are in series. The total drop across the light circuit is 12 volts, and this is equal to the applied voltage. If you place another resistor in series with the light, then there must be an additional drop across that resistor. If the battery voltage is 12 volts, and some voltage is dropped across a series resistor, then there cannot be a drop of 6 volts across each light. To answer the question correctly, note that the current ratings of the lights are different. You cannot have the rated current flowing through each light. Study Fig. 13-18.* Then proceed to block 3.

23.   *Your answer to the question in block 11 is **A**. This answer is wrong. The power dissipated (that is, radiated) in a resistor circuit is always found*

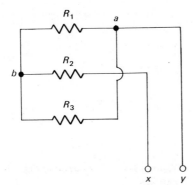

Figure 13-31.   A resistive circuit.

by adding the powers dissipated by each resistor. This is true regardless of whether the resistors are in series, parallel, or series-parallel. Proceed to block 19.

24. *Your answer to the question in block 12 is **A**. This answer is wrong. The total circuit current is* I_b, *or 1 ampere. This current divides, so that part flows through light a* (L_a) *and the rest flows through light b* (L_b). *The currents through these lights must add to equal 1 ampere. Note that 0.5 ampere is flowing through* L_a, *so you must determine how much current must flow through* L_b *so that, when it is added to* I_a, *the total is 1 ampere.* Proceed to block 5.

25. *Your answer to the question in block 20 is **B**. This answer is wrong. Each resistor in a circuit dissipates a certain amount of power. In every case, the power is supplied by the battery (or power supply). It is reasonable to expect, then, that the total power dissipated in a series-parallel circuit is found by adding all the powers dissipated by each resistor.* Proceed to block 4.

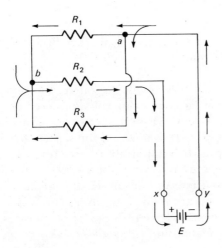

Figure 13-32.   Setup for tracing the circuit of Fig. 13-31.

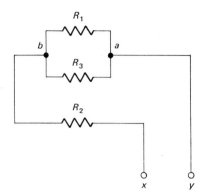

Figure 13-33. The circuit of Fig. 13-31 redrawn.

26. *The circuit of Fig. 13-31 is connected in series-parallel. The arrows in Fig. 13-32 show the electron current path with an assumed voltage E. Note that the current divides at point a, with part of it flowing through* R_1 *and the rest flowing through* R_3. *This means that* R_1 *and* R_3 *are in parallel. The currents recombine at junction b, and the total current flows through* R_2. *This indicates that* R_2 *is in series with the parallel combination* R_1 *and* R_3.

The circuit of Fig. 13-31 is redrawn in Fig. 13-33. You should assume a battery across terminals x and y of this circuit, with the negative terminal at y so that the result will be similar to that obtained in the circuit of Fig. 13-31. Trace the electron current path and note that the circuits of Figs. 13-32 and 13-33 are the same as far as electrical connections are concerned.

This problem demonstrates the importance of being able to trace a circuit. You have now completed the programmed questions. The next step is to put some of these ideas to work in laboratory experiments. Proceed to the "Experiments" section that follows.

Experiments

(The experiments described in this section may be performed on the circuit board described in Appendix C or on a similar laboratory setup.)

EXPERIMENT 1

Purpose—The purpose of this experiment is to demonstrate the effect of large resistance values on series and parallel circuits.

Theory—A transformer can be considered to be a voltage source. The output of a transformer can only be an alternating current or voltage, never dc. In order to make this into a dc voltage or current, it is necessary to use a rectifier. However, in circuits which contain only pure resistance, all the theory that has been discussed for dc will also apply for ac. It is important to note that you will be measuring ac voltages in this experiment, and you must use the ac portion of your meter to obtain the readings required. As you will see, the voltage, current, and resistance values in the ac circuits are related by Ohm's law, just as in dc circuits. This is only true for ac circuits that have resistors. If other types of components are added, the Ohm's law relationship *may not hold true.*

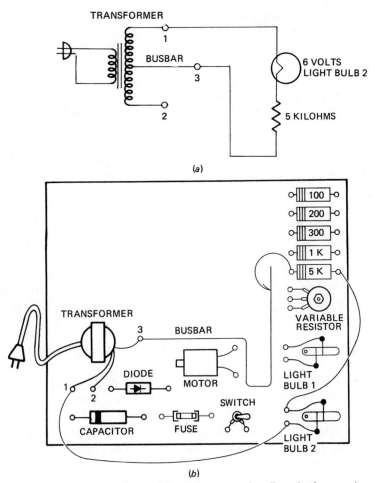

Figure 13-34. *This experiment setup demonstrates the effect of a large resistance in series with a small resistance.* (a) *The circuit for Experiment 1.* (b) *This is the way your circuit board will look if you have wired the circuit correctly.*

When a small resistance value, such as the resistance of a filament in a light bulb, is placed in series with a very large resistance value, for all practical purposes you can ignore the resistance of the smaller resistor. Naturally, the smaller resistance will affect the amount of current somewhat, but the amount of effect is so small that it cannot be measured with most types of troubleshooting instruments. As a general rule, if the ratio of large resistance to small resistance is greater than 10:1 (read "ten to one"), you can ignore the small resistance as far as measurements in a series circuit are concerned.

When resistors having different resistance values are placed in parallel, the greatest amount of current flows through the smallest resistance value. If a large resistance is placed in parallel with a small resistance, then for all practical purposes the effect of the large resistance can be ignored. As a general rule, this will be true if the large resistance is 10 or more times greater than the small resistance.

Test Setup—For the first part of the experiment wire the circuit board as shown in Fig. 13-34. The schematic representation of this circuit is also shown in Fig. 13-34. The power transformer is shown schematically, and the output voltage of this transformer is taken from pins 1 and 3.

Procedure—
Part 1—

*Step 1—*Using the ac voltmeter, measure the voltage between pins 1 and 3 of the transformer in the circuit of Fig. 13-34. Record the voltage here.

_____ volts

*Step 2—*Measure the ac voltage across the light bulb in the circuit of Fig. 13-34. Record the voltage here. (If you cannot obtain a reading, record 0 volt.)

_____ volts

*Step 3—*Measure the voltage across the 5,000-ohm (5-kilohm) resistor. Enter the value here.

_____ volts

If you have performed this part of the experiment correctly, the voltages that you entered in Steps 2 and 3 should add to equal the voltage that you entered in Step 1.

Procedure—
Part II—

*Step 1—*You are now going to demonstrate a short circuit. Wire the circuit as shown in Fig. 13-35. Note that two 6-volt lights are placed across a 6-volt source. The lights should glow with partial brightness.

*Step 2—*When the switch is closed, a 5,000-ohm resistor is placed in parallel with the low resistance of the light circuit. Does the 5,000-ohm resistor seriously affect the brightness of the lights?

_____
yes or no

Procedure—
Part III—

*Step 1—*In this part of the experiment you are going to demonstrate that a very small resistance value in parallel with another resistance value presents a *short circuit*. Wire the circuit as shown in Fig. 13-36. With the switch open, notice that the lights glow at about half their normal brightness due to the fact that two 6-volt lights are connected across a 6-volt source. Since the voltage across each light is only about 3 volts, they cannot glow at full brightness.

*Step 2—*Close the switch and note that light 2 goes out and light 1 increases to full brightness. This is because the closed switch contacts have offered a very low resistance path to electrons around the light. For all practical purposes, all the electrons flow through the switch and none flow through light 2. A short circuit, then, is a low-resistance circuit path that allows electrons to flow around a desired load resistance.

Conclusion—If you have performed these experiments correctly, it will be apparent that whenever a large resistance is connected *in series* with a small resistance, the small resistance can be ignored. Also, a large resistance *in parallel* with a small resistance does not seriously affect the circuit operation.

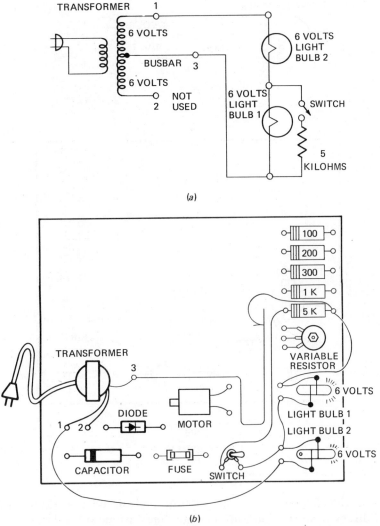

Figure 13-35. *This experiment setup demonstrates the effect of a large resistance in parallel with a small resistance.* (a) *The circuit for Part II of Experiment 1.* (b) *This is the way your circuit board will look if you have wired the circuit correctly.*

When the large resistor is in series with the small resistor, all the voltage can be considered to be dropped across the larger resistance value. However, if the large resistance is in parallel with the small resistance, as shown in Fig. 13-35, all the current can be considered to be flowing through the small resistance, and the effect of the 5,000-ohm resistor can be ignored.

These are important points that you should remember when trouble-shooting in a circuit.

EXPERIMENT 2

Purpose—In this experiment you are going to verify the calculation of series-parallel resistance values. You will also learn how to determine the resistance of a parallel circuit by voltage measurement and Ohm's law.

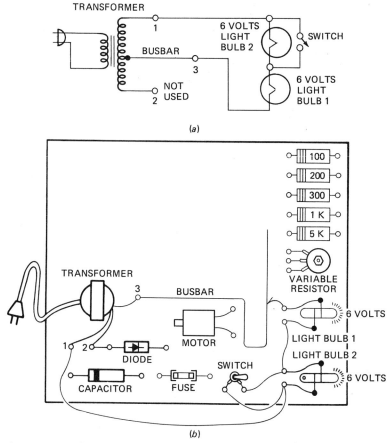

Figure 13-36. *This experiment demonstrates the meaning of a short circuit.* (a) *The circuit for Part III of Experiment 1.* (b) *This is the way your circuit board will look if you have wired the circuit correctly.*

Theory — In the chapter you learned to find the resistance of series-parallel circuits. In this experiment you will calculate the resistance of a series-parallel circuit, and then you will verify your results by measurements and the use of Ohm's law.

Test Setup — Wire the circuit as shown in Fig. 13-37. Both the schematic and the physical wiring layout are shown in this illustration. The letters representing the resistors R_1, R_2, and R_3 are arbitrarily chosen. Resistors on the circuit board are not numbered or lettered. It is convenient to use letters so that specific resistances can be described in the writeup.

Procedure —
Part I —
Step 1 — Calculate the value of parallel resistance R_p for R_2 and R_3. Enter your calculated value here.

$$R_p = \underline{\hspace{2cm}} \text{ohms}$$

Step 2 — Redraw the circuit with R_1 in series with the resistance value as calculated for the parallel resistance value.

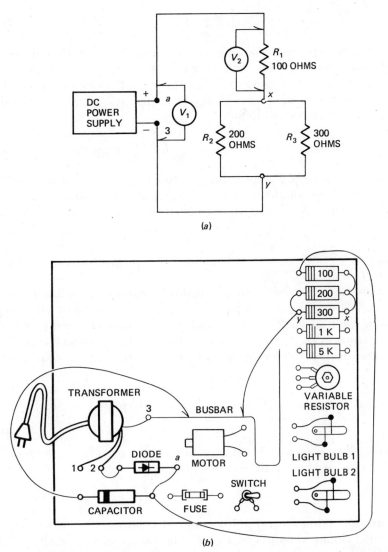

Figure 13-37. Circuit for experiment in series-parallel circuits. (a) *The circuit for Experiment 2.* (b) *This is the way your circuit board will look if you have wired the circuit correctly.*

Step 3 — Add the two series resistances of the circuit that you have drawn to determine the complete circuit resistance in Fig. 13-37. Enter that value here.

$$R_1 + R_p = R_T = \underline{\hspace{3cm}} \text{ ohms}$$

Step 4 — Measure the applied voltage across the complete circuit. This will be voltage V_1 (a dc voltage!) between terminals *a* and 3 of Fig. 13-37. Enter the voltage value here.

$$V_1 = \underline{\hspace{3cm}} \text{ volts}$$

Step 5 — The amount of current that flows in a circuit should be equal to the applied voltage (V_1) divided by the series-parallel circuit resistance. Calculate the current by the equation

$$I = \frac{V_1}{R_T} = \underline{\hspace{2cm}} \text{ amperes}$$

Step 6—To verify your calculations, it is necessary only to find the amount of total circuit current. This can be done by measuring the voltage (V_2) across R_1. Knowing the voltage across the resistor and the resistance value, find the amount of current flowing through R_1. This current is equal to the total circuit current that you calculated in Step 5.

$$I = \frac{V_2}{R_1} = \frac{?}{100} = \underline{\hspace{2cm}} \text{ amperes}$$

If you have properly performed this experiment, the value of current that you entered in Step 6 should be the same as the value of current in Step 5. This can be true only if you have properly calculated the series-parallel resistance value of the circuit.

Procedure—

Part II—

Step 1—You can verify your calculation of parallel resistance by making a voltage measurement between points *x* and *y* in the circuit of Fig. 13-37, and then using Ohm's law ($R = E/I$) to find the resistance between points *x* and *y*. This resistance value should be equal to the parallel resistance that you calculated. Measure the voltage between points *x* and *y*. Enter your value here.

$$V_{xy} = \underline{\hspace{2cm}} \text{ volts}$$

Step 2—Enter the value of current that you calculated in Step 6 of Part I.

$$I = \underline{\hspace{2cm}} \text{ amperes}$$

Step 3—The resistance between points *x* and *y* must equal the total voltage drop (V_{xy}) divided by the circuit current (I).

$$R = \frac{E}{I} = \frac{?}{?} = \underline{\hspace{3cm}}$$

Compare this resistance with the amount of resistance that you calculated for the parallel circuit in Step 1 of Part I. The values should be close. However, remember that there are some inaccuracies to be expected due to the fact that the resistors have tolerances—that is, they are not manufactured exactly to the value specified. Also, the meter that you used for making measurements may be slightly inaccurate. This is normal and is expected in troubleshooting procedures.

Conclusion—One obvious conclusion is that the series-parallel resistance value can be calculated by the methods shown in this chapter.

Another important conclusion is that the total circuit current can be obtained simply by measuring the voltage across one of the resistors through which all the circuit current flows, and then by Ohm's law determining the current through it. Technicians often prefer using this method rather than using an ammeter (or a microammeter or a milliammeter) to measure current. The reason is that, in order to measure the current with such a meter, it is necessary to open the circuit and insert the ammeter in series with it. However, with the voltmeter method, you can measure the voltage across the resistor, and you do not have to open the circuit in order to make such measurements. Of course, you do need to make an Ohm's law calculation, but many technicians still prefer this method because it is much quicker.

Self-Test with Answers

(Answers with discussions are given in the next section.)

1. Which of the following is *true?* (*a*) Batteries are never connected in a series-parallel arrangement; (*b*) Batteries may be connected in a series-parallel arrangement to obtain a desired voltage and current-delivering ability.
2. To trace a circuit, (*a*) start at the negative terminal of the source, and follow the path of electron current flow through the circuit; (*b*) start at the smallest resistance value in a parallel branch and determine the path that the current takes to get to the battery.
3. A resistor having a resistance value of 500 ohms is connected in parallel with a 5-kilohm resistor. The parallel resistance value is (*a*) about 500 ohms; (*b*) about 5,000 ohms.
4. A 500-ohm resistor is connected in series with a 10-kilohm resistor. This series circuit is connected across a 10-volt source. The voltage drop across the 10-kilohm resistor is about (*a*) 0 volt; (*b*) 10 volts.
5. The total current in a series-parallel circuit is (*a*) the sum of all the currents through every resistor; (*b*) the current delivered by the battery.
6. You wish to trace a series-parallel circuit. It is a resistance circuit, and there is no source of voltage. The correct procedure is to (*a*) assume a voltage; (*b*) start with the smallest value of resistance and work toward the largest.
7. The symbol Ω stands for (*a*) ohms; (*b*) amperes.
8. A single value of resistance that can be used to replace a number of resistances in parallel (for the purpose of simplifying the circuit) is called the (*a*) important resistance; (*b*) equivalent resistance.
9. When tracing a circuit by following the electron current path, you come to a point where the current divides into two parts, with each part flowing through a resistor. The resistors are (*a*) in series; (*b*) in parallel.
10. Which of the following will normally cost more? (*a*) A 100-ohm resistor rated at 100 watts; (*b*) A 100-ohm resistor rated at 25 watts.

Answers to Self-Test

1. (*b*)—Figure 13-29 shows an example of batteries connected in series-parallel.
2. (*a*)—When tracing an ac circuit, the power source does not have a positive or negative terminal. In such cases, start at one terminal and return to the other.

3. (*a*) — In cases where the resistances are in a 10:1 ratio, the parallel resistance value is approximately equal to the smaller resistance value. The actual parallel resistance value is over 454 ohms. (A 500-ohm resistor with a tolerance of 20 percent may have a resistance of only 400 ohms.)

4. (*b*)

5. (*b*)

6. (*a*)

7. (*a*)

8. (*b*)

9. (*b*) — This is an important way to determine if resistors are in series or parallel.

10. (*a*)

14.
How Do DC Generators Work?

Introduction

An automobile has a *dc* electrical system because the lead-acid battery is the most practical and most economical portable power supply that can provide enough current to run the dc starter motor. Lead-acid batteries must be *recharged* (by forcing a current to flow backward through the battery), and a dc generator run by the fan belt of the engine can be used for this purpose.

There are many other applications for dc generators. For example, they are used as portable power supplies for radio transmitters, and for supplying dc power to industrial machines.

There are certain disadvantages of dc generators, and the use of *alternators* — a special type of generator for producing a dc voltage — in newer cars is now common practice. An alternator produces an ac voltage which is rectified by semiconductor diodes that operate on the same principle as the one used on the experiment board. A diode rectifier converts ac to dc, and the output of the complete alternator system is a dc voltage.

In this chapter you will learn about dc generators and alternators. You will be able to answer these questions after studying this chapter:

How is voltage induced in a conductor?

What does the amount of induced voltage depend upon?

What is the left-hand generator rule?

How is voltage generated in a single rotating conductor?

How is direct current produced by a conductor that is rotating in a magnetic field?

What is a self-excited generator?

How does an automobile alternator work?

What is a dynamotor?

Instruction

HOW IS VOLTAGE INDUCED IN A CONDUCTOR?

In an earlier chapter you learned that there are six ways of generating a voltage. They are by use of chemical action, heat, light, pressure, friction, and mechanical motion. Mechanical motion involves moving a conductor through a magnetic field.

Mechanical motion is employed for generating voltages in both dc and ac generators. Figure 14-1 shows the basic principle involved. A permanent bar magnet is shown inserted into a coil. Both the coil and the magnet are at rest in Fig. 14-1*a*. The voltmeter shown in the illustration is designed so that 0 volt is at the center of the meter scale.

As long as the magnet and coil remain at rest, there is no voltage being generated. Remember that there are magnetic field lines around the bar magnet at all times, and these magnetic field lines will actually exist throughout the turns of the coil. However, the *law of induction* states that the conductor must be moving through the magnetic field for a voltage to be generated.

In Fig. 14-1*b* the magnet is being pulled out from the center of the coil. Now the magnetic field around the bar magnet is cutting across the conductors of the coil, and during this period of time a voltage is generated. The polarity of the generated voltage depends upon the direction of the motion, and upon which side of the bar magnet is a north pole and which side is a south pole. It also depends upon the direction that the coil is wound.

As long as the magnet is moving, the voltage will continue to be generated.

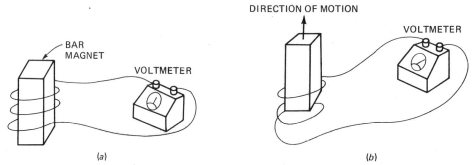

Figure 14-1.   *The principle of induced voltage.* (a) *When the magnet is not moving, no voltage is generated.* (b) *Moving the magnet out of the coil causes a voltage to be generated.*

If you stop moving the magnet, the voltage will drop to zero again. If you reverse the direction of motion of the magnet—that is, push it into the coil instead of pull it out of the coil—the voltage will again be generated. However, the polarity of the voltage will be opposite to that produced when the magnet is pulled out. The voltmeter needle will deflect to the left of zero, instead of to the right of zero.

WHAT DOES THE AMOUNT OF INDUCED VOLTAGE DEPEND UPON?

If you performed the simple experiment of Fig. 14-1, you would soon learn that the amount of voltage induced depends upon a number of things. For example, if you move your magnet quickly, the induced voltage will be larger than if you move it slowly. We can make a basic rule from this observation: *The amount of induced voltage depends upon the rate at which a magnetic field cuts across the conductor wires.*

Another way of increasing the voltage is to increase the number of turns in the coil. This is because the greater the number of turns, the greater the number of conductors cut by the magnetic field. We can make a basic rule based on this observation as follows: *The amount of voltage induced depends upon the number of conductors being cut by the magnetic field.*

The greater the strength of a magnet, the greater the number of magnetic flux lines surrounding it. If you change the bar magnet in Fig. 14-1 to one that is a stronger magnet—that is, one that has a greater number of flux lines—you will find that the amount of induced voltage increases. A method of increasing the strength of the magnetic field is illustrated in Fig. 14-2. In this drawing the magnetic field is being held motionless, and the conductor is moving through the field of the magnet. The only difference between this setup and the one shown in Fig. 14-1 is that in one case the conductor is moved and in the other case the magnetic field is moved. Either method of generating a voltage may be used in ac and dc generators. Always keep in mind that it is only necessary that there is motion between the conductor and the field in order to generate a voltage.

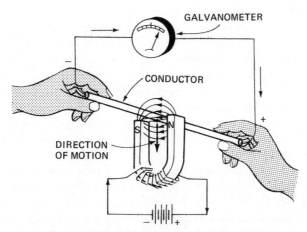

Figure 14-2. The amount of induced voltage depends upon the strength of the magnetic field that the conductor is moving through.

In order to increase the strength of the magnetic field in the system of Fig. 14-2, a battery and a coil are added to the permanent magnet. The greater the amount of turns of the coil, the stronger the magnetic field. This in turn leads to an increase in the amount of voltage for a given amount of speed that the conductor is moving.

To summarize then, the amount of induced voltage is dependent upon three things:

The number of turns in the conductor
The strength of the magnetic field
The speed of the relative motion between the conductor and the magnetic field

SUMMARY

1. The electrical system of an automobile is dc. The reason is that the lead-acid battery is a portable dc supply that can supply enough power to start a car. It is the most convenient power source capable of supplying that much power.
2. A dc generator operates on the principle that a voltage is always induced in a conductor when it is moved through a magnetic field.
3. The amount of voltage induced in conductors moving through a magnetic field depends upon the speed of the motion, the strength of the magnetic field, and the number of conductors.
4. Reversing the direction of the motion of a conductor through a magnetic field will cause the polarity of the induced voltage to reverse.
5. An alternator is an ac generator. The output voltage of an alternator is rectified to make a dc voltage that can be used in the electric circuits of automobiles.

WHAT IS THE LEFT-HAND GENERATOR RULE?

The relationship between a moving conductor, a magnetic field, and an induced voltage was discussed in the previous section. We noted that the polarity of the induced voltage depends upon the direction of motion between the conductor and the magnetic field.

The relationship between the polarity of the induced voltage and the direction of motion is stated in a law which says that the direction of the induced voltage is always such that it will produce a magnetic field which will react with the magnetic field producing it in such a way that the motion producing it will be opposed. (Remember that like poles repel, and the opposition is produced by like poles of magnetism.) This is a complicated law that can be simplified as follows: *Whenever a current flows as a result of a conductor moving through a magnetic field, the magnetic field of that current will be such that it will oppose the motion by reacting with the magnetic field already present.*

The idea of this law can be demonstrated physically by having a dc generator that can be turned by a crank. The generator will turn quite easily while it is not delivering a current to a load. However, if you connect a low resistance across the generator terminals so that it is supplying current, it will be observed that the generator is very hard to turn. This is because the current flowing through the generator has a magnetic field that opposes the motion.

The left-hand rule for generators is based on the law just described. It

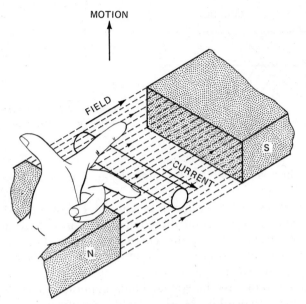

Figure 14-3. The left-hand rule for generators.

simplifies the method of determining the direction of induced voltage, or the direction that the induced current will flow. It should be pointed out that it is not theoretically possible for a generator to induce current. Only a voltage can be induced. However, when we speak of induced current what we mean is the current that will flow in a complete circuit as a result of induced voltage. The direction of induced current, of course, depends upon the polarity of the induced voltage.

Figure 14-3 illustrates the left-hand rule for generators, which defines the direction of induced current flow. The three fingers of the left hand are extended so that they are at right angles to each other. The first finger points in the direction of the magnetic field. Remember that the magnetic field always has a direction presumed to be north to south. The thumb points in the direction that the conductor is moving in the magnetic field. In the illustration the conductor is moving up. When the thumb and forefinger are pointed in these directions, the second finger will point in the direction of the induced current. This is clearly shown in the illustration.

The left-hand rule for generators is used for determining the direction of the electron current flow. The second finger of the left hand points in the direction that electrons will flow as a result of induced voltage, assuming, of course, that there is a complete circuit path. In some textbooks this is stated as the right-hand rule for generators, because they are talking about conventional current flow. For conventional current flow the electricity is presumed to flow from plus to minus, rather than from minus to plus as in electron flow.

HOW IS VOLTAGE GENERATED IN A SINGLE ROTATING CONDUCTOR?

We will now discuss the voltage generated in a single conductor being rotated in a magnetic field. Figure 14-4 shows the conductor and the magnetic poles that produce the magnetic field. Note that the poles are curved so that the

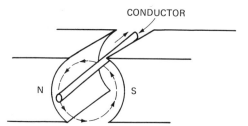

Figure 14-4. A single conductor rotating in a magnetic field.

conductor can be maintained at the same distance from the pole face as it is rotated.

The arrow beside the conductor in Fig. 14-4 shows the direction of induced current flow at the instant the conductor is in that position. The left-hand rule can be used to verify this current direction. Only a section of the conductor is shown, and it is presumed that the ends of this conductor are connected to a complete circuit.

We are going to discuss the voltage being generated in the conductor as it is rotated. In order to do this, we will simplify the illustration so that the conductor is being viewed from the end. In Fig. 14-5 a cross indicates current flowing into the paper and a dot indicates current flowing out of the paper. Again, in order for current to flow the circuit must be complete. Therefore, you should assume that you are looking at a cross section of the rotating conductor which is connected into a complete circuit.

Figure 14-6 shows the voltage generated in the rotating conductor. In position *A* the conductor is moving down through the flux. Remember that the flux lines are between the north and south magnetic poles, so the conductor is moving at right angles to the flux. Under this condition, the amount of voltage generated is maximum. The cross shows that the induced current will flow into the paper when the conductor is in position *A*. We are going to make a graph of the voltage produced as the conductor rotates in the magnetic field. The first point on the graph (in Fig. 14-6*a*) shows that the voltage is positive at this instant.

Figure 14-6*b* shows the voltage generated after the conductor has moved past the position that it was in (position *A*). Note that the voltage is still induced in the same direction, as indicated by the cross in the conductor. However, at this moment the conductor is no longer moving at right angles to the flux, but rather it is moving more toward the south pole. The voltage induced at this in-

CURRENT FLOWING
INTO PAPER

CURRENT FLOWING
OUT OF PAPER

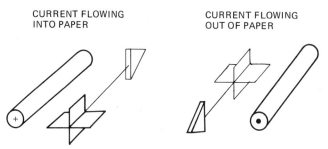

Figure 14-5. A cross indicates that the current is flowing into the paper. A dot indicates that the current is flowing out of the paper.

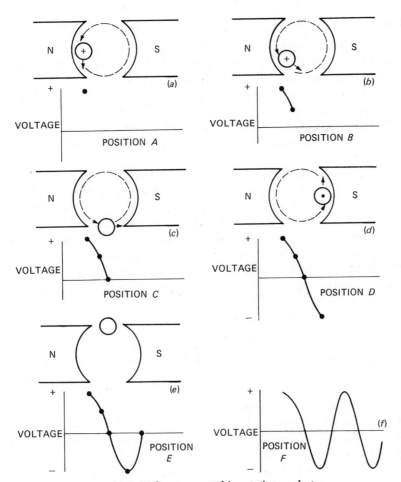

Figure 14-6. *Voltage generated in rotating conductor.*

stant is lower than it was in position *A*. This is indicated by the graph in Fig. 14-6*b*.

Figure 14-6*c* shows what happens at the instant when the conductor is moving along the direction of the flux lines between the north and south magnetic poles. Note that for this instant the arrow shows that the conductor is moving parallel to the flux. Since the conductor is not cutting any flux lines at this instant, no voltage is induced. This is indicated on the graph in Fig. 14-6*c*.

In Fig. 14-6*d* the conductor is moving upward through the flux. At the instant shown, the conductor is moving at right angles through the flux lines and the maximum amount of voltage is induced. The graph shows that the voltage is maximum and has a negative polarity. The dot on the conductor indicates that the induced current is moving out of the conductor.

Figure 14-6*e* shows what happens when the conductor has moved all the way to the top position. At this instant, it is moving parallel to the flux lines. Since the conductor is not cutting *across* any flux lines, the voltage induced at this instant is zero. The graph of Fig. 14-6*e* shows this condition.

After several rotations the voltage will appear as shown in the graph of Fig. 14-6*f*. Note that the voltage moves from positive to negative and back to positive. This is an ac voltage, and alternating current will flow as a result of this generated voltage.

SUMMARY

1. The left-hand generator rule makes it possible to predict the polarity of voltage generated, or the direction of induced current flow, for a conductor that is moving through a magnetic field.
2. The direction of induced current flow is always such that its magnetic field will oppose the motion that produced it.
3. When a conductor is moved through magnetic flux in such a way that its motion is at right angles to its flux lines, the maximum amount of voltage is induced.
4. When a conductor is moved through magnetic flux in such a way that its direction of motion is parallel to the flux lines, no voltage will be induced.
5. As a conductor rotates in a magnetic field, an alternating voltage is produced. This is due to the fact that part of the time the conductor is moving in one direction through the flux lines, and part of the time the conductor is moving in the opposite direction through the flux lines.
6. Application of the left-hand rule to a rotating conductor in Fig. 14-6 shows that an alternating voltage is produced by the rotating conductor.

HOW IS DIRECT CURRENT PRODUCED BY A CONDUCTOR THAT IS ROTATING IN A MAGNETIC FIELD?

We have just shown that a single conductor moving around a magnetic field produces an alternating voltage. In order to produce a dc voltage it is necessary to periodically switch the connections to the conductor so that the output voltage always has the same polarity.

Figure 14-7 is a simplified illustration of a generator which provides the necessary switching to convert the generated ac voltage to a dc voltage. It is important to understand the purpose of each part of the generator. The permanent magnet creates a magnetic field through which the rotating conductor

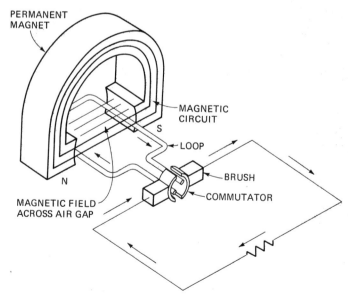

Figure 14-7. Construction details of a simple dc generator. Arrows show path of induced current flow.

moves. Instead of moving a single conductor through the field, the conductor has been formed into a loop which is technically known as the *armature* of the generator. The left-hand rule will show that at any instant the induced current flowing in one side of the loop is in the opposite direction to the current on the other side of the loop. The result is that the two currents add.

As shown in Fig. 14-7, the ends of the loop are connected to two half-ring conductors called the *commutator*. As the loop rotates the commutator segments rotate with it.

The voltage induced in the armature loop must be taken away from the commutator and delivered to the external circuit. This is done by the use of *brushes*. The brushes are usually made of a carbon-type material; they rub against the commutator as it turns.

Figure 14-8 shows how the dc voltage is produced by a generator as its armature turns in the magnetic field. In order to simplify the discussion, one side of the armature is shown as black and the other side is shown as white. This will make it easier to identify the armature conductors as they are rotated.

In Fig. 14-8a the armature loop is positioned so that the conductors are moving parallel to the magnetic flux lines. At this instant the voltage induced in the armature is 0 volt. This is indicated in the graph of Fig. 14-8a. Note that at this instant the brushes are at the commutator slots. Thus, no voltage is being generated and connection is made to the generator brushes.

In Fig. 14-8b the armature has turned so that the conductors are moving at right angles to the magnetic flux. Now the voltage induced will be the maximum value. This is also indicated in the graph at position B.

In position C the loop has again turned so that the conductors are moving parallel to the flux lines of the magnetic field and no voltage is generated. Note that the commutator is in a position so that the black and white wire connections will be interchanged on the brushes. During the half cycle of rotation just described, the black wire was connected to the black brush and the white wire was connected to the white brush. Now, the positions are being reversed. This is clearly seen in Fig. 14-8d. The black side of the loop goes to the white brush, and the white side of the commutator goes to the black brush. What has happened is that the commutator has switched the loop connections. The result is that the induced current will flow in the same direction. The left-hand rule will show that the white wire of the armature is producing a voltage in the same direction in Fig. 14-8d that the black wire was producing in Fig. 14-8b.

At the instant shown in Fig. 14-8d the loop is moving at right angles to the flux and the maximum voltage is being produced. This is shown in the graph.

In position E the armature has again turned so that its position is identical to that of Fig. 14-8a. Note that the slots of the commutator are opposite the brushes, and the connections of the black and white wires of the loop are again being reversed at the instant shown.

To summarize, the commutator actually serves as a switch to periodically reverse the connections of the armature. The generator output voltage will always have the same polarity, and the induced current will always flow in the same direction through the load.

Figure 14-8 shows that the dc voltage is produced in pulses rather than as a constant value of dc. The voltage drops to zero every instant when the open part of the segment of the commutator is opposite the brushes and the conductors are moving parallel to the flux lines.

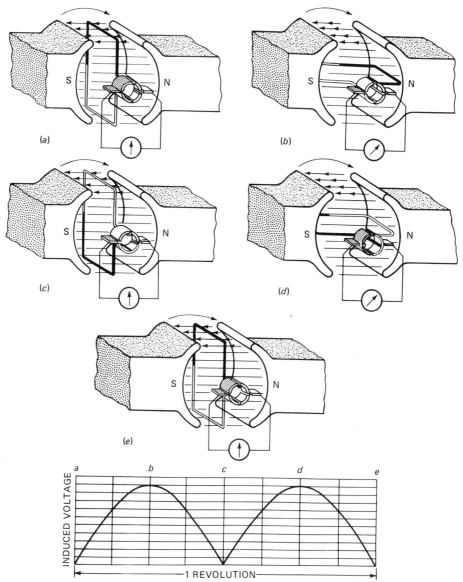

Figure 14-8. *Voltage generated by a single loop of wire turning in a magnetic field. The commutator converts the alternating voltage (ac) to dc.*

When you are trying to generate a dc voltage, it is better to have a generator output voltage that does not drop to 0 volt periodically. This can be accomplished by using more than one loop in the armature. (See Fig. 14-9.) Figure 14-9a shows how it is accomplished. Instead of rotating a single loop in the magnetic field, three loops are being rotated. Each loop will deliver a voltage to the segmented commutator. The brushes are not shown, but they are similar to the ones shown in Fig. 14-8.

In the generator of Fig. 14-9a, a voltage is delivered to the brushes at all times. After one loop has passed its maximum voltage position, the next loop moves into place. The commutator switches each loop to the brushes as it

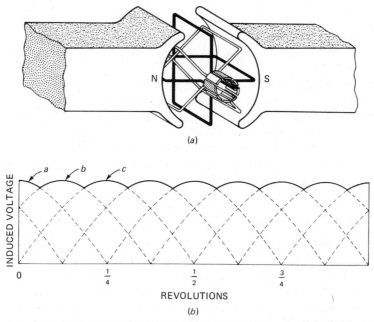

(a)

(b)

Figure 14-9.   By using more loops of wire in the armature, a steadier dc voltage is generated. (a) *If more loops of wire are used, the generated voltage will not drop to 0 volt. Three loops are shown here.* (b) *This is a graph of the voltage generated by the rotating armature in Fig. 14-9a. The dark line shows the actual output voltage. Waveforms a, b, and c are generated by different loops.*

moves to its maximum-voltage position. The voltage generated by this type of arrangement is shown in the graph of Fig. 14-9*b*. Note that this voltage never drops completely to zero.

   In actual practice a large number of turns may exist on a single armature. One example is illustrated in Fig. 14-10. It shows an actual armature and commutator from a dc generator. Note that the commutator consists of a large number of segments, and each segment is connected to the ends of one loop of coil. The coil wires lie in slots in the armature, and the complete assembly is wound on a single shaft which turns special coils through the magnetic fields.

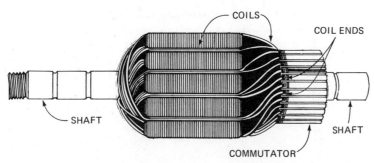

Figure 14-10.   This is the way the armature and commutator of a dc generator actually look.

SUMMARY

1.   A single conductor rotating in a magnetic field will produce an alternating voltage, which can be converted to a dc voltage by a mechanical switching arrangement.
2.   The rotating wire in a generator is called the *armature*. It is usually a complete loop of wire or a number of loops of wire.
3.   A mechanical switching arrangement in a dc generator is called a *commutator*. Its purpose is to periodically switch the voltage so that the current flowing through the load is always in the same direction.
4.   By using more than one single turn of wire for the armature, it is possible to produce a dc voltage which does not drop to 0 volt periodically.

WHAT IS A SELF-EXCITED GENERATOR?

You have learned that the strength of the magnetic field has a direct effect on the amount of voltage induced in a wire that is moving through the field. One way to increase the output voltage of a generator is to increase the strength of its magnetic field. Figure 14-11 shows one way to accomplish this.

The generator of Fig. 14-11 is said to be *separately-excited.* This means that the current for the electromagnetic field is obtained from a source external to the generator. In this case, the source is a battery. The rheostat controls the amount of current through the field winding. With this arrangement, varying the adjustment of the rheostat varies the current through the field coil. As the current in the field coil changes, the strength of the magnetic field is changed, and the overall result is a change in the output voltage from the armature. Figure 14-12 shows the circuit for the separately-excited generator.

Since a dc generator delivers current to a load, it is reasonable to expect it to be able to also deliver current for the magnetic field winding. When a generator supplies the current for its own field winding, it is said to be *self-excited.*

There are several different ways to connect the generator field winding so that it will receive current from its own armature. Figure 14-13 shows the three most commonly used connections. The terms *series, shunt,* and *com-*

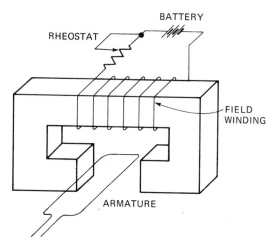

Figure 14-11. A separately-excited generator.

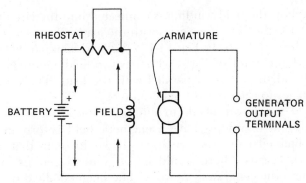

Figure 14-12.   Circuit for a separately-excited generator.

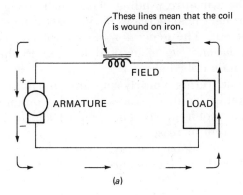

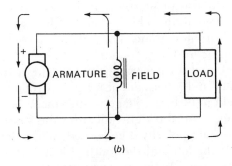

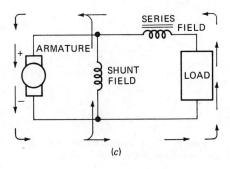

Figure 14-13.   Three types of self-excited generators. Arrows show paths for electron current flow. (a) *A series generator.* (b) *A shunt generator.* (c) *A compound generator.*

pound describe how the field winding is connected into the circuit.

In Fig. 14-13a the field is in series with the load. This means that the same current flows through both. In Fig. 14-13b the field is in parallel with the load. The compound generator of Fig. 14-13c has two field windings—one in series with the load and the other in parallel with the load. There are advantages and disadvantages to each type of generator.

Self-excited generators need current from the armature to produce the magnetic field, and they must have a magnetic field before an induced current can be obtained from the armature. It would seem that such a system could not work because there would be no way to get the system started. Actually, though, the generators *do* work. The magnetic field does not actually drop to zero when the field current stops flowing. A small amount of magnetism—called the *residual magnetism*—remains in the iron. When the generator shaft starts to turn, this small amount of magnetism in the iron is enough to induce a small amount of voltage in the armature winding. This small voltage starts the field current flowing. The stronger field increases the induced voltage, which further increases the field current. The generator output voltage increases in this process. (The name *generator buildup* is used to refer to the process of starting the generator action described above.)

The series generator is capable of delivering a nearly constant current to a load. The shunt generator, on the other hand, produces a nearly constant output voltage for different load current requirements. The terminal voltage of a shunt generator drops as the load current increases.

Compound generators can deliver varying currents to a load with a more constant terminal voltage than possible with shunt generators.

HOW DOES AN AUTOMOBILE ALTERNATOR WORK?

An alternator is an ac generator. In order to use an alternator in an automobile it is necessary to convert its ac output voltage to dc. This is done by the use of *silicon diodes* which are mounted on the alternator frame. These diodes serve as *rectifiers*—that is, components that convert ac to dc.

Figure 14-14 shows a simple method used for generating an ac voltage. In this generator the coils are held stationary and a permanent magnet is rotated near the coils. The moving magnetic field induces a voltage in the coil. When the north pole of the permanent magnet is on the right side, and the south pole is on the left side, a voltage with a certain polarity is generated. When the magnet turns so that the north pole is on the left and the south pole is on the right, the polarity of generated voltage reverses. As the magnet turns, the volt-

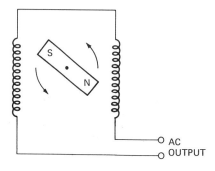

Figure 14-14. A rotating magnet induces an ac voltage in the coils.

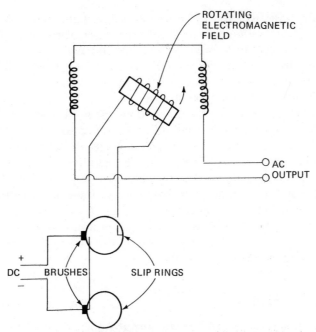

Figure 14-15. A simplified drawing of an automobile alternator circuit.

age across the output terminals periodically reverses polarity. In other words, the output voltage is ac rather than dc.

If the ac voltage is applied across a load resistor, an alternating current will flow. In other words, the current will flow back and forth in the load. Since the currents in an automobile are dc, meaning that the current always flows in the same direction, the ac output of the alternator is not useful. As mentioned before, however, rectifiers can be used to convert the ac to dc.

A rotating electromagnet will have a stronger magnetic field than the permanent magnet shown in Fig. 14-14. Figure 14-15 shows how an alternator with a rotating magnetic field is made. The dc voltage for the rotating winding is supplied through slip rings. The slip rings permit the winding to rotate while it is receiving a direct current from the battery. The output ac voltage is taken from the coils. The output ac voltage is then rectified, as shown in Fig. 14-16. to produce a dc voltage.

Instead of using a single-coil arrangement as shown in Fig. 14-15, the actual alternators used in cars have a number of windings. This makes it possible to generate a voltage that does not drop to 0 volt periodically. However, the basic theory of operation is no different from that illustrated in Fig. 14-15.

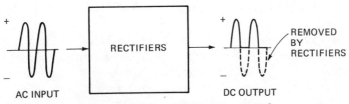

Figure 14-16. A rectifier converts ac to dc.

One of the basic differences between an alternator and a dc generator is that semiconductor rectifiers are used to convert generated ac to dc in an alternator circuit, but a commutator is used for this purpose in a generator circuit. The alternator discussed in this section has a rotating field rather than a rotating armature, but this is not true of all alternators. Some have rotating armatures, as you will learn in the next chapter.

A dc generator will run like a motor when voltage is fed to its terminals. In an automobile circuit a *cutout* is needed for the generator circuit. Whenever the generator voltage drops below the battery voltage, the generator tries to turn like a motor. The cutout prevents this from happening. An alternator, unlike a dc generator, will not turn like a motor when a voltage is applied to its terminals. Thus, the circuit of an alternator is different than the circuit for a dc generator in an automobile.

Alternators are more efficient than dc generators—especially at low rpms (revolutions *per* minute). This is largely due to the fact that the output current does not have to flow through brushes.

WHAT IS A DYNAMOTOR?

In order for a generator to produce an output voltage its shaft must be turned. In the system used with automotive electrical systems the generator (or alternator) is turned by a fan belt connected to the engine.

Occasionally a student of basic electricity develops the idea for a perpetual-motion machine using a dc motor and a dc generator. The idea is to use the dc motor to turn the generator, and to use the output of the generator to provide the voltage for running the motor.

Of course, the perpetual-motion system will not actually work because of the friction and the power loss in the electric wires. However, there are systems in which electric motors are used for turning generators. These are called *motor-generator* systems, and the voltage for operating the motor comes from some source other than the generator being driven.

A special type of motor-generator system is made with the motor and generator mounted on the same shaft. This eliminates the need for gears or belts and pulleys. When the motor and generator are both mounted on a single shaft, it is called a *dynamotor*.

SUMMARY

1. An excited generator is one in which the magnetic field is obtained from current flowing in a field coil.
2. The current through the coil, which produces the magnetic field required for generator operation, is called the *excitation current,* or *field current.*
3. If the field current flows through the load *and* the field, it is a series generator.
4. In a shunt generator the load and the field winding are in parallel.
5. A compound generator has both a series and a shunt winding.
6. An alternator is an ac generator. In automobile electrical systems the output ac voltage of the alternator is converted to a dc voltage by silicon rectifiers mounted on the alternator frame.

7. The alternator used in cars has a rotating electromagnetic field. Current for the electromagnet is fed through slip rings.
8. Alternators can produce a higher output voltage at lower rpms.

Programmed Review Questions

(Instructions for using this programmed section are given in Chap. 1.)

We will now review the important concepts of this chapter. If you have understood the material, you will progress easily through this section. Do not skip this material, because some additional theory is presented.

1. Figure 14-17 shows a conductor moving through a magnetic field. The direction of motion is upward, as indicated by the arrow. Using the left-hand rule, determine which of the following is correct.
   A. The induced electron current will flow from point *a* to point *b*. (Proceed to block 7.)
   B. The induced electron current will flow from point *b* to point *a*. (Proceed to block 17.)

2. *The correct answer to the question in block 17 is **A**. As the conductor rotates, its direction of motion through the magnetic flux continually changes. This causes the polarity of the generated voltage to change periodically. Thus, an ac voltage is generated.*
   Here is your next question.
   A mechanical switching device in a dc generator (sometimes called the *mechanical rectifier*) which converts the ac voltage generated in the rotating loop into a dc voltage is called the
   A. commutator. (Proceed to block 13.)
   B. armature. (Proceed to block 21.)

3. *Your answer to the question in block 18 is **C**. This answer is wrong. If the brushes were made of hard steel, they would wear the commutator very quickly.* Proceed to block 24.

4. *The correct answer to the question in block 19 is **A**. Silicon diodes are used on alternators to convert the generated ac to dc. The diode on the experiment board is a silicon diode.* Here is your next question.
   Which of the following statements is correct?
   A. The magnetic field around an induced current always helps the motion that produced it. (Proceed to block 15.)
   B. The magnetic field around an induced current always opposes the motion that produced it. (Proceed to block 25.)

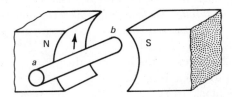

Figure 14-17. In which direction is electron current flowing?

5. *Your answer to the question in block 17 is **B**. This answer is wrong. When the conductor moves up through the flux, the polarity of the generated voltage will be opposite to the polarity of the voltage generated when the conductor moves down. As the conductor moves in a circle, the polarity of the voltage reverses periodically, and this is an ac voltage.* Proceed to block 2.

6. *The correct answer to the question in block 23 is **B**. A dynamotor is a system in which both the motor and the generator are mounted on the same shaft.* Here is your next question.
   What is the purpose of a cutout in an automobile generator circuit? (Proceed to block 14.)

7. *Your answer to the question in block 1 is **A**. This answer is wrong. Study Fig. 14-3 carefully. Note that it is very similar to Fig. 14-17. Use your left hand to show that the electron current flow in Fig. 14-17 will be from point b to point a.* Proceed to block 17.

8. *The correct answer to the question in block 24 is **B**. The direction of magnetic flux is presumed to be from the north magnetic pole to the south magnetic pole. This information is needed if you want to use the left-hand generator rule.* Here is your next question.
   The electrical system in an automobile is
   A.   dc. (Proceed to block 23.)
   B.   ac. (Proceed to block 16.)

9. *Your answer to the question in block 25 is **B**. This answer is wrong. The amount of induced voltage does **not** depend upon the type of insulation on the conductor. This assumes, of course, that the insulation is sufficient to prevent the conductor from being short-circuited.* Proceed to block 18.

10. *Your answer to the question in block 13 is **A**. This answer is wrong. The commutator delivers the generated voltage to the brushes. Study the different parts of the generator in Fig. 14-7.* Then proceed to block 19.

11. *Your answer to the question in block 18 is **A**. This answer is wrong. In fact, brushes are used in dc generators but not in alternators.* Proceed to block 24.

12. *Your answer to the question in block 19 is **B**. This answer is wrong. A relay is an electrically operated switch. When a voltage is applied to the relay coil, the switch contacts close. A relay does not convert ac to dc.* Proceed to block 4.

13. *The correct answer to the question in block 2 is **A**. The commutator switches the connection of the rotating loop to the load. This causes the current to always flow in the same direction through the load. Since the commutator converts the ac of the loop to dc, it is sometimes called a "mechanical rectifier."* Here is your next question.
   The rotating conductor in a dc generator is called the
   A.   commutator. (Proceed to block 10.)
   B.   armature. (Proceed to block 19.)

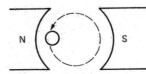

***Figure 14-18. Is the voltage generated
ac or dc?***

14. *The cutout disconnects the generator from the battery circuit whenever the
generator voltage is lower than the battery voltage. This prevents the generator from "motoring" whenever the battery voltage is greater than the generator voltage.* Here is your next question.
What is the advantage of using more than one loop of wire for the
armature of a dc generator? (Proceed to block 26.)

15. *Your answer to the question in block 4 is* ***A***. *This answer is wrong. In order to
generate a current to a load you must do work. If you turn the generator
when it is not connected into a circuit, you will find that it turns easily. However, as soon as a complete circuit is connected to the generator terminals, it
becomes difficult to turn. The reason is that the magnetic field around the
generated current reacts with the magnetic field in the generator and opposes
the motion.* Proceed to block 25.

16. *Your answer to the question in block 8 is* ***B***. *This answer is wrong. All automobiles have dc power supply systems.* Proceed to block 23.

17. *The correct answer to the question in block 1 is* ***B***. *Electron current will flow
from point b to point a.* Here is your next question.
Figure 14-18 shows the end view of a conductor being rotated
through a magnetic field. If the conductor follows the path of the
dotted arrow, the voltage generated in it will be
A.   ac. (Proceed to block 2.)
B.   dc. (Proceed to block 5.)

18. *The correct answer to the question in block 25 is* ***A***. *The amount of voltage
induced in a conductor is determined by the speed at which it moves through
the magnetic field and the strength of the magnetic field. The amount of induced voltage also depends upon the number of conductors being moved
through the magnetic field.* Here is your next question.
Which of the following statements is correct?
A.   Brushes are used in alternators but not in dc generators. (Proceed to block 11.)
B.   Brushes are used for picking the voltage off the commutator.
(Proceed to block 24.)
C.   Brushes are made of very hard steel. (Proceed to block 3.)

19. *The correct answer to the question in block 13 is* ***B***. *The armature rotates in
the magnetic field as the generator shaft is turned.* Here is your next question.
Which of the following is a semiconductor device that converts ac
to dc?
A.   A silicon diode (Proceed to block 4.)
B.   A relay (Proceed to block 12.)

20.   *Your answer to the question in block 24 is **A**. This answer is wrong. Magnetic flux lines do not actually move, but their direction is presumed to be from north to south.* Proceed to block 8.

21.   *Your answer to the question in block 2 is **B**. This answer is wrong. The armature is the rotating loop of wire in the generator.* Proceed to block 13.

22.   *Your answer to the question in block 23 is **A**. This answer is wrong. A transformer is used in ac circuits to step the voltage up or down. There are no moving parts in a transformer.* Proceed to block 6.

23.   *The correct answer to the question in block 8 is **A**. Dc is used because the lead-acid battery which produces a dc voltage is a very convenient, compact, portable source of voltage compared to ac sources.* Here is your next question.
When a dc motor and generator are both mounted on the same shaft it is called
A.   a transformer. (Proceed to block 22.)
B.   a dynamotor. (Proceed to block 6.)

24.   *The correct answer to the question in block 18 is **B**. The purpose of the brushes is to take the generated voltage from the commutator and deliver that voltage to the load.* Here is your next question.
The direction of the flux lines of a magnetic field is presumed to be
A.   from the south magnetic pole to the north magnetic pole. (Proceed to block 20.)
B.   from the north magnetic pole to the south magnetic pole. (Proceed to block 8.)

25.   *The correct answer to the question in block 4 is **B**. The magnetic field around an induced current always opposes the motion that produced it.* Here is your next question.
Whenever a conductor is moved through a magnetic field, a voltage is induced. Which of the following determines the amount of voltage generated?
A.   The speed at which the conductor moves through the magnetic flux (Proceed to block 18.)
B.   The type of insulation on the conductor (Proceed to block 9.)

26.   *A smoother output voltage is obtained when there are more conductors in the armature and more sections in the commutator.*
You have now completed the programmed questions. The next step is to put some of these ideas to work in laboratory experiments. Proceed to the "Experiment" section that follows.

Experiment

(The experiment described in this section may be performed on the circuit board described in Appendix C or on a similar laboratory setup.)

EXPERIMENT 1

Purpose — To demonstrate a dc motor-generator system.

Theory — A dc motor can be used to turn a dc generator, and the combination is called a *motor-generator system.* If the motor and the generator are both on the same rotating shaft the combination is called a *dynamotor.*

By using a motor-generator or dynamotor it is possible to convert a dc voltage from one value to another. For example, a 12-volt dc motor can be operated from a car battery, and the motor shaft can be coupled to a 250-volt dc generator. This is a practical way to get a high dc voltage from a low dc voltage.

In this experiment you are going to use a dc motor to turn the shaft of another dc motor. It is an interesting and important fact that a dc motor will generate a voltage if its shaft is turned. Furthermore, if you apply a dc voltage to a dc generator, it will run like a motor. (There are exceptions to this. For example, some dc generators have an automatic mechanically operated *off* switch that opens the field windings or commutator connection whenever the shaft is not turning. Thus, applying a voltage to such a generator cannot cause it to *motor* — that is, run like a dc motor.)

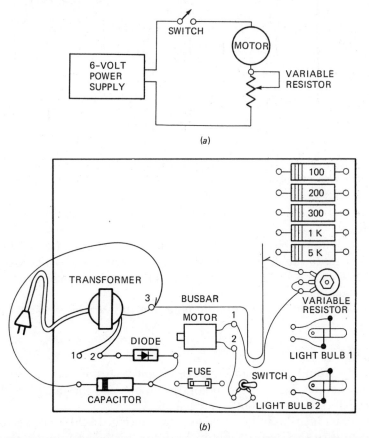

Figure 14-19. Test setup for Part I of Experiment 1. (a) *The circuit for Part I of Experiment 1.* (b) *This is the way your circuit board will look if you have wired the circuit correctly.*

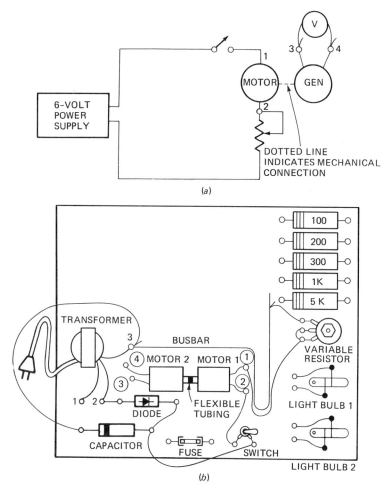

Figure 14-20. *Test setup for Experiment 1.* (a) *The circuit for Experiment 1.* (b) *This is the way the circuit board will look if you have wired the circuit correctly.*

Procedure—

Part I—Wire the circuit board as shown in Fig. 14-19. The variable resistor controls *current* to the dc motor; therefore, it is connected as a rheostat.

Step 1—Apply voltage to the circuit. Note that the switch turns the motor circuit on and off.

Step 2—With the motor running note that the variable resistor controls its speed.

Procedure—

Part II—Mount two motors on the circuit board as shown in Fig. 14-20. Make sure that the shafts of the motors are in line and the short length of flexible tubing fits securely on both of the shafts.

The circuit board is to be wired as shown in Fig. 14-20. In this setup motor 1 is used to turn the shaft of motor 2. Motor 2 is being used as a generator.

Step 1—Apply voltage to the circuit and note that the motor turns the generator.

Step 2 — Vary the resistance of the circuit by turning the variable resistor shaft. Note that this controls the speed of the motor-generator system.

Step 3 — With the motor-generator system running, use a voltmeter to measure the generator output voltage across terminals 3 and 4. As with all voltage measurements, you should start with the highest voltmeter scale and work down until you get a meter reading.

Step 4 — Vary the speed of the motor and note that the dc output voltage at terminals 3 and 4 changes with changes in motor speed.

Conclusion — A dc motor will generate a voltage when its shaft is turned. The greater the speed of the generator shaft, the greater its output voltage. A motor-generator system can be used for changing a dc voltage from one value to another.

Self-Test with Answers

(Answers with discussions are given in the next section.)

1. Reversing the direction that a conductor is moving through a magnetic field will (*a*) not affect the voltage generated in the conductor; (*b*) reverse the polarity of the voltage generated in the conductor.
2. Increasing the speed at which a conductor is moving through a magnetic field will (*a*) increase the voltage generated in the conductor; (*b*) decrease the voltage generated in the conductor.
3. Increasing the strength of the magnetic field through which a conductor moves will (*a*) increase the frequency of the ac voltage generated in the conductor; (*b*) increase the voltage generated in the conductor.
4. A rotating loop of wire in a magnetic field will generate an ac voltage. In a dc generator the ac voltage is converted to dc at the (*a*) field; (*b*) commutator.
5. A smoother output voltage is obtained in a dc generator by using (*a*) a weaker magnetic field; (*b*) more than one loop of wire in the armature and more segments in the commutator.
6. A generator or an alternator is used in the electrical system of a car to (*a*) recharge the battery when the engine is running; (*b*) increase the dc voltage in the electrical system.
7. The magnetic field of an induced current always reacts with the field in a generator so as to (*a*) oppose the motion; (*b*) aid the motion.
8. An alternator is basically an ac generator. In order to use an alternator in the dc circuit of an automobile, its output voltage must be (*a*) decreased; (*b*) rectified.
9. It is harder to turn a dc generator when (*a*) it is delivering current to a load; (*b*) it is not delivering current to a load.
10. When a conductor moves parallel to magnetic flux lines, the voltage generated in it is (*a*) 0 volt; (*b*) greater than when it moves at right angles to the flux lines.

Answers to Self-Test

1. (b) – In electricity the term "polarity" means the arrangement of positive and negative terminals. When you reverse the direction of motion, the voltage reverses so that the terminal which was positive will become negative, and the terminal which was negative will become positive.
2. (a) – The amount of voltage generated is directly related to the speed at which a conductor moves through a magnetic field.
3. (b) – The amount of voltage generated is directly related to the strength of the magnetic field through which the conductor moves.
4. (b) – The commutator is sometimes thought of as being a mechanical rectifier that converts the ac to dc.
5. (b) – This is apparent from the voltage waveform in Fig. 14-9.
6. (a)
7. (a)
8. (b) – Semiconductor diodes are usually used for rectifying the ac voltage.
9. (a) – The reason for this is related to Question 7.
10. (a) – This is clear from the graphs of Fig. 14-6. Note that the voltage is 0 volt at positions C and E.

15.
How Do DC Motors Work?

Introduction

A dc generator will turn like a motor when a dc voltage is applied to its terminals. This is the reason that a cutout circuit must be used with the generator in automobiles. When the car engine turns the generator slowly, there is actually more voltage at the terminals of the battery than there is at the generator. This causes the battery to feed voltage to the generator. Under this condition the generator tries to turn like a motor. The effect is known as *motoring*.

Motoring can quickly destroy a generator, because when it is acting as a motor it tries to turn in the opposite direction that the fan belt is rotating its pulley. The result is that you are moving a dc motor backward while the voltage applied is trying to move it forward.

Just as a dc generator can be used as a motor, it is also true that a dc motor acts like a generator when its shaft is turned. As a matter of fact, dc motors and dc generators have the same basic construction.

Dc motors are used extensively in modern industry and other applications. In your car, for example, dc motors operate the windshield wipers, electric windows, electric seats, and many other components placed in automobiles for convenience. In industry, dc motors are sometimes preferred over ac motors because it is easier to control their speed and direction of rotation.

You will be able to answer these questions after studying this chapter:

How is a current-carrying wire affected when it is placed in the field of a strong magnet?

What is the right-hand rule for motor action?

Why is a commutator needed for dc motors?

How are the speed and direction of rotation of a simple motor controlled?

What are the characteristics of a series motor?

What are the characteristics of a shunt motor?

What are the characteristics of a compound motor?

Instruction

HOW IS A CURRENT-CARRYING WIRE AFFECTED WHEN IT IS PLACED IN THE FIELD OF A STRONG MAGNET?

In order to better understand the relationship between a current-carrying wire and a magnetic field, it is important to first review some basic electrical facts.

Every time an electric current flows, there is an accompanying magnetic field. The magnetic field circles the electric current. Usually the current is confined to some type of conducting wire, but in some cases the current travels through a vacuum or in another manner. Regardless of the manner in which it flows, the basic rule regarding the magnetic field surrounding the current always holds true.

Another important basic electrical fact is that like poles of a magnet repel, and unlike magnetic poles attract. It is natural to assume, then, that the magnetic field surrounding a wire will be affected by the presence of a magnetic field that is external to the wire.

Consider the simple arrangement of Fig. 15-1. Here we have a conductor between the north and south poles of a magnet. It is connected in a circuit with a battery and switch. You know from your study of electricity that such a circuit *should not be connected* because it could cause a short circuit across the battery when the switch is closed. However, this illustration is only for the purpose of demonstrating the relationship between a current-carrying conductor and a magnetic field. If you were actually going to conduct this experiment, you would place a resistor in the battery circuit to limit the amount of current from the battery.

When the switch of the circuit is closed, as shown in Fig. 15-2, the conductor moves rapidly between the poles of the magnet. The direction of motion is shown in the illustration. From this simple experiment you can conclude

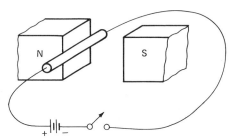

Figure 15-1. A conductor in a strong magnetic field.

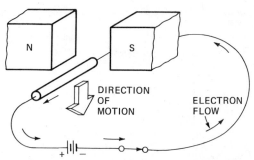

Figure 15-2. When the switch is closed, the conductor moves.

that in the presence of a strong magnetic field, a current-carrying conductor will move if free to do so.

In the circuit of Fig. 15-3 the battery has been turned around so that the electron current flows in the opposite direction. When the switch is closed, the conductor moves again, but in this case its direction of motion is opposite to the way it moved in the arrangement of Fig. 15-2. From this simple experiment you can conclude that the direction of the motion of a current-carrying conductor in a magnetic field is related to the direction of electron current flowing through that conductor.

WHAT IS THE RIGHT-HAND RULE FOR MOTOR ACTION?

By "motor action" we mean the motion of the current-carrying conductor when in the presence of a strong magnetic field. In the last chapter you learned that the polarity of an induced voltage or an induced current flow can be found by the use of the left-hand rule. In order to determine the direction of motion of the wire in Figs. 5-2 and 5-3, another rule—called the *right-hand motor rule*—is used.

Figure 15-4 illustrates the use of the right-hand motor rule. The forefinger of the right hand is extended in the direction of the magnetic field—that is, pointing from north to south along the flux lines. The second finger of the right hand is pointed in the direction of electron current flow. When the hand is

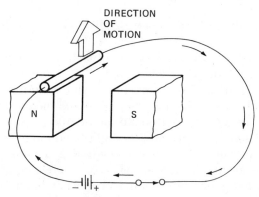

Figure 15-3. When the direction of current reverses, the direction of motion also reverses.

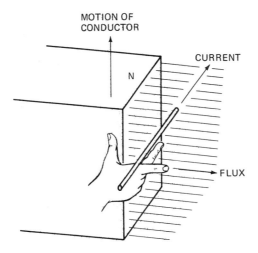

MOTION OF
CONDUCTOR

CURRENT

N

FLUX

Figure 15-4. The right-hand rule shows the direction that a current-carrying conductor will move when it is placed in a magnetic field.

held in this position, the thumb will point in the direction that the conductor will move when current is flowing through it.

You should use the right-hand motor rule to verify that the directions of conductor motion shown in Figs. 15-2 and 15-3 are correct.

The left-hand generator rule and the right-hand motor rule stated in this book apply only if you are talking about electron current. Conventional current, which is opposite in direction to electron current, was used by the early experimenters in electricity. Therefore, the rules were originally given as the left-hand motor rule and the right-hand generator rule, since they were based on conventional current flow. You may find these rules in other textbooks, and they may be opposite to the ones given in this book. That is the reason.

The left-hand generator rule and the right-hand motor rule are sometimes referred to as "Fleming's rules."

SUMMARY

1. Dc motors are constructed like dc generators.
2. If you apply a dc voltage to the terminals of a dc generator, it will turn like a motor. (This assumes that there is no cutout to prevent the generator from motoring.)
3. If you turn the shaft of a dc motor, it will generate a voltage across its terminals.
4. When a current-carrying conductor is placed in a strong magnetic field, it will move if free to do so.
5. The direction of motion for a current-carrying conductor in a strong magnetic field can be determined by the right-hand motor rule.
6. Figure 15-4 illustrates the right-hand motor rule.
7. The left-hand generator rule and the right-hand motor rule as described in this book apply when electron current flow is assumed. If conventional current flow is assumed, the rules must be interchanged so that the right hand is used for generator action and the left hand is used for motor action.

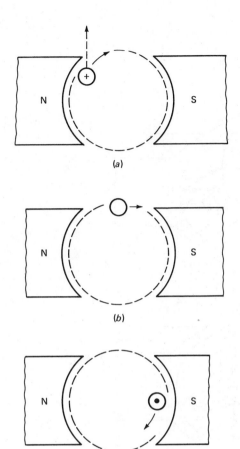

Figure 15-5. The force on a current-carrying conductor must be reversed periodically by reversing the direction of current flow. Unless this is done, the conductor will not move in a circle. (a) The force on the conductor is along the direction of the arrow. (b) At this point the force on the conductor is up, but the conductor continues to move in a circle. (c) The direction of current flow has reversed. The force on the conductor is downward.

WHY IS A COMMUTATOR NEEDED FOR DC MOTORS?

You will remember that the commutator in a dc generator serves as a mechanical switch to periodically reverse the connections to the armature wire. In a dc motor the commutator accomplishes the same thing. However, in the motor there is a current being delivered *to* the armature, instead of being taken *from* it as in the case of a generator.

Figure 15-5 shows why the commutator is needed. (You should use the right-hand motor rule to verify the forces shown on the conductors in this illustration.) As before, a + sign indicates that current is flowing into the conductor, and a dot indicates that current is flowing out of the conductor.

In Fig. 15-5a the current is flowing into the conductor and the force—as determined by the right-hand motor rule—is upward. This is shown by the dotted arrow. The conductor is free to move only in the circle shown by the dotted lines, so it follows the path of the solid arrow.

In Fig. 15-5b the conductor has moved to the top center position. If the direction of current flow at this point was into the paper, the force on the conductor would be straight upward, and if the current flow was out of the paper the force would be straight downward. Neither of these forces would move the

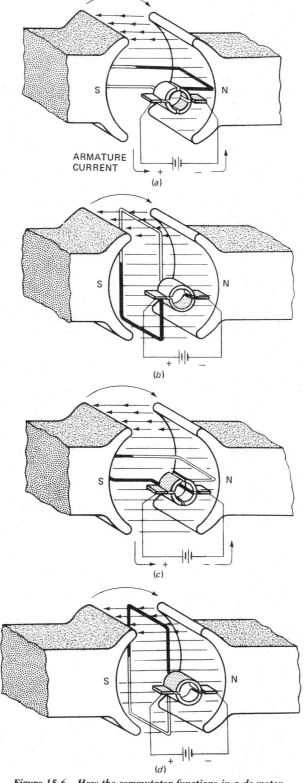

Figure 15-6. How the commutator functions in a dc motor.

conductor along the curved path. However, the conductor has enough inertia to coast past this point.

When the conductor is at the point shown in Fig. 15-5c, the current is flowing out of the paper. The right-hand motor rule shows that the force is downward, forcing the conductor to move along the curved path.

To summarize, in order for the right force to be exerted on the conductor for the positions shown in Fig. 15-5a and 15-5c, it is necessary to reverse the direction of current. Also, regardless of the direction of current flow, no force will be exerted on the conductor when it is in the position shown in Fig. 15-5b. This means that it is not necessary to supply current to the conductor at this time.

Figure 15-6 shows the action of the commutator in a simple dc motor. In this illustration, the armature is moving clockwise as viewed from the commutator. In Fig. 15-6a, the armature is positioned so that the electron current enters the black side of the commutator, goes through the black side of the armature wire, and then goes back through the white side, which is connected to the positive terminal of the battery. The right-hand motor rule can be used to show that the black wire of the armature will be thrust downward and the white wire of the armature will be forced upward at this moment. This causes the armature to move clockwise.

When the armature has moved to the position shown in Fig. 15-6b, the slots of the commutator are opposite to the brushes. Therefore, no current is being delivered to the armature wire at this moment. However, the armature does not stop at this point because it has inertia. Therefore, the armature continues to move past the point shown in Fig. 15-6b.

At the same time the brushes are switched to opposite segments of the commutator. The black brush will now be connected to the white commutator segment, and the white brush will be connected to the black segment. This is shown in Fig. 15-6c. Electron current again flows through the armature. The right-hand motor rule will show that the white side of the armature is now being forced downward, while the black side is being thrust upward.

When the armature reaches the position shown in Fig. 15-6d, the brushes are again switched to different segments of the commutator. Now the black brush will be connected to the black segment and the white brush connected to the white segment, as shown in Fig. 15-6a.

The process continues as described above. The overall effect is that the armature rotates clockwise in the magnetic field when a voltage is applied. The armature is connected to the motor shaft (not shown in the illustration), and this shaft can be used to turn a *load*. When we talk about the load in a motor system, we mean the opposition to turning. If the motor is turning a machine, then the machine becomes the load.

HOW ARE THE SPEED AND DIRECTION OF ROTATION OF A SIMPLE MOTOR CONTROLLED?

Figure 15-7 shows the effect of reversing either the magnetic field or the armature current in a simple dc motor.

In Fig. 15-7a the connections are the same as shown for Fig. 15-6. The right-hand rule will again show that the direction of rotation for this system is clockwise as viewed from the commutator.

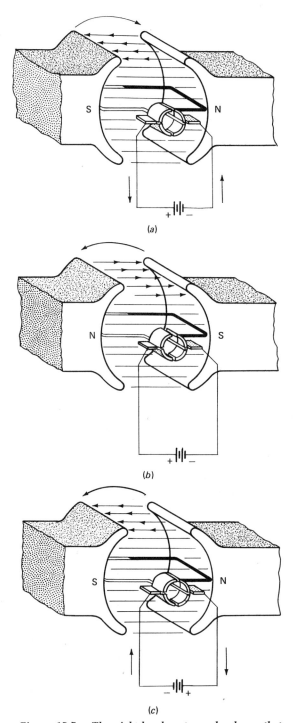

Figure 15-7. The right-hand motor rule shows that changing either the direction of current or the direction of the field will reverse the direction of rotation. (a) The direction of rotation in this circuit is clockwise. (b) Reversing the field reverses the direction of rotation. (c) The field is now the same as in Fig. 15-7a, but the direction of current is reversed. This also reverses the direction of rotation.

In Fig. 15-7*b* the magnetic field of the motor has been reversed. Thus, the flux lines are moving from left to right instead of from right to left. It would be very difficult to reverse the flux if a permanent magnet were used for the motor. However, permanent magnets are usually used for establishing the field in very small motors. In larger motors, the magnetic field is established by an electromagnet, and the field is easily reversed by simply reversing the current through the coil of the electromagnet.

In Fig. 15-7*b*, the right-hand rule will now show that the direction of rotation of the armature has been reversed. Now, the black wire is being thrust upward and the white wire is being thrust downward, thus causing the armature to rotate in a counterclockwise direction.

From this you can make a very important rule regarding simple dc motors: *The direction of rotation of the armature can be reversed by reversing the direction of the magnetic field in which the armature turns.*

Figure 15-7*c* shows what happens when the voltage connections to the motor—that is, to the armature—are reversed. In this illustration you should note that the magnetic field has now been returned to its original position of Fig. 15-7*a*. The only thing that has changed, then, is that the direction of current is reversed in the armature. The right-hand rule applied to this system shows that the white wire is being thrust downward and the black wire is being thrust upward. This causes the armature to turn in a counterclockwise direction.

From this illustration you can make another very important rule for basic dc motors: *The direction of rotation of a simple dc motor can be reversed by reversing the voltage leads to the motor* (also known as "permanent-magnet motors" or "PM motors").

If you reverse *either* the field *or* the polarity of the voltage, you will reverse the direction of dc motor rotation. However, if you reverse both the field and the applied voltage polarity, you will *not* change the direction of rotation. To prove this, refer again to Fig. 15-7*b*. In this case the field has been reversed (with reference to the field in Fig. 15-7*a*). Assume that the polarity of the applied voltage is also reversed, so that the positive terminal of the battery is connected to the black segment of the commutator and the negative terminal is connected to the white segment. With the polarity of the applied voltage reversed, both the field and voltage polarity have been changed from the condition shown in Fig. 15-7*a*. If you apply the right-hand rule, you will see that the motor turns clockwise as it did before.

From this you can make a third important rule for dc motors: *Reversing both the field and the polarity of applied voltage will not affect the direction of rotation.*

There are many occasions in which it is desirable to be able to reverse the direction of rotation of a motor. As mentioned before, this is one of the advantages of dc motors over ac motors. It is easier to reverse their rotation. An example of a practical application in which a dc motor is used is for electric windows in automobiles. Pushing the window switch in one direction causes the motor to turn in such a way that the window goes up. Pushing the switch in the opposite direction reverses the polarity of the voltage to the dc motor and causes the window to go down.

Before we discuss the methods by which motor speed is controlled, it is necessary to review a very basic principle regarding electric current flow. Whenever a current flows through a conductor, there is always a magnetic field surrounding the conductor. The strength of the magnetic field depends directly upon the amount of current flowing in that wire. For small currents the magnetic field is weak, and for large currents the magnetic field is strong.

If you have ever experimented with simple magnets, you know that weak magnets have a very light attraction to metal and strong magnets have a strong attraction. The same is true for electromagnets. If the amount of current flowing through the armature of a simple dc motor is weak, then the magnetic field is weak. This means that the armature will not be turned with as great a force as if a heavy current were flowing through it. In order to control the amount of speed of a simple dc motor, it is necessary to control only the amount of current flowing through the armature. This, in turn, controls the strength of the magnetic field around the armature wires.

Figure 15-8 shows two different ways of controlling armature current. A rheostat is shown in the circuit of Fig. 15-8a. A rheostat is a variable resistor connected in such a way that it controls the amount of circuit current. In this illustration the rheostat is connected directly in series with the armature of the motor. Adjustment of the rheostat will change the amount of current flow, and thus change the amount of thrust of the armature in the field. This, in turn, changes the speed of the motor.

The dotted lines in Fig. 15-8 indicate a mechanical connection to a load. The motor speed will change whether or not there is a load connection, but this illustration shows how a load can be represented in a diagram.

Figure 15-8b shows another way of controlling motor speed. Instead of using a variable resistor, three different resistance values—R_1, R_2, and R_3—are

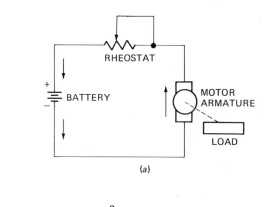

(a)

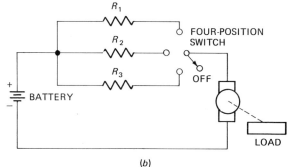

(b)

Figure 15-8. Two ways to control the speed of a dc motor. (a) *This circuit gives continuous control of motor speed.* (b) *This circuit gives step control of motor speed.*

connected to a four-position switch. The fourth position of the switch is not connected to anything. It opens the circuit and stops the motor. When the switch is put into any other position, motor current flows. If R_2 is a smaller resistance than R_3, then tuning to that resistor from R_3 will cause a larger current to flow and cause the armature to turn faster.

The important thing to note from this is that a switch and resistors can be used for controlling the speed of the motor. In many models of cars the windshield wiper is operated by a small dc motor. A so-called *three-speed* windshield wiper is connected in a circuit like the one shown in Fig. 15-8*b*.

SUMMARY

1. A commutator is used in a dc motor to switch current flowing to the armature.
2. As the armature turns, it is necessary to switch the dc voltage periodically so that the force on the armature wires always produces the same direction of rotation.
3. The direction of rotation of a dc motor can be reversed by reversing its magnetic field.
4. The direction of rotation of a simple dc motor can be reversed by reversing the polarity of the applied voltage.
5. Reversing both the magnetic field and the polarity of the applied voltage will not change the direction of rotation of a motor.
6. Changing the amount of current flowing in the armature will change the speed of a dc motor.
7. A rheostat can be used for varying the armature current, and thus varying the motor speed.
8. A switch can be used to insert different values of resistance into the armature circuit in order to control the motor speed.

WHAT ARE THE CHARACTERISTICS OF A SERIES MOTOR?

Series-wound motors, like electric generators, are made in such a way that the field winding is in series with the armature. A *shunt-wound motor* has the field winding in parallel with the armature. A *compound-wound motor* is made with both series and shunt field windings.

Figure 15-9 shows the circuitry for a series-wound motor. Note that the field coil, which produces the electromagnetic field within the motor, is in series with the armature winding. You will remember that whenever components are in series, the same current flows through them. With the series-wound motor it

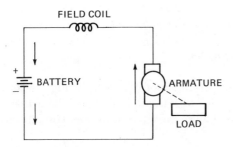

Figure 15-9. In a series-wound motor the field current flows through the armature.

is necessary to make the field coil so that it has a relatively small resistance. This is necessary so that it will not seriously limit the amount of current flow in the armature.

When the armature moves through the magnetic field of a motor, its motion is opposed by the resistance of the motor bearings. It is also opposed by a *countervoltage* which is produced in the armature winding. To understand why this occurs, remember that any time a conductor moves through a magnetic field, a voltage is generated. The armature is a conductor, and it is moving through the magnetic field of the motor. Therefore, it is reasonable to expect that it will have a voltage generated in it. This voltage is called a *countervoltage* because it is opposite to the voltage that produces the rotation of the armature. Of course, the countervoltage can never be as large as the applied voltage, or the armature would not turn.

If you connect a load to a series-wound motor, it turns more slowly. When this happens, the countervoltage decreases, and the armature current increases. This is a very important characteristic of series-wound motors: the armature current increases when the load increases.

A series-wound motor can be damaged if it is operated without a load. The reason is evident from Fig. 15-9. Since the field coil has low resistance and the armature has low resistance, there is practically no opposition to current flow. The high current in the armature causes the motor to turn faster and faster until the motor reaches such a high speed that it will fly apart. It does not do this when the motor is connected to a load, because the countervoltage opposes the applied voltage, and the load acts to further limit the speed.

The fact that the motor speed varies with changes in load is very important when you are trying to decide which motor to use in a certain application. Motors are rated according to their *speed regulation*. When motors have good speed regulation, their speed does not vary much with changes in load. If they have poor speed regulation, the speed varies considerably with small changes in load. According to this method of rating, a series motor is one that has very poor speed regulation.

Another important feature of the series motor can be understood from studying Fig. 15-9. When the motor is first started with a load, there is practically no countervoltage generated. This means that a high field-coil current and armature current flows. The result is a very strong turning effect of the motor. In other words, we say that the series-wound motor has a *high starting torque*. (Torque is defined as the turning force created by the motor. A motor with a low torque rating cannot turn a heavy load, but a motor with a high torque rating can turn a very heavy load.)

We can summarize this discussion on the characteristics of series-wound motors by writing the following rules:

1. *The speed of the series-wound motor varies considerably with changes in motor load.*
2. *A series-wound motor will destroy itself if permitted to operate without a load.*
3. *A series-wound motor has a high starting torque.*

To control the speed of the series-wound motor shown in Fig. 15-9, it is necessary only to insert a rheostat in series with the field coil and the armature. Since they are in series, whenever you change the current in one, you automatically change the current in the other.

WHAT ARE THE CHARACTERISTICS OF A SHUNT MOTOR?

Figure 15-10 shows the circuitry for a shunt-wound motor. The field winding is in parallel with the armature winding. Since the field winding is connected directly across the battery, it must have a high resistance. Otherwise, it would be a short across the battery, and this would seriously limit the battery life.

The amount of field current in the shunt-wound motor does not depend upon the amount of armature current, and therefore, changes in load for this type of motor do not affect its speed of rotation. This type of motor has a much better speed regulation than the series-wound type, and it is sometimes referred to as a *constant speed motor* for this reason. However, the term "constant" is misleading. Although the speed for this type of motor varies very little with changes in load, it does vary a small amount.

Large shunt-wound motors are sometimes used with a special starting circuit. The need for this starting circuit is apparent from Fig. 15-10. Note that the armature is connected directly across the battery. Therefore, a very high armature current will start to flow the minute the circuit is energized. The countervoltage is so small that it can be ignored when the motor is at rest, or when it is turning very slowly. The result is that the armature winding can be burned out. After the armature is turning at full speed, this does not happen because the armature generates a countervoltage, which opposes the battery voltage and thus limits the amount of current. Excessive armature current occurs only when the motor is first starting. To eliminate the problem of high current when starting a shunt-wound motor, some industrial motor circuits have a resistor which can be inserted in series with the armature until the motor has reached its running speed.

The shunt-wound motor does not have a large starting torque like the series motor, and this is a serious disadvantage of this type of motor in industrial applications. The speed of a shunt-wound motor can be varied by changing either the field current or the armature current. Since the field current is usually much smaller, it is usually more sensible to put the rheostat in series with the field winding. You know that the amount of power dissipated by a resistor depends upon the amount of current flowing through it. Therefore, when you put the rheostat in series with the field winding, it will dissipate only a small amount of power due to the small current. If the rheostat were placed in series with the armature, the larger current would cause the rheostat to dissipate, or waste, more power. We can summarize the important characteristics of the shunt-wound motor as follows:

1. *It has a poor starting torque.*
2. *It has a relatively constant speed with changes in load. In other words, it has good speed regulation.*

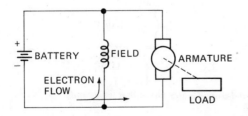

Figure 15-10. In a shunt-wound motor the field current flows in parallel with the armature current.

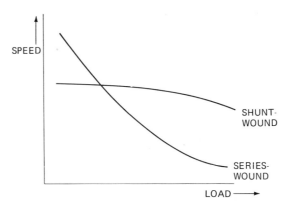

Figure 15-11.   This graph shows that a shunt motor is less affected by changes in speed than a series motor.

3.   Its speed can be varied by controlling either the speed current or the armature current.

Figure 15-11 shows a graph that compares the speed regulation of series-wound and shunt-wound motors. This graph shows the speed of the motor for different amounts of load. As you go from left to right on the horizontal line, the load is increasing, and as you go from bottom to top on the vertical line, the speed is increasing. You can see that with a low load the speed is high for the series-wound motor, and with a high load the speed is low. This means that the series-wound motor has a speed that varies widely with changes in load. In contrast, note that the speed of the shunt-wound motor drops off only slightly as the load is increased.

WHAT ARE THE CHARACTERISTICS OF THE COMPOUND MOTOR?

The circuit for a compound-wound motor is shown in Fig. 15-12. It is a combination of both the series-wound and shunt-wound motors. It has a field winding in series with the armature, so the motor has some of the characteristics of a series-wound motor such as a high starting torque. At the same time the motor has a shunt field winding, and therefore, the motor has better speed regulation than a series-wound motor. In this type of motor the speed can be varied by changing either the shunt motor or the series field and armature current.

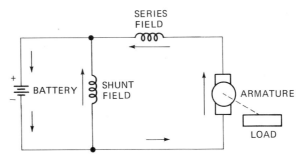

Figure 15-12.   A compound-wound motor has both a series and a shunt field.

We can summarize the characteristics of the compound motor as follows:

1. *It has a better starting torque than a shunt-wound motor, but not as good as a series-wound motor.*
2. *It has better speed regulation than a series-wound motor, but not as good as a shunt-wound motor.*

In other words, the compound-wound motor is a compromise between the characteristics of series- and shunt-wound motors.

SUMMARY

1. In a series motor, the same current flows through the armature and the field windings.
2. The field windings in a series motor have low resistance.
3. Series motors have a high starting torque but poor speed regulation.
4. In a shunt motor, the field current does not flow through the armature.
5. A shunt motor has good speed regulation but poor starting torque.
6. With a shunt motor the speed can be varied by varying *either* the field current *or* the armature current, but it is better to vary the field current because less power is wasted in the rheostat.
7. A compound motor is a compromise between the series motor and shunt motor. It has good starting torque and good speed regulation.

Programmed Review Questions

(Instructions for using this programmed section are given in Chap. 1.)

We will review the important concepts of this chapter. If you have understood the material, you will progress easily through this section. Do not skip this material, because some additional theory is presented.

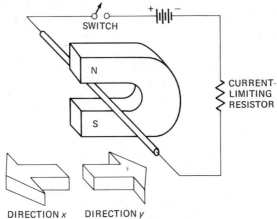

Figure 15-13. Will the conductor move in direction x *or direction* y *when the switch is closed?*

1. In Fig. 15-13, when switch *SW* is closed the conductor will move in
   A. direction *x*. (Proceed to block 7.)
   B. direction *y*. (Proceed to block 17.)

2. *The correct answer to the question in block 16 is **B**. The cost of a motor is more directly related to its "horsepower rating." This is a method of rating a motor according to how much work it can do.*
   It is easier to control the speed and direction of rotation for a dc motor than it is for an ac motor. Here is your next question.
   In Fig. 15-14 electron current is flowing into the conductor as indicated by the cross. This conductor can move only along the direction shown by the dotted line. In this case, the conductor will move
   A. counterclockwise. (Proceed to block 15.)
   B. clockwise. (Proceed to block 18.)

3. *Your answer to the question in block 5 is **B**. This answer is wrong. A shunt-wound motor can be operated with no load.* Proceed to block 13.

4. *Your answer to the question in block 17 is **A**. This answer is wrong. The load of a battery or generator is the amount of current flow. Thus, if a battery or generator must deliver a high current, you would say that it "has a heavy load." The load of a motor is the opposition that it turns against.* Proceed to block 12.

5. *The correct answer to the question in block 15 is **B**. The left-hand motor rule is used with positive-to-negative conventional current flow.* Here is your next question.
   Which type of dc motor would destroy itself if operated without a load?
   A. Series-wound (Proceed to block 13.)
   B. Shunt-wound (Proceed to block 3.)

6. *The correct answer to the question in block 19 is **A**. Dc generators and dc motors are constructed the same way.* Here is your next question.
   If you reverse the direction of current through the armature *and* reverse the direction of the field, what will be the effect on the direction of rotation? (Proceed to block 20.)

7. *Your answer to the question in block 1 is **A**. This answer is wrong. Study the right-hand rule as illustrated in Fig. 15-4. Apply this rule to Fig. 15-13.* Then proceed to block 17.

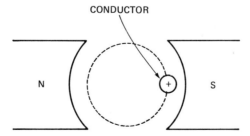

CONDUCTOR

N   S

Figure 15-14. Will the conductor move clockwise or counterclockwise?

8. *Your answer to the question in block 12 is **A**. This answer is wrong. Ohm's law defines the relationship between voltage, current, and resistance in an electric circuit.* Proceed to block 16.

9. *Your answer to the question in block 15 is **A**. This answer is wrong. The right-hand motor rule is used for electron current flow.* Proceed to block 5.

10. *Your answer to the question in block 16 is **A**. This answer is wrong. Since dc motors must have a commutator, they may be more expensive than an ac motor.* Proceed to block 2.

11. *Your answer to the question in block 13 is **A**. This answer is wrong. A potentiometer is used for varying the **voltage** in a circuit.* Proceed to block 19.

12. *The correct answer to the question in block 17 is **B**. The load of a motor is the physical opposition that it turns against. If the motor is running but it is not turning against an opposition, it is said to be "operating with no load."* Here is your next question.
    The right-hand motor rule and the left-hand generator rule are sometimes called
    A. Ohm's laws for motors and generators. (Proceed to block 8.)
    B. Fleming's rules. (Proceed to block 16.)

13. *The correct answer to the question in block 5 is **A**. A series-wound motor is characterized by a high starting torque and poor speed regulation. If operated without a load, this type of motor runs faster and faster until it is destroyed.* Here is your next question.
    To make the speed of a simple dc motor continuously variable, you would use
    A. a potentiometer. (Proceed to block 11.)
    B. a rheostat. (Proceed to block 19.)

14. *Your answer to the question in block 19 is **B**. This answer is wrong. An ac generator does not use a commutator, so it will not work as a dc motor.* Proceed to block 6.

15. *The correct answer to the question in block 2 is **A**. The right-hand rule shows that the conductor moves upward. However, since it must follow the dotted path, it will move upward and to the left. This is opposite to the direction that the hands on a clock move, thus it is **counterclockwise**.* Here is your next question.
    When you are using conventional current flow, the direction of motion of a current-carrying conductor in a magnetic field is determined by the
    A. right-hand motor rule. (Proceed to block 9.)
    B. left-hand motor rule. (Proceed to block 5.)

16. *The correct answer to the question in block 12 is **B**. Sir John Ambrose Fleming (1849–1945) was an English electrical engineer.* Here is your next question.

In industry, dc motors are sometimes preferred over ac motors because

A. they are much cheaper. (Proceed to block 10.)

B. it is easier to control their speed and direction of rotation. (Proceed to block 2.)

17. *The correct answer to the question in block 1 is* **B**. *Application of the right-hand rule shows that the conductor will move in direction* y. Here is your next question.

The load of a motor is

A. the current that flows through it. (Proceed to block 4.)

B. the mechanical opposition that it turns. (Proceed to block 12.)

18. *Your answer to the question in block 2 is* **B**. *This answer is wrong. Use the right-hand motor rule to determine the direction of motion.* Then proceed to block 15.

19. *The correct answer to the question in block 13 is* **B**. *In order to control the speed of a motor, you want to vary the field* **current**. *This is accomplished by using a rheostat.* Here is your next question.

The construction of a dc motor is the same as the construction of

A. a generator. (Proceed to block 6.)

B. an ac generator. (Proceed to block 14.)

20. *There will be no effect on the direction of rotation if you reverse both the armature current and the field. In order to reverse the direction that a dc motor turns, you must reverse either the armature current or the field, but not both.* You have now completed the programmed question. The next step is to put some of these ideas to work in laboratory experiments. Proceed to the "Experiments" section that follows.

Experiments

(The experiments described in this section may be performed on the circuit board described in Appendix C or on a similar laboratory setup.)

EXPERIMENT 1

Purpose — To show how the speed of a dc motor is controlled by a rheostat, and by a resistor and switch.

Theory — A rheostat is a variable resistor connected into a circuit in such a way that it controls the amount of current flow. If you lower the amount of current flowing through the armature, then you also reduce the strength of the magnetic field surrounding the armature wires. This, in turn, reduces the amount of force on the wires and lowers the motor speed.

In this experiment you will use a rheostat to control the speed of a simple dc motor.

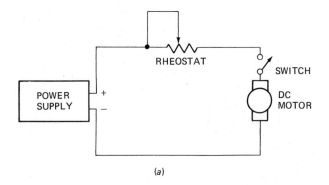

(a)

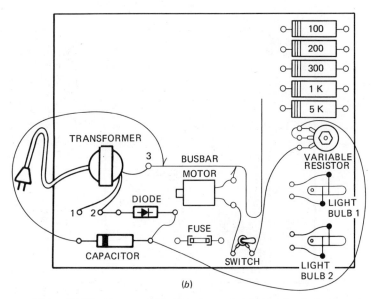

(b)

Figure 15-15. Test setup for Experiment 1. (a) *The circuit for Experiment 1.*
(b) *This is the way the circuit board will look if you have wired it correctly.*

Procedure—

Part I—Wire the circuit as shown in Fig. 15-15. This test setup was used in Chap. 14, and is repeated here to emphasize the importance of varying current in motor speed control.

Step 1—Apply voltage to the circuit by closing the switch.

Step 2—With the motor running note that adjusting the rheostat varies the motor speed.

Procedure—

Part II—Instead of using a variable resistor to change the motor speed, a number of different values of resistance can be switched into the motor circuit. Each different value of resistance causes the motor to turn at a different speed.

Wire the circuit as shown in Fig. 15-16. In this circuit you have two values of resistance: 0 ohm and the resistance of the light.

Step 1—Apply power to the circuit. Close the switch (*SW*). The arrows of Fig. 15-17 show that the current is limited only by the motor. With this setup your motor should turn at full speed.

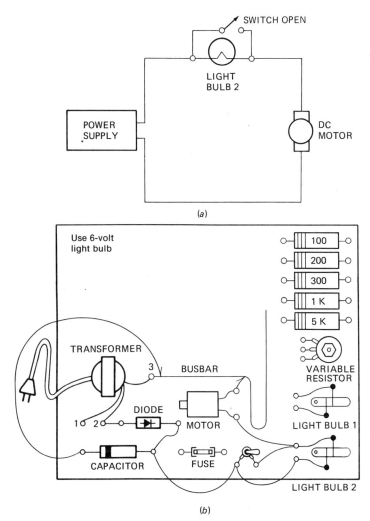

(a)

(b)

Figure 15-16. *Test setup for Experiment 2.* (a) *The circuit for Experiment 2.* (b) *This is the way the circuit board will look if you have wired it correctly.*

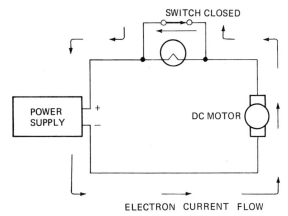

Figure 15-17. *Arrows show the path of current with the switch closed. The switch offers a short circuit around the lamp.*

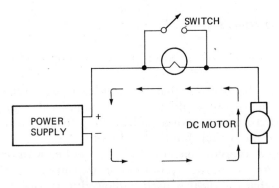

Figure 15-18. With the switch open, current must flow through the light. This reduces the amount of current through the motor.

Step 2—Open the switch (*SW*). Figure 15-18 shows that current must now flow through the resistor (that is, the light). Since the resistance of the circuit has increased, the current has decreased. Note that this causes the motor to turn more slowly. (You can tell by the sound of the motor when it is going faster and when it is going slower.)

EXPERIMENT 2

Purpose—To show that reversing the voltage leads to a dc motor causes it to reverse its direction of rotation.

Theory—If you reverse the plug for an ac electric fan, it does not have any effect on the direction of rotation. However, if you reverse the leads to a simple dc motor (also called a "permanent-magnet" or "PM motor"), the motor reverses its direction of rotation.

Procedure—
Step 1—Wire the circuit shown in Fig. 15-15. With the switch (*SW*) closed, adjust the rheostat until the motor is turning very slowly and you can tell its direction of rotation.
Step 2—Reverse the connections of the circuit to the power supply. (The lead that was connected to the positive terminal will now go to the negative terminal, and the lead that was connected to the negative terminal will now go to the positive terminal.)

Conclusion—Changing the amount of current flowing through a dc motor will change its speed.
Reversing the leads of a dc motor to its voltage source causes the direction of rotation to reverse.

Self-Test with Answers

(Answers with discussions are given in the next section.)

1. To determine the direction that a current-carrying conductor will move in a magnetic field, you would use (*a*) Fleming's rule; (*b*) Kirchhoff's law.
2. If you needed a motor with a high starting torque, you would use a (*a*) series-wound motor; (*b*) shunt-wound motor.
3. If you needed a motor that could be used either with or without a load, you would use a (*a*) series-wound motor; (*b*) shunt-wound motor.
4. A dc motor that will destroy itself if allowed to run without a load is (*a*) a series-wound motor; (*b*) a shunt-wound motor.
5. To vary the speed of a dc motor, you would use (*a*) ac instead of dc for a power supply; (*b*) a rheostat.
6. In some applications, dc motors are preferred over ac motors because (*a*) they are able to operate for a longer period of time without the need for maintenance; (*b*) it is easier to control their speed and direction of rotation.
7. The electric starter motor used for starting a car is (*a*) a dc motor; (*b*) an ac motor.
8. The type of dc motor that has both series and shunt field coils is called a (*a*) complex-wound motor; (*b*) compound-wound motor.
9. How are dc motors rated? (*a*) According to the size of their armature windings; (*b*) By horsepower.
10. If you apply an ac voltage to a dc motor, it will (*a*) not turn; (*b*) turn, but at a slower-than-normal rate.

Answers to Self-Test

1. (*a*)
2. (*a*)
3. (*b*) — A series-wound motor should *never* be operated without a load!
4. (*a*)
5. (*b*) — You could also use different values of resistors and a selector switch. However, this was not one of the choices.
6. (*b*) — This does not mean that it is not possible to control the speed or direction of rotation of an ac motor.
7. (*a*) — There is no ac power available in a car to run an ac motor. You *could* convert the car's dc to ac by using an *inverter*, but this is not done for the starter motor.
8. (*b*)
9. (*b*) — A dc motor has the same parts as a dc generator.
10. (*a*) — It was mentioned in the "Experiments" section that an ac voltage cannot be used for operating a dc motor.

16.
What Is Alternating Current and How Is It Generated?

Introduction

When a direct current flows in a wire, electrons leave the negative terminal of the voltage source, flow through the circuit, and return to the positive terminal of the source. The direction of electron motion is always the same—from negative to positive.

When an alternating current flows, the voltage source periodically reverses its polarity. For one short period of time the electrons flow through the circuit in one direction, and then the polarity of the source is reversed and the electrons flow in the opposite direction. Alternating current, then, is a back-and-forth motion of electrons in a circuit.

In an automobile, dc sources are used for producing electric current. However, the power generated for use in homes and industry is usually ac. There are advantages and disadvantages to using dc and ac in any application. However, you should not believe that one is better than the other. This just simply is not true. Dc power is used in automobiles because of the need for portable electricity to go along with the car. The lead-acid battery is the only portable source of electricity that can be economically produced and that can generate enough electricity to run the starter motor of a high-compression engine. Alternating current, on the other hand, is usually generated for applications where the electricity must be sent over long distances. As you will see later, the amount of power loss in a long-distance dc line is prohibitive; but for an ac line, it is possible to step the voltage up at regular intervals so that all the houses in a neighborhood can receive approximately the same amount of voltage at their ac outlets.

Both types of electricity (dc and ac) are in common use. In some applications there is a need for converting from one type of electricity to the other. You have already seen that the diode rectifier on the experiment board can be used to convert an ac voltage to a dc voltage. That is the purpose of rectifiers. On the other hand, components that convert dc to ac so that ac appliances can be operated from a 12-volt dc source are called *inverters*.

You will be able to answer these questions after studying this chapter:

What is meant by the term "frequency," and how is it measured?
What is meant by the term "amplitude"?
What is meant by the expressions "average voltage" and "average current"?
What is meant by the expressions "rms voltage" and "rms current"?
How do you convert an rms value into an average value?
What is meant by the term "phase"?
How does an ac motor work?
How does an ac generator work?

Instruction

WHAT IS MEANT BY THE TERM "FREQUENCY,"
AND HOW IS IT MEASURED?

If you were able to watch the electrons flowing in an alternating-current circuit, you would see that they move in one direction for a period of time and then stop. Next you would see the current move in the opposite direction for a period of time and then stop again. This back-and-forth motion continues as long as the alternating current is flowing.

Figure 16-1 shows a graph that can be used to represent the ac voltage or the flow of alternating current in the circuit of your home. This graph has a particular wave shape known as a *sine wave*. It is a fairly accurate representation

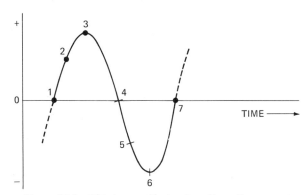

Figure 16-1. This is a graph showing alternating current.

of the relationship of the voltage or current used in homes and in industries. This type of ac voltage or alternating current is known as a *sinusoidal voltage* or *sinusoidal current.*

We will now discuss the graph to make sure its meaning is clearly understood. The discussion will be with respect to current flow, but the graph is equally related to voltage. At point 1 on the graph the value of current is considered to be 0 ampere. Point 1 is on a time line, which is often called the *time base.* So we can say that at this particular time—that is, the time corresponding to point 1 on the graph—the current is zero.

At a later time the current has increased to a positive value indicated at point 2. You should understand that the terms "positive" and "negative" are relative—they have no actual physical meaning. If you were standing in the circuit watching electrons go from left to right, you might say they are going in a positive direction; when they travel from right to left, you could say they are going in a negative direction. The terms "positive" and "negative" do not have a real meaning but are only a convenience.

At point 3 the current has reached its maximum possible positive value. After this instant, the current begins to decrease until it reaches 0 ampere at point 4.

If you were watching this current flow through a wire for the period of time between 1 and 4, you would say that the electrons are moving from left to right (using the above definition of positive and negative). After the current has stopped at point 4 it begins to flow in the opposite direction. Between points 4 and 5 the current has increased in the negative direction. It continues to increase until it reaches its maximum negative value at point 6. After this point the current begins to decrease again until it finally reaches a value of 0 ampere at point 7.

If you were watching the electron flow for the period of time between points 4 and 7, you would say the electrons are moving from right to left. After passing point 7, the electrons are beginning to move in a positive direction again—that is, they would move from left to right.

A complete *cycle* of current occurs between points 1 and 7. For a cycle the current starts to flow in one direction, stops, and then starts flowing in the opposite direction.

If you were able to watch the electrons flowing back and forth in the wires in your home, you would find that there are 60 complete cycles of current every second. Of course, no one could really count these by watching the electrons, but there are methods of indirectly determining the number of electrons per second.

If you think about it for a moment, 60 cycles per second represents a very rapidly changing current. However, by the standards of electricity and electronics, this is a rather low number of cycles per second. The number of cycles per second is known as the *frequency* of the current. The frequency of the voltage and current in your house wiring system is 60 *cycles per second.*

Frequency used to be measured in cycles per second. Unfortunately, the terms "cycle" (the variations between points 1 and 7 of Fig. 16-1) and "cycles per second" (which is the frequency) were often interchanged and confused by writers and technicians alike. To get around this problem, the term "cycles per second" is no longer used. It has been replaced by the term *hertz.* One hertz is one cycle per second. (The time duration of 1 second is always the unit of time for measuring frequencies.) A frequency of 60 cycles per second is now called a frequency of 60 hertz.

It has been noted that 60 hertz represents a rapidly changing current, but this frequency is low compared to the frequency of other systems. To give you an idea of how low it is by comparison with other systems, we will compare it with the frequency of a radio station operating at 1500 on your radio dial. If you could watch the electrons flowing in the antenna circuit of this radio station, you would find that the current has a frequency of 1,500,000 hertz (1,500,000 cycles per second). Such a rapidly changing current staggers the imagination, but it is by no means the highest frequency in use in electronics. In an X-band radar system, for example, the current changes at a rate of ten-thousand million hertz (10,000,000,000 cycles per second). So you can see that 60 cycles per second alternating current in your home is a low frequency for alternating current.

Although the sine wave is the most important of all the waveforms of alternating currents and ac voltages, it is not the only one that is in use. The ac waveform may be a square wave, a sawtooth wave, or any other wave shape. Figure 16-2 shows some examples of the wave shapes that are used in electronic systems.

The horizontal line in Fig. 16-1 is sometimes called the *time base*. This time base may be marked in seconds, milliseconds, or microseconds.

There is another way of illustrating the time base, as shown in Fig. 16-3. In this drawing the time base is marked in degrees rather than in seconds, or divisions of seconds. The reason for the degree markings can be understood from the conductor rotating in a magnetic field, shown to the left of the graph. As the conductor rotates, an ac voltage will be generated.

The conductor starts at point *a* and rotates in a counterclockwise direction. We will consider point *a* to be 0° on the voltage graph. The voltage generated at this point is 0 volt.

When the conductor has rotated 45° to position *b*, the voltage generated has increased to a positive value. This is represented by point *b* on the graph.

When the conductor is at point *c*, the amount of voltage generated is max-

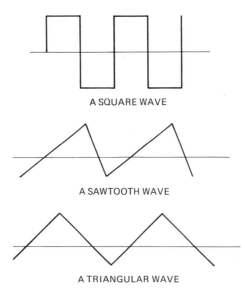

A SQUARE WAVE

A SAWTOOTH WAVE

A TRIANGULAR WAVE

Figure 16-2. Not all alternating currents and ac voltages have a sinusoidal waveform. These are examples of other waveforms in use.

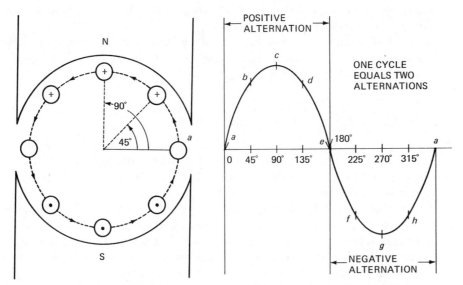

Figure 16-3. The time base may be marked in degrees rather than seconds.

imum, and this is indicated at the 90° mark on the graph. Note that at this time the conductor has rotated 90° from its zero position between the magnetic poles.

If you follow the path of the rotating conductor through 360°, you will find corresponding points on the graph related to the voltage induced in the conductor at each angle. Thus, instead of marking the graph base in seconds, it can be marked in degrees, as shown in Fig. 16-3.

SUMMARY

1. When direct current is flowing in a wire, the electrons always flow in the same direction. With alternating current the electrons flow back and forth in the wire.
2. There are some applications where dc is needed, and some applications where ac is needed. It is not a matter of which is better, but rather, it is a matter of which is better for each particular application.
3. Dc is used in automobile systems because a high-power compact supply (the lead-acid battery) is available for starting the engine.
4. Ac is used for power distribution to homes because transformers can be used to step up the voltage. This compensates for voltage loss along the power lines. Transformers do not work for dc.
5. A rectifier can be used to convert ac to dc. An inverter can be used to convert dc to ac.
6. A cycle is a complete alternation of voltage or current.
7. Frequency was measured in cycles per second at one time, but it is now measured in hertz. One hertz is one cycle per second.
8. The ac voltage and alternating current used in the home are said to be "sinusoidal." The name comes from the shape of the graph used to represent the ac voltage or alternating current of the power lines.
9. The frequency of the ac voltage and alternating current from the power

company is 60 hertz. This is a low frequency when compared to the frequency of some electronic systems.

10. Instead of marking the time base for the graph of an ac voltage or alternating current in seconds, microseconds, or milliseconds, it is also a common practice to mark it in degrees.

WHAT IS MEANT BY THE TERM "AMPLITUDE"?

There are several important terms related to the measurement of ac voltage or alternating current. They are illustrated in Fig. 16-4. Although this waveform shows a sinusoidal voltage, the terms that will be defined apply also to an alternating sinusoidal current.

The portion of the graph from 0 point to point *b* on the horizontal line is one-half cycle of waveform. From *b* to *d* is also a half cycle. You will notice that for each half cycle there is one point (marked E_M) at which the voltage reaches its maximum value. This is known as the *maximum voltage*, or *peak voltage*. It is also known as the *amplitude* of the ac waveform. In other words, the amplitude is the maximum value of voltage (or current) during one-half cycle of waveform. During a complete cycle, then, there are two values of maximum voltage, and the amplitude of the wave is considered to be either one of these values when the wave is symmetrical. The word "symmetrical," as applied to waveforms, means that there is as much positive voltage on one-half cycle as there is negative voltage on the next half cycle. If you are asked what is the amplitude of a certain waveform, the answer is always the maximum value during a half cycle.

Another value of importance is the *peak-to-peak voltage*, which is also marked in Fig. 16-4. Note that the peak-to-peak value in this symmetrical waveform is actually twice the value of the amplitude.

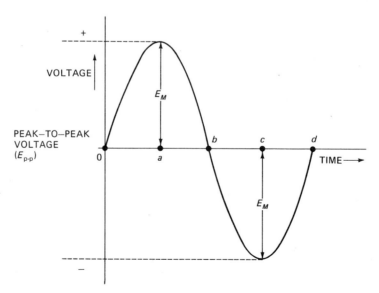

*Figure 16-4. The maximum value of voltage occurs at points **a** and **c**.*

WHAT IS MEANT BY THE EXPRESSIONS "AVERAGE VOLTAGE"
AND "AVERAGE CURRENT"?

Let's review the meaning of the word "average." Suppose you take a series of tests, and your scores are as follows: 95 percent, 90 percent, and 85 percent. Your average grade is easy to determine. You simply add the three grades and divide by the total number of grades involved. Here is how it is done mathematically:

$$95$$
$$90$$
$$\underline{85}$$
$$\text{TOTAL OF 3 GRADES} = 270$$

$$\text{AVERAGE} = 270 \div 3 = 90 \text{ percent}$$

In this case, the average value of the three grades is 90 percent. The average grade is always some value between the minimum grade and the maximum grade that you received.

A sinusoidal voltage or current is actually a combination of a large number of values. When we speak of *average voltage* or *average current,* we are always referring to the average of all the values for a half cycle. This is very important, because if you took the complete average for a complete cycle using both positive and negative values, you would find that the average is zero. However, if you take the average of the values over a half cycle, the situation is quite different.

Figure 16-5 shows a simple way of finding the average value of a sinusoidal waveform. We simply divide the half cycle into a number of vertical segments. For this graph, the length of each segment represents a value of voltage. If you measured the length of these segments and took the average value of them, you would find that it comes out to be about 6.4 volts. In other words, the average value is a little over $^6/_{10}$ of the maximum value.

If you increased the number of segments and took the average value, you would find that the average value comes closer to 6.36. In fact, mathematicians have a very precise way of figuring the average value, and they tell us that the exact value is 0.636 times the maximum value when we are talking about a half cycle of pure sine wave.

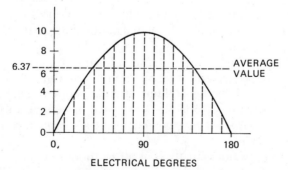

Figure 16-5. How a half cycle of sine wave is divided into segments for finding the average value.

We can write an important rule for alternating current circuits: *The average value of sine-wave voltage or current is 0.636 times the maximum value.*

You should always keep this very important point in mind when we are talking about average values: the average is always the average for one-half cycle of waveform.

The average value is 0.636 times the peak value only when we are talking about a sine wave. For any other waveform the average value is quite different. We will give a sample problem to show you how the average value is found.

Example 16-1 – The peak value of voltage for a certain sine wave is 50 volts. What is the average value of the wave?

Solution – Average value is equal to 0.636 times the peak value.

$$\text{AVERAGE VOLTAGE} = 0.636 \times \text{PEAK VOLTAGE}$$
$$= 0.636 \times 50$$
$$= 31.8 \text{ volts} \qquad \textit{Answer}$$

As you progress through your studies of electricity and electronics you will find occasion to use the average value of the waveform. For the time being it is necessary only for you to understand how it is found.

SUMMARY

1. For a sinusoidal waveform the maximum or peak value is known as the *amplitude.*
2. For a sinusoidal waveform, the peak-to-peak value is twice the peak value. In other words, it is twice the amplitude.
3. The terms "average voltage" and "average current" mean the average of all the values that occur during a half cycle.
4. The average value of sinusoidal voltage or current is 0.636 times the maximum value.

WHAT IS MEANT BY THE EXPRESSIONS "RMS VOLTAGE" AND "RMS CURRENT"?

If you apply a sinusoidal voltage to a resistance, then the current flow in the resistor will also have a sine waveform. Suppose, for example, that you apply the sine-wave voltage of Fig. 16-4 to a light bulb. You can see that when the waveform is at points 0, *b*, and *d*, the actual value of applied voltage to the light bulb is 0 volt, and no current will flow through the light-bulb filament at those instants. On the other hand, when the waveform is at points *a* and *c*, the maximum voltage will occur, and the maximum amount of current will flow through the filament of the light. Thus, the filament current varies from zero to maximum and back to zero as the waveform of voltage varies. This means that the filament is actually being heated, and then cooled, continuously when an ac voltage is applied.

The filament does not actually cool off to room temperature at the points where the voltage is zero. It takes a while for the filament to cool, just as it takes a while for a pan to cool after it has been removed from the stove. At the same time, the filament does not become as hot as it would be if the maximum voltage were applied at all times. It is reasonable to expect, then, that the filament is heated to some value between what it would have been if 0 volt were applied and the heat that would occur if the maximum voltage were applied at all times.

We have a special name for the heating effect of an alternating current, or for the ac voltage that produces the current. It is called the *effective voltage* or the *rms voltage*. The name "rms" comes from the mathematical procedure for finding the rms value. If you took all the voltage values for a half cycle, as shown in Fig. 16-5, squared them, took the mean (or average), and then took the square root, you would have the effective or rms value. Thus, the rms value is the *root* of the *mean* of the *squares*. It is sometimes called the *root-mean-square value*.

Unless you are mathematically inclined, the process described for finding the rms value is tedious. Figure 16-6 shows a method by which we can find the rms voltage experimentally. Of course, this method is not as accurate as mathematical methods, but it is a very good way to learn the meaning of rms voltage and current. Here we have a switch which connects a light to either an ac voltage source or a dc source. The ac voltage source has a maximum (peak) value of E_M, and the dc voltage (V) applied to the light depends on the setting of the variable resistor R. This dc voltage is called V.

(You will not be able to measure E_M directly with most voltmeters, since voltmeters are calibrated to read the rms value. However, E_M is 1.414 times greater than the rms value, so it is a simple matter to take a voltage reading and multiply it by 1.414, and thus you will have the peak value.)

When the switch is in position A, so that the light filament is being heated by the ac voltage source, a certain amount of light will be given off. When you switch to point B, the amount of light is dependent upon the amount of dc voltage applied. You can adjust the variable resistor until the light gives off the same amount of light as occurred for the ac source.

It is difficult to compare the brightness of the light with the naked eye, but photographers use a little instrument called a *light meter*, which can accurately measure the amount of light given off. The experiment works better in a darkened room, and the light should be glowing at some brightness that is less

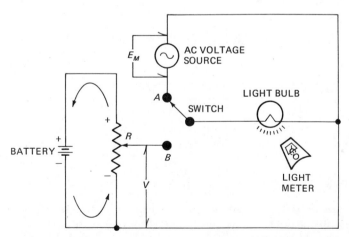

Figure 16-6. This circuit can be used to find the approximate value of rms voltage.

than its rated value. For our purposes we will assume that a light meter is being used to determine how much light is being given off. Once you have variable resistor R set, you should be able to adjust the switch from point B to point A without having any difference in light.

After the experiment is completed, if you measure the value of E_M and V, you will find that the dc voltage is only about $7/10$ of the peak value of ac voltage. In other words, the heating *effect* of the dc equals the heating effect of the ac whenever the value of dc is about $7/10$ of the peak value of ac. The heating effect of the ac voltage, then, is only about $7/10$ of the peak value. We can express this relationship mathematically as follows:

EFFECTIVE VOLTAGE = 0.7 × PEAK VOLTAGE

Mathematicians tell us that the exact multiple should be 0.707, but for most practical purposes it is not necessary to find the value with an accuracy of three decimal places.

When you measure an ac voltage with a voltmeter, or an alternating current with an ammeter, you are actually measuring the effective value. This is the value in which you are usually more interested. The readings of voltmeters and ammeters are based on the effective value of a sine wave. If you are measuring a voltage or current with a different waveform, you cannot use the reading from such a meter. Thus, if you are using an ac voltmeter or an ac ammeter to measure an ac current or voltage, *you must be sure that you are measuring in a circuit that has a pure sine-wave voltage or current!*

SUMMARY

1. When you apply a sinusoidal voltage to a resistor, a sinusoidal current flows.
2. A sinusoidal current through a light-bulb filament causes it to heat to some value between room temperature and the temperature that it would have if the maximum current flowed at all times.
3. The effective value, or rms value, is the value of dc voltage or current that would be needed to produce the same heating effect as a given ac voltage or alternating current.
4. For a pure sine wave,

EFFECTIVE VOLTAGE = 0.707 × PEAK VOLTAGE

and

EFFECTIVE CURRENT = 0.707 × PEAK CURRENT

5. Most ac voltmeters and ac ammeters are designed to read the rms value. These instruments will give an accurate reading only when measuring voltages and currents having sinusoidal waveforms.

HOW DO YOU CONVERT AN RMS VALUE INTO AN AVERAGE VALUE?

We have shown that you can find the average voltage or current by taking 0.636 times the peak. We have also shown that you can find the rms value of voltage or current by taking 0.7 times the peak value. It is possible to learn the peak value if you know the average or rms value. Furthermore, the average value can be obtained from an rms value, and the rms value can be determined from the average value.

Table 16-1 is given for your convenience in converting from one value to another.

TABLE 16-1. CONVERSIONS FOR SINE-WAVE VOLTAGES AND CURRENTS

Multiply the	*By*	*To Get the*
Peak-to-peak value	½	peak value
Peak value	2	peak-to-peak value
Peak value	0.707	rms (effective) value
Rms (effective) value	1.414	peak value
Peak value	0.636	average value
Average value	1.57	peak value
Rms (effective) value	0.9	average value
Average value	1.1	rms (effective) value

Example 16-2 — A voltmeter indicates that the voltage across a certain load is 30 volts. What is the peak voltage across this load?
Solution — A voltmeter measures the effective value of voltage, so 30 volts is the rms voltage across the load. According to Table 16-1, you can convert from the rms value to the peak value by using the equation

$$\text{RMS VALUE} \times 1.414 = \text{PEAK VALUE}$$

Therefore,

$$30 \text{ volts} \times 1.414 = 42.4 \text{ volts}$$

The peak voltage across the load is 42.4 volts.

Example 16-3 — The current through a light bulb is measured with an ac ammeter and found to be 0.83 ampere. What is the average value of this current?
Solution — An ac ammeter measures the rms value of current. According to Table 16-1 this value can be converted to the average current as follows:

$$\text{RMS VALUE} \times 0.9 = \text{AVERAGE VALUE}$$
$$0.83 \text{ ampere} \times 0.9 = 0.747 \text{ ampere}$$

The average value of current through the light bulb is 0.747 ampere.

It is a bother to write out "average value of voltage" and "effective value of voltage" each time you want to use these terms, so their symbols are usually used. Table 16-2 shows the symbols for the quantities that have been discussed in this chapter.

TABLE 16-2.   SYMBOLS FOR AC VOLTAGE AND ALTERNATING-CURRENT VALUES

	Voltage	Current
Peak-to-peak value	E_{p-p}	I_{p-p}
Peak value	E_{max}	I_{max}
Rms (or effective) value	E	I
Average value	E_{ave}	I_{ave}

WHAT IS MEANT BY THE TERM "PHASE"?

When you apply an ac voltage across a resistor, an alternating current flows through it. When the voltage reaches its maximum value, the current also reaches its maximum value at the same instant. When the voltage reaches the zero value, the current also reaches its zero value. In other words, the voltage and current are in step, or as we say in electricity, they are *in phase*.

Figure 16-7 shows what is meant by the term "phase." In Fig. 16-7a you will note that the voltage and current reach their maximum and minimum values at the same time. Under this condition, the voltage and current are in phase.

It is not always necessary in ac circuits for the voltage and current to be in step. In some cases the voltage reaches its maximum value before the current reaches its maximum value, and the voltage arrives at the zero value before the current reaches zero. When this happens, we say that the voltage is leading the current. Another way of saying this is that the current is lagging the voltage. You can think of it this way: if two people are walking and one person is ahead of the other, you can say that he is ahead, or you can say that the other man is behind. Likewise, it does not matter if you say that the voltage is leading the current, or if you say that the current is lagging the voltage.

Figure 16-7b shows the relationship between the current and voltage when

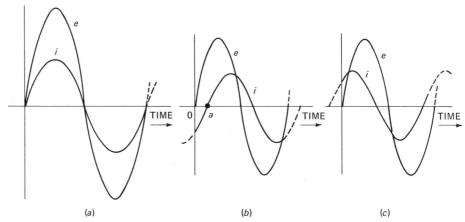

Figure 16-7.   One method of illustrating phase. (a) *The voltage and current are shown in phase.* (b) *The current is lagging behind the voltage.* (c) *The current is leading the voltage.*

the current is lagging behind the voltage. You will note that only one cycle of waveform is shown, but the dotted lines indicate that you are only looking at one cycle of a number of cycles that have been generated.

To clearly understand the phase relationship shown in Fig. 16-7*b*, you must understand that time on the graph becomes later as you move from left to right. Point *a* is to the right of point 0, and therefore, point *a* occurs at a later time. Of course, this time is very short, but nevertheless, it is later than the time at zero. We sometimes refer to the 0 point as the *origin* or the point at which we decided to start measuring time.

Instead of having the voltage ahead of the current, it is also possible for the voltage to lag behind the current. This situation is illustrated in Fig. 16-7*c*. The current reaches its maximum value before the voltage, and the current reaches its zero value before the voltage. Therefore, the current is leading the voltage. Another way of saying this is that the voltage is lagging behind the current.

Drawing the waveforms, as shown in Fig. 16-7, is one way of illustrating phase. There is another method, which is shown in Fig. 16-8. You could refer to the arrows shown in this illustration as *vectors*, but their correct name is *phasors*. Regardless of which name you call them, you can see that the current and the voltage in Fig. 16-8*a* are on the same line. This is an illustration of zero phase angle.

Figure 16-8*a* represents the same situation as shown in Fig. 16-7*a*. Figure 16-8*b* shows a representation of the current lagging behind the voltage. The angle by which the current lags the voltage is marked *a* in this illustration. To be technically correct, you would say the phase angle between the voltage (*e*) and current (*i*) is *a*, or the current lags the voltage by an angle of *a* degrees. (Of course, you could say the voltage leads the current by an angle of *a* degrees also.) The method of showing phase in Fig. 16-8*b* represents the same relationship between voltage and current as shown in Fig. 16-7*b*.

Figure 16-8*c* shows the current leading the voltage by an angle *b*. This is the same situation as shown graphically in Fig. 16-7*c*. Again, you can say that the current is leading the voltage, or you can say that the voltage is lagging the current.

Although both illustrations show the same thing, the one in Fig. 16-8 is easier to draw than the one in Fig. 16-7.

HOW DOES AN AC MOTOR WORK?

Figure 16-9 shows the relationship between the electromagnetic flux and the electric circuit that produces it. A current-carrying coil is used for producing a strong magnetic field.

When the battery is connected as shown in Fig. 16-9*a*, the magnetic flux

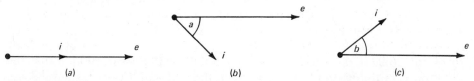

Figure 16-8. Another way of illustrating phase. (a) *Voltage and current in phase.* (b) *The current is lagging the voltage.* (c) *The current is leading the voltage.*

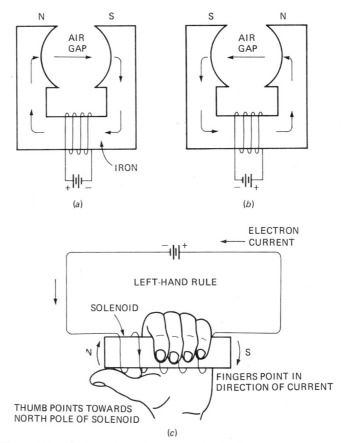

Figure 16-9. Flux lines in an electromagnetic circuit. (a) *The coil sets up a flux in the magnetic circuit.* (b) *Reversing the current causes the flux to reverse.* (c) *The left-hand rule for coils shows the direction of electromagnetic flux.*

lines of the coil are set up in the iron and across the air gap as shown by arrows. The flux lines do not actually move, but rather, the arrows represent the north-to-south direction and the path of the flux. The flux path is called a *magnetic circuit.*

Note that the flux path in Fig. 16-9*a* is from left to right across the air gap. Since the direction of flux is north to south, the north pole is on the left and the south pole is on the right.

In Fig. 16-9*b* the connection of the battery has been reversed. This causes the direction of the flux to reverse, as shown by the arrows. Now the north pole is on the right, and the south pole is on the left.

There is a simple method of determining the actual direction of the flux lines that are produced by a current-carrying coil. This method is shown in Fig. 16-9*c*, and it is called *the left-hand rule for coils.* When the fingers are circled around the coil in the same direction as the electrons are flowing, the thumb points in the direction of the flux—that is, north to south. If you grasped the coil of Fig. 16-9*a* so that your fingers circled the coil in the direction of the current flow, your thumb would be pointing to the left; and if you grasped the coil in Fig. 16-9*b* so that your fingers circled the coil in the direction of electron current flow, your thumb would point to the right.

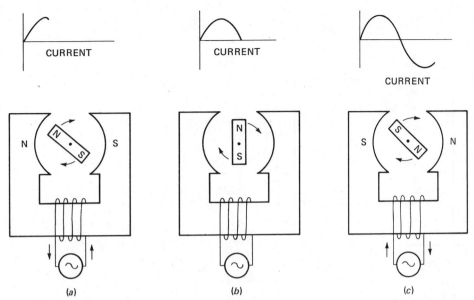

Figure 16-10. A very simple type of ac motor. (a) **When the current flows during this half cycle, the permanent magnet turns.** (b) **No current flows at this instant, but the bar magnet continues to turn.** (c) **The coil current has reversed, and the permanent magnet is again being forced to turn.**

The important thing about the magnetic circuits in Fig. 16-9 is that the direction of the magnetic flux depends directly upon the direction of current flow through the coil that produced it. You can reverse the flux simply by reversing the coil current. (You can also reverse the flux by reversing the directions of coil windings around the magnetic circuit, but in practical circuits this is not possible.) You can conclude from Fig. 16-9 that an alternating current flowing through the coil will cause the polarity of the magnetism at the air gap to periodically reverse.

Figure 16-10 shows the simple magnetic circuit with an ac voltage applied to the coil. A small bar magnet, which is free to rotate, has been inserted between the poles in the illustration. Figure 16-10*a* shows what happens during the first half cycle of alternating current. Since the south pole of the bar magnet is adjacent to the south pole of the electromagnet, and the north pole is adjacent to the north pole of the electromagnet, the bar magnet will rotate. This is because like poles repel.

When the current in the coil has reached zero at the end of the first half cycle, the amount of magnetic flux in the circuit also reaches zero. However, the permanent magnet continues to rotate during this period because it has inertia (Fig. 16-10*b*).

During the negative half cycle of current, shown in Fig. 16-10*c*, the magnetic poles across the air gap have reversed. At the same time the permanent magnet has moved into the position shown. Note that the like poles are again near to each other. This produces a turning effect in the bar magnet.

In order for the simple motor of Fig. 16-10 to work, it is necessary for the permanent magnet to turn at exactly the right speed so that the like poles of the permanent magnet and electromagnet are adjacent at just the right moment. By increasing the number of electromagnetic poles, you can alter the *speed* of the rotating permanent magnet.

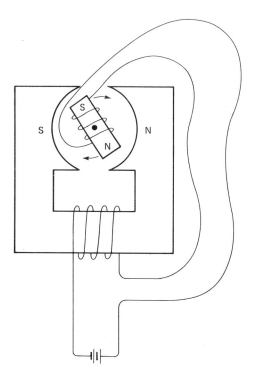

Figure 16-11. When an electromagnet instead of a permanent magnet is used, the turning force is greater.

To increase the *torque*—that is, the turning force—of the simple ac motor, an electromagnet is used instead of a permanent bar magnet for the air gap. This is illustrated in Fig. 16-11. Note that the coil that produces the flux in the air gap and the electromagnet that is turning in the air gap are actually two coils connected in series, and the same current flows through both of them.

Of course, this motor could turn through only a few revolutions before the wire would become completely wrapped in the motor, so in actual practice it is necessary to use *slip rings* for connection to the rotating coil. These are metal conductors used with brushes. (Figure 16-13 shows the type of slip rings used in both ac motors and ac generators.) Slip rings are not shown in Fig. 16-11. We are concerned here only with the theory that causes the armature to turn. If you employ the left-hand rule for coils in Fig. 16-11, you will see that the two north poles and the two south poles are correctly labeled on the illustration.

If you reverse the battery current, the direction through both coils reverses at the same time. This causes the two north poles to be at the left side and the two south poles to be at the right side. However, the situation will be the same. In other words, the like poles will be repelling and causing the armature to turn.

From the simple illustration of Fig. 16-11 you can conclude that an ac voltage will cause the motor to turn if there is a slip-ring arrangement to deliver current to the rotating coil.

A simple series-wound dc motor will operate on alternating current. A shunt-wound motor will not run on ac simply because the field coil has a very high resistance and opposition to alternating-current flow.

Figure 16-12 shows how an alternating current can cause the armature to rotate in a simple series-wound dc motor. In Fig. 16-12*a* the current is flowing through the field coils in such a way as to cause the north pole at the air gap to be on the right and the south pole to be on the left. The current flows through

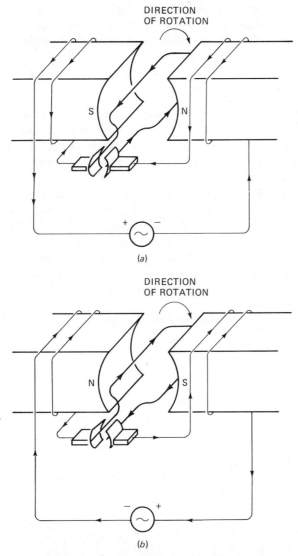

Figure 16-12. *A series-wound dc motor will run on ac.* (a) *During this half cycle of ac the armature is caused to turn clockwise.* (b) *During this half cycle the field and armature currents are reversed. The armature continues to turn in the same direction.*

the commutator and through the armature wire in the direction of the arrows. Use of the left-hand rule for coils will verify that the north and south poles of the field are correctly drawn, and use of the right-hand motor rule will show that at the instant shown in Fig. 16-12*a* the armature is being forced to turn clockwise as viewed from the commutator.

The next half cycle of alternating current is shown in Fig. 16-12*b*. The direction of current through both the field-coil windings and also the armature have reversed. Again, the left-hand rule for coils will verify that the north and south poles are correctly drawn, and the right-hand motor rule will show that the direction of rotation is again clockwise.

When a series-wound motor is designed to run on both dc and ac, it is called a *universal motor*. In actual practice slight changes are made in the magnetic circuitry to reduce the losses when an alternating current flows through the field coil. These changes are minor, and the basic theory of operation is not altered.

HOW DOES AN AC GENERATOR WORK?

In the previous chapter you learned that a wire rotating in a permanent magnetic field will cause a voltage to be induced which periodically reverses its polarity. The rotating inductor will normally induce an ac voltage, and in order to have a dc generator a commutator is needed to periodically switch the output of the armature.

In order to produce an ac voltage the same rotating armature setup can be used. However, instead of using a commutator, two slip rings are used to remove the voltage from the rotating conductor. This is illustrated in Fig. 16-13. With the slip-ring arrangement shown, one of the generator output leads is always connected to one half of the armature, and the other generator lead is always connected to the other half of the armature. Therefore, the voltage across the output will continually reverse direction when the armature rotates in the magnetic field.

An ac generator will not necessarily turn like a motor when an ac voltage is applied to its leads.

SUMMARY

1. When the voltage and current of a circuit are in phase, they both reach zero and their maximum values at the same instant.

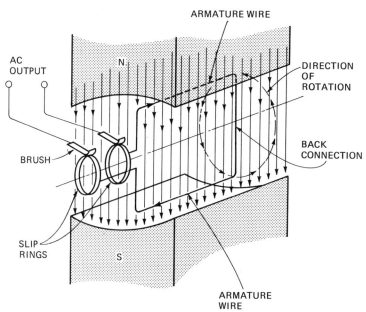

Figure 16-13.   A basic ac generator. It is similar to a dc generator except that slip rings are used instead of a commutator.

2. Vector-type arrows that represent the phase difference between voltage and current are called *phasors.*
3. A series-wound motor will run when an ac voltage is applied.
4. A series-wound motor designed to operate on either ac or dc is called a *universal motor.*
5. Ac generators have their armatures connected to slip rings rather than to a commutator.

EQUATIONS RELATED TO THIS CHAPTER

We have discussed average, effective, and peak values of ac voltage and alternating current. Although Table 16-1 gives the relationship between these values of voltage and current, we have not shown their relationship in equation form. The related equations are given here for your convenience.

RMS Voltage and RMS Current

$$E = 0.707\ E_{max} \qquad I = 0.707\ I_{max}$$
$$E = 1.1\ E_{ave} \qquad I = 1.1\ I_{ave}$$

Average Voltage and Average Current

$$E_{ave} = 0.636\ E_{max} \qquad I_{ave} = 0.636\ I_{max}$$
$$E_{ave} = 0.9\ E \qquad I_{ave} = 0.9\ I$$

Peak Voltage

$$E_{max} = \frac{E_{p-p}}{2} \qquad I_{max} = \frac{I_{p-p}}{2}$$
$$E_{max} = 1.414\ E \qquad I_{max} = 1.414\ I$$

Programmed Review Questions

(Instructions for using this programmed section are given in Chap. 1.)

We will review the important concepts of this chapter. If you have understood the material, you will progress easily through this section. Do not skip this material, because some additional theory is presented.

1. An inverter is used for the purpose of
   A. changing ac to dc. (Proceed to block 7.)
   B. changing dc to ac. (Proceed to block 17.)

2. *The correct answer to the question in block 9 is* **B**. *The unit of measurement for frequency is named for Heinrich Hertz (1857–1894) a German physicist.* Here is your next question.
   In Fig. 16-14,
   A. phasor *a* leads phasor *b*. (Proceed to block 13.)
   B. phasor *b* leads phasor *a*. (Proceed to block 8.)

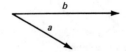

Figure 16-14. Which phasor is leading?

3. *Your answer to the question in block 17 is* **B**. *This answer is wrong. There are many components and circuits that will change a small voltage into a large one. An example is the transformer which you will study in a later chapter. A rectifier is* **not** *a component that will change a small voltage into a larger voltage.* Proceed to block 9.

4. *The correct answer to the question in block 8 is* **A**. *However, it is not a good practice to operate a series-wound dc motor on ac, because the losses in the magnetic circuit are too high. A series-wound motor designed for operation on either dc or ac is called a* **universal motor**. *Here is your next question.*
    In Fig. 16-15,
    A. waveform *a* leads waveform *b*. (Proceed to block 14.)
    B. waveform *b* leads waveform *a*. (Proceed to block 12.)

5. *Your answer to the question in block 14 is* **A**. *This answer is wrong. Table 16-1 shows how the rms voltage can be converted to the peak voltage. Rework the problem.* Then proceed to block 11.

6. *The correct answer to the question in block 11 is* **170 volts**. *(By actual multiplication the value is slightly less, but it rounds off to 170 volts.)*

    $$RMS \ VOLTAGE \times 1.414 = PEAK \ VOLTAGE$$
    $$120 \ volts \times 1.414 = 170 \ volts \qquad Answer$$

    Here is your next question.
    What is another name for rms current? (Proceed to block 20.)

7. *Your answer to the question in block 1 is* **A**. *This answer is wrong. A* **rectifier** *converts ac to dc.* Proceed to block 17.

8. *The correct answer to the question in block 2 is* **B**. *It is easy to determine which phasor is leading and which is lagging. Just imagine that the phasors are turning together in a counterclockwise direction, and ask yourself which phasor is ahead. Here is your next question.*
    Which of the following will run with an ac voltage applied?
    A. A series-wound dc motor (Proceed to block 4.)
    B. A shunt-wound dc motor (Proceed to block 18.)

9. *The correct answer to the question in block 17 is* **A**. *A rectifier allows current*

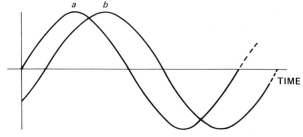

Figure 16-15. Which waveform is leading?

to flow through it easily in one direction, but will not allow it to flow in the opposite direction. Since current can flow through it only in one direction, it is a form of dc. A diode is an example of a rectifier. Here is your next question.

The frequency of an ac voltage or an alternating current is now measured in

A. cycles per degree. (Proceed to block 15.)
B. hertz. (Proceed to block 2.)

10. *The correct answer to the question in block 11 is* **C**. *The waveform is non-sinusoidal, and the method of finding the rms value involves calculations not given in Table 16-1. It is very important to understand that you could not measure the rms value with an ordinary ac voltmeter because it is not a sinusoidal waveform.* Here is your next question.

The ac power line delivers 120 volts to your home. This is the voltage as measured by an ac voltmeter. What is the peak value of this voltage? (Proceed to block 6.)

11. *The correct answer to the question in block 14 is* **B**. *According to Table 16-1, the peak value is determined by multiplying the rms voltage times 1.414.*

$$RMS\ VALUE \times 1.414 = PEAK\ VALUE$$
$$115\ volts \times 1.414 = 162.6\ volts$$

This peak voltage of 162.6 volts exceeds the maximum allowable peak value of 130 volts that can be placed across the diode, so the correct answer for the question in block 14 is **B**. Here is your next question.

The peak-to-peak voltage of the waveform in Fig. 16-16 is 10 volts. The rms value of the voltage

A. is 7.07 volts. (Proceed to block 19.)
B. is 3.58 volts. (Proceed to block 16.)
C. cannot be determined from the information given in Table 16-1 of this chapter. (Proceed to block 10.)

12. *Your answer to the question in block 4 is* **B**. *This answer is wrong. Study Fig. 16-7. Then proceed to block 14.*

13. *Your answer to the question in block 2 is* **A**. *This answer is wrong. Study Fig. 16-8. Then proceed to block 8.*

14. *The correct answer to the question in block 4 is* **A**. *Since time is later as you move to the right on the time line, it follows that waveform b is later than*

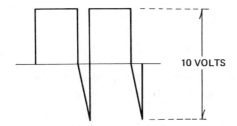

10 VOLTS

Figure 16-16. Waveform for the question in block 11.

(*behind*) *waveform* a. *Thus,* b *lags* a. *Another way of saying this is that* a *leads* b. Here is your next question.

Some components, such as diodes, are rated according to the peak voltage they can tolerate without being destroyed. If the voltage across the component exceeds the peak rating, the component will be destroyed.

If a certain diode has a peak voltage rating of 130 volts, can it be placed across an ac sine-wave voltage having an rms value of 115 volts?

A. Yes (Proceed to block 5.)

B. No (Proceed to block 11.)

15. *Your answer to the question in block 9 is* **A**. *This answer is wrong. There is no standard measurement of frequency in units of "cycles per degree."* Proceed to block 2.

16. *Your answer to the question in block 11 is* **B**. *This answer is wrong. You cannot use Table 16-1 to find the rms value of a nonsinusoidal waveform.* Proceed to block 10.

17. *The correct answer to the question in block 1 is* **B**. *Sometimes the manufacturers of inverters call them "converters." To be technically correct, though, you should use the name "converter" for a circuit that changes dc from one value of voltage to another value, and you should use the name "inverter" for circuits that change dc to ac.* Here is your next question.

A rectifier is used for the purpose of

A. changing ac to dc. (Proceed to block 9.)

B. changing a small voltage to a larger voltage. (Proceed to block 3.)

18. *Your answer to the question in block 8 is* **B**. *This answer is wrong. The field winding of a shunt-wound motor has a large number of turns. Its opposition to the flow of alternating current is high, and the motor will not work very well (if at all) on ac.* Proceed to block 4.

19. *Your answer to the question in block 11 is* **A**. *This answer is wrong. You cannot use Table 16-1 to find the rms value of a nonsinusoidal waveform.* Proceed to block 10.

20. *Another name for rms current is "effective current."*

You have now completed the programmed questions. The next step is to put some of these ideas to work in laboratory experiments. Proceed to the "Experiment" section that follows.

Experiment

(The experiment described in this section may be performed on the circuit board described in Appendix C or on a similar laboratory setup.)

EXPERIMENT 1

Purpose – The purpose of this experiment is to show that an rms voltage across a light will produce the same amount of brightness as a dc voltage across the same light, providing the voltage values are the same.

Theory – When you measure an ac voltage with an ac voltmeter, you are actually measuring the rms value. A dc voltage of the same value should produce the same amount of brightness when placed across the same light. This follows because the rms value is the effective value of ac. In other words, the rms value tells us what heating effect it will have on a light filament. An ac voltage of 6 volts (rms) should produce the same amount of light as a dc voltage of 6 volts when they are placed across identical light bulbs.

In this experiment you will be operating 12-volt light bulbs with only 6 volts across their filaments. The filaments will glow, but not at their full rated brightness. This is desirable because the human eye can detect smaller changes

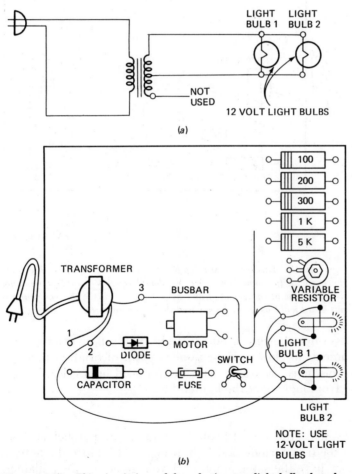

Figure 16-17. *This circuit is used for selecting two light bulbs that glow with the same brightness for a given voltage.* (a) *Circuit for matching light bulbs.* (b) *This is the way your circuit board will look if you have wired the circuit correctly.*

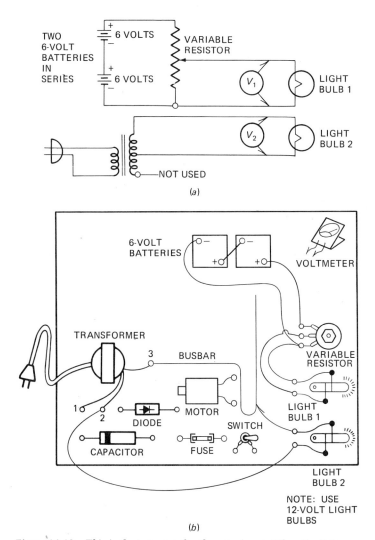

Figure 16-18. This is the test setup for the experiment. When the lights are glowing with the same brightness, V_1 should equal V_2. (a) Test setup for measuring the dc equivalent of an rms voltage. (b) This is the way your circuit board will look if you have wired the circuit correctly.

in brightness when the light is not as bright. You will be making adjustments to get two filaments to glow with the same amount of brightness, and you will be able to do this more accurately with the reduced light; also, there will be less annoyance from looking at the lighted bulbs.

Procedure —

Step 1 — Wire the two 12-volt light bulbs in parallel across the 6-volt winding of the transformer as shown in Fig. 16-17. Since the light bulbs are in parallel, they have the same amount of voltage across their filaments. Therefore, they should glow with the same brightness.

If the lights do not glow with the same brightness, try substituting other bulbs.

Step 2 — Wire the circuit as shown in Fig. 16-18. The ac voltage from the transformer is used for light 2, and two 6-volt batteries are used for light 1. The variable resistor is used as a potentiometer to control the dc voltage across light 1.

Step 3 — Adjust the variable resistor so that light 1 glows with the same brightness as light 2. Make this adjustment very carefully.

Step 4 — Measure the ac voltage across light 2. Make this measurement with the lights glowing. Record the rms value of voltage here.

$$\text{Ac voltage across light 2 } (V_2) = \underline{\hspace{2cm}} \text{ volts}$$

Step 5 — Measure the dc voltage across light 1. Make this measurement with both lights glowing. Record the dc value here.

$$\text{Dc voltage across light 1 } (V_1) = \underline{\hspace{2cm}} \text{ volts}$$

Conclusion — If you have performed this experiment carefully, the voltage measurements taken in Steps 4 and 5 should be the same. This shows that the rms value and the dc values of voltages produce the same heating effect.

Self-Test with Answers

(Answers with discussions are given in the next section.)
1. The rms value of voltage is also known as the (*a*) average value; (*b*) effective value.
2. If you know the peak value of voltage, you can find the rms value by multiplying by _____.
3. What is the average value of voltage of the 120-volt power line? (*a*) 117 volts; (*b*) 108 volts.
4. Frequency is measured in (*a*) hertz; (*b*) microseconds.
5. The time base for the graphical representation of a sine wave may be marked in seconds or (*a*) degrees; (*b*) hertz.
6. An ac ammeter normally measures (*a*) average current; (*b*) rms current.
7. If you know the direction of electron current in a coil, you can tell the direction of electromagnetic flux by using the (*a*) right-hand rule; (*b*) left-hand rule.
8. The maximum value of voltage during a half cycle of sine-wave voltage or current is called the (*a*) frequency; (*b*) amplitude.
9. A circuit that changes dc to ac is called (*a*) an inverter; (*b*) a rectifier.
10. A component that changes ac to dc is called (*a*) an inverter; (*b*) a rectifier.

Answers to Self-Test

1. (*b*)
2. 0.707
3. (*b*) — The voltage rating of the power line is 120 volts. This is the rms

value of the voltage. The rms value times 0.9 gives the average value.

$$120 \times 0.9 = 108 \text{ volts}$$ *Answer*

4. (*a*)—A microsecond is a millionth of a second. It is a unit of *time* measurement, not frequency measurement.
5. (*a*)—Hertz is a unit of frequency, or cycles per second. The time base is not marked in hertz.
6. (*b*)
7. (*b*)—See Fig. 16-9.
8. (*b*)
9. (*a*)
10. (*b*)

17.
How Do AC Measuring Instruments Work?

Introduction

In Chapter 10 you learned how a simple dc measuring instrument is constructed. The operation of this instrument is based on the fact that electron current flowing through a coil always produces a magnetic field. The field of the coil reacts with a permanent magnetic field, causing the coil to turn.

Figure 17-1 shows the simple dc meter. In Fig. 17-1a the current is flowing through the coil and produces a north-to-south pull as indicated on the drawing. The permanent-magnet north pole and the coil north pole are close together. Likewise, the two south poles are close together. Since *like magnetic poles repel*, the coil will turn. This moves the needle (that is, the pointer) up scale in the direction shown by the arrow. The amount that the needle turns is dependent upon the strength of the current flowing through the coil.

If the current in the coil is reversed, as shown in Fig. 17-1b, the coil will try to turn in a counterclockwise direction. This is because the unlike poles attract. The coil is prevented from moving counterclockwise by a needle rest. However, if the current is strong enough, it will push the needle so hard against the stop that the pointer will bend. For this reason, it is important that the current through the meter movement always be in the proper direction.

If the current through the coil in Fig. 17-1 varies rapidly, but always flows

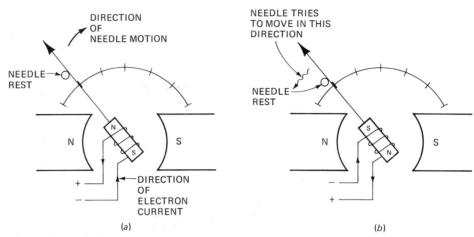

Figure 17-1.   The basic moving-coil meter movement used for making electrical measurements. (a) *When electron current flows in the direction shown, the meter pointer will move up scale.* (b) *When electron current flows in the direction shown, the pointer will press against the needle rest.*

in the same direction, the needle will not be able to move up and down scale to follow the rapid variations in current.

Figure 17-2 shows what happens when a rapidly varying current flows through the meter movement. A direct current through the coil causes the needle to deflect as shown. Varying resistance R slowly causes the needle to move back and forth between the extremes shown in dotted lines.

If you could vary the resistance R very rapidly, say 60 times per second, the needle could not follow the rapid changes in current. Instead, it would stay at the center, or *average*, point indicated by the solid black needle. This is because the meter movement has inertia.

The average value of an alternating current over a full cycle is zero. Thus, if you applied an alternating current to the movement shown in Fig. 17-1, the needle would stay at the zero mark. (Actually, you may be able to see the

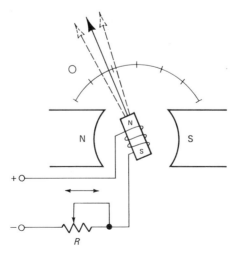

Figure 17-2.   When the current is varied rapidly, the needle cannot follow the variations, so it deflects to the average value.

needle vibrating rapidly in a very sensitive instrument, but it would vibrate around the zero mark.)

In order to measure an alternating current it is necessary to rectify it—that is, convert it to direct current—so that it can produce an upward motion of the needle on the meter scale. In this chapter you will learn about the circuits used for rectifying an alternating current in order to measure it with a simple instrument like the one shown in Fig. 17-1. These rectifier circuits are also used in many other applications.

You will be able to answer these questions after studying this chapter:

How is current rectified?
What is a full-wave rectifier?
What is a bridge rectifier?
How is a meter calibrated to read rms voltage?
How is a multimeter constructed?
How can power be measured?

Instruction

HOW IS CURRENT RECTIFIED?

A rectifier is a component that will allow current to flow through it in one direction, but not in the reverse direction. There are a number of components on the market (such as vacuum-tube diodes and semiconductors) that can rectify current. For the purpose of instrumentation, the rectifier normally used is the semiconductor diode, like the one shown in Fig. 17-3.

Figure 17-3*a* shows the symbol for a semiconductor diode. There are two ways that a voltage can be placed across a diode. When the voltage is placed as shown in Fig. 17-3*b*, so that the negative polarity of the voltage is at the cathode and the positive voltage is at the anode, electron current will flow through the diode as shown by the arrow. The voltage across a diode is called the *bias voltage,* and when the diode is biased as shown in Fig. 17-3*b*, it is said to be *forward-biased.*

If you reverse the voltage across the diode as shown in Fig. 17-3*c,* so that the positive voltage is applied to the cathode and the negative voltage is applied to the anode, current will *not* flow through the diode. Electrons simply cannot pass through the component in the direction of the wavy arrow. When a voltage is applied to a diode in this manner, it is said to be *reverse-biased.*

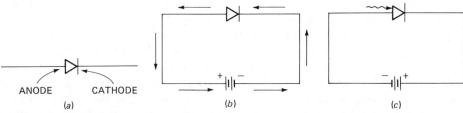

Figure 17-3. Comparison of forward and reverse bias for a diode. (a) *Symbol for a semiconductor diode.* (b) *Forward bias for a diode.* (c) *Reverse bias for a diode.*

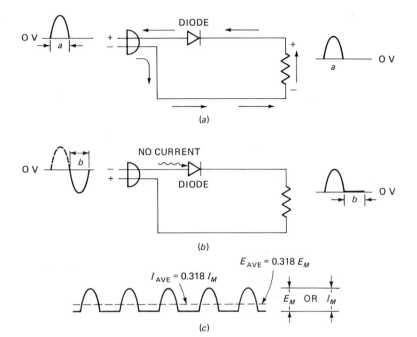

Figure 17-4. *The half-wave rectifier circuit produces a waveform with a low average value. (a) Current flows in the circuit during this half cycle of power. The current through the resistor causes a voltage drop across it. (b) No current flows during this half cycle of input power, and therefore, there is no voltage drop across the resistor. (c) This is the waveform of both the current and the voltage.*

Figure 17-4 shows how a diode can be connected into a circuit to convert the ac power line voltage into a form of dc voltage. In Fig. 17-4a the half cycle of input voltage is such that a positive voltage is applied to the anode side of the diode. The diode is forward-biased. This causes the diode to conduct, and electron current flows in the direction shown by the arrows.

When the electron current flows through the resistor in the direction shown, there will be a voltage drop across that resistor. This is true because there is *always* a voltage drop when an electron current flows through a resistor. The negative side of the voltage drop is always at the side of the resistor where the current enters the resistor, and the positive polarity is always at the point where the electrons leave the resistor. The polarity of the voltage across the resistor during this half cycle is shown in Fig. 17-4a.

During the half cycle shown in Fig. 17-4a, the current starts at zero, rises to a maximum value, and then drops down to zero again. Since the current is varying at all times, the voltage across the resistor will also vary. This voltage is always directly related to the current. Note that the waveform of the voltage across the resistor is the same as the waveform of the input voltage. (It is also the same as the current waveform, which is not shown.)

In Fig. 17-4b the power line voltage has reversed so that the negative voltage is applied to the anode of the diode. The diode cannot conduct under this condition, and therefore, there is no current flowing through it. Since the resistor and diode are in series, the same current must flow through both components. If there is no current through the diode, then there is obviously no current through the resistor. Therefore, even though there is an input voltage

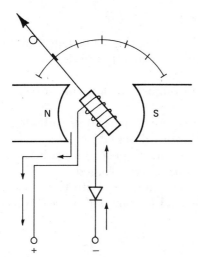

Figure 17-5. This diode circuit prevents current from going through the meter coil in the wrong direction.

applied at the plug, there is no current and no output voltage waveform during this half cycle. This is illustrated in the waveform of the voltage across the resistor for the period marked *b.*

If you were to observe the waveform of the voltage across the resistor in the circuit of Fig. 17-4 for several cycles, you would see that it consists of a series of impulses as shown in Fig. 17-4*c.* This is a series of half waves with the other half wave being eliminated by the diode.

You will remember that the needle of the meter deflects to the average value of the voltage and to the average value of the current flowing through the coil. If the waveform of Fig. 17-4*c* were applied to the coil in the meter movement, it would deflect to a value that is 0.318 times the maximum value. You should not expect the value to be 0.636 times the maximum. That does not apply here, since it was only for pure sine waveforms. The waveform in Fig. 17-4 is definitely not a pure sine wave.

The half-wave rectifier circuit shown in Fig. 17-4 is a very popular one, and it is used in many radios and other electronic equipment. It has the definite advantage of being simple and inexpensive. The disadvantage is that current flows in the load resistance for only one-half cycle of every input cycle. It would be better if current flowed through the load resistance at all times. In terms of measuring alternating current, if current flowed at all times, it would raise the average value of the current and produce a greater deflection of the meter.

The diode could be used for measuring alternating current with a simple moving-coil meter movement if connected into the circuit as shown in Fig. 17-5. If you apply the left-hand rule to the coil, you will see that current flowing in the direction shown will produce a magnetic field of the coil that reacts with the permanent magnetic field to produce an upward movement of the pointer. Current can flow in the circuit only in the direction shown by the arrows. During the half cycle of input in which the current tries to reverse, the diode prevents current from flowing. Therefore, when an ac voltage is applied to the terminals of the meter, the current flowing in the coil will be a form of direct current.

SUMMARY

1. A dc meter movement can be used for measuring ac voltage and alternating current. However, the ac must first be converted to dc.
2. A rectifier is used for converting ac to dc.
3. A diode will allow current to flow through it in only one direction. It is useful as a rectifier.
4. When the anode of a diode is positive with respect to its cathode, current flows through it, and it is said to be "forward-biased."
5. When the anode of a diode is negative with respect to its cathode, no current flows through it, and it is said to be "reverse-biased."
6. A half-wave rectifier converts ac to a pulsating dc. With this type of rectifier circuit, current flows through the load for one half of each full cycle of input power.

WHAT IS A FULL-WAVE RECTIFIER?

It has been shown that a half-wave rectifier circuit allows current to conduct through the load for only one-half of each input cycle. This causes the average value of the current, and of the voltage across the load, to be low. A small amount of deflection results when a half-wave rectifier is used (with a moving-coil meter movement) to measure ac voltage or alternating current. It would be better if current flowed through the meter coil at all times. This can be accomplished by the use of a *full-wave rectifier.*

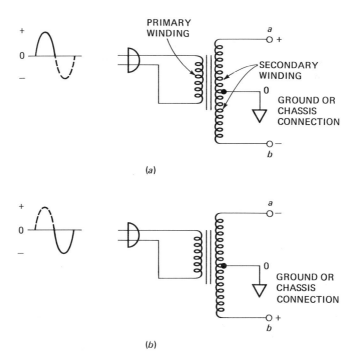

Figure 17-6. *Action of a transformer that has a grounded center tap on the secondary winding.* (a) *The polarity of voltage across the secondary during one-half cycle of input power.* (b) *On the next half cycle of input power the polarity of the voltage across the secondary changes as shown here.*

In order to understand how a full-wave rectifier works, we will first look at the circuit shown in Fig. 17-6. The transformer consists of two windings—the *primary winding* in which the power is applied, and the *secondary winding* from which the power is taken.

The secondary winding of this transformer has three leads. It is similar to the transformer used in the circuit board for the experiments described in this book. The center lead is called the *center tap* of the transformer. If this center lead is grounded, the voltage at the center will be zero at all times. This is true because ground is always considered to be 0 volt.

On one-half cycle of input power to the transformer, as shown in Fig. 17-6a, the output voltage across the secondary will be as shown. Note that there is a positive voltage at point a and a negative voltage at point b during this half cycle of input power. It is important to understand that voltages are relative. In other words, if the voltage at a is positive and the voltage at the center tap is zero, it is just another way of saying that the voltage at the center tap is negative with respect to the voltage at a. Likewise, if the voltage at b is negative with respect to the center tap, then the center tap is actually positive in respect to the voltage at point b.

On the next half cycle of input power, as shown in Fig. 17-6b, the voltage across the secondary of the transformer reverses. Now the voltage at point a is negative and the voltage at point b is positive. Again, this is just another way of saying that the voltage at a is negative with respect to the center tap, or the center tap is positive with respect to the voltage at point a. Likewise, the voltage at the center tap is negative with respect to the voltage at point b.

The significant thing about the transformer shown in Fig. 17-6 is that the voltage at point a and at point b alternates between positive and negative depending upon the polarity of the input voltage.

Figure 17-7 shows how the transformer of Fig. 17-6 can be used to make a full-wave rectifier circuit. Notice that the center tap transformer is connected to ground, and it is also connected to one side of the load resistor. The other side of the load resistor is connected to the cathode side of both diodes. The anode of one diode goes to point a, and the anode of the other diode goes to point b.

On one-half cycle, shown in Fig. 17-7a, point a is positive and point b is negative. Diode D_2 cannot conduct during this half cycle because of a negative voltage on its anode. In order for a diode to conduct, its anode must by positive with respect to its cathode. However, diode D_1 has a positive voltage on its anode at this time, and it is able to conduct. The path of electron flow is shown by the arrows in Fig. 17-7a. You have been instructed to trace circuits by starting at the negative terminal of the voltage. Actually, the center tap is marked 0 volt, and that makes it negative with respect to point a. Therefore, the center tap is the starting point for tracing electron current.

Starting at the center tap, the current flows through the load resistor, through diode D_1, ending at positive point a. Note that current does not flow through D_2 at this time because that diode has a negative voltage on its anode. In other words, D_2 is reverse-biased.

Figure 17-7b shows what happens during the next half cycle of input power. Note that the voltage across the secondary is reversed. Now the voltage at point a is negative. With a negative voltage on the anode of D_1, that diode is reverse-biased and cannot conduct. However, the positive voltage at point b means that there is a positive voltage on the anode of D_2, and it can conduct.

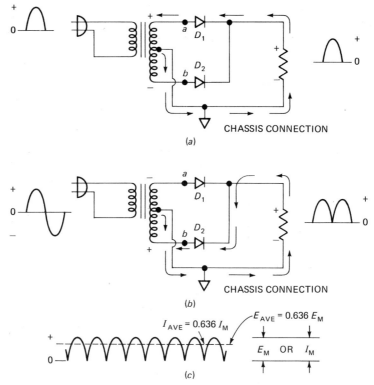

Figure 17-7. Operation of a full-wave rectifier. (a) *Output waveform for one-half cycle of power.* (b) *Output waveform for one complete cycle.* (c) *Output waveform for several cycles of input power.*

The conduction path is again shown by the arrows in this illustration. Starting at the center tap, which is actually negative with respect to the voltage at point *b*, and following the arrows, you see that the current flows through the resistor, through diode D_2, and back to the transformer. Current does not go through D_1 because that diode is reverse-biased.

Note that the current flowing in Fig. 17-7*b* is in the same direction as was shown in Fig. 17-7*a*. Therefore, there is a voltage drop across the resistor during both half cycles, and the voltage drop always has the same polarity.

Figure 17-7*b* shows two half cycles of output waveform for one cycle of input power. If you were to observe the waveform for several cycles of input power, it would look like the one shown in Fig. 17-7*c*. Each of the impulses shown in the waveform in Fig. 17-7*c* is produced during one-half cycle of the primary power.

The average value of voltage or current for a half cycle of a sine wave is 0.636 times the maximum value. Since the output waveform consists of a series of half cycles, its average value is 0.636 times the maximum value. This is illustrated by the dotted line in Fig. 17-7*c*.

You can readily see the advantage of having full-wave rectification if you want to use a rectified voltage for operating a meter movement. With full-wave rectification the average value is twice the average value for half-wave rectification. Since the needle deflects to an average value, it means that for a given input waveform the needle will deflect to twice as much deflection for full-wave rectification. This means that the meter will be more sensitive, and it can be used for measuring smaller voltages.

The disadvantage of the full-wave rectifier shown in Fig. 17-7 is that a transformer is needed. You have not studied transformers in detail at this time, so this may not appear to be much of a disadvantage, but transformers have several disadvantages that must be taken into consideration. *First,* they are rather bulky and expensive, especially if the transformer has to be large enough for a wide variety of input voltages. *Second,* transformers have a tendency to slightly distort the ac waveform. This distortion might be quite small for small amounts of currents, but it becomes considerable when the current increases. Distorting the waveform will change its average value, and that in turn will produce a reading error. Fortunately, it is possible to obtain full-wave rectification without the use of a transformer, although it is necessary to use two additional diodes.

SUMMARY

1. With a half-wave rectifier, current flows in the circuit for only one-half cycle of each input cycle. This means that the average value of current is low.
2. With a full-wave rectifier, current flows in the circuit for two half cycles of each input cycle.
3. The average value of a full-wave rectified current is higher than the average value of a half-wave rectified current.
4. When a transformer has a grounded center tap, the voltage across the secondary periodically reverses polarity. This type of secondary can be used with two diodes to make a full-wave rectifier.
5. The transformer-type full-wave rectifier has the disadvantage of being bulky and expensive, and it may distort the ac waveform.

WHAT IS A BRIDGE RECTIFIER?

Figure 17-8 shows a bridge rectifier circuit. As shown in Fig. 17-8*a* it is made with four diodes and a resistor. This rectifier will produce full-wave rectification of an ac voltage directly from the power line (or any other ac source) without the need for a transformer.

In Fig. 17-8*b* the input power is shown to be positive at one power input terminal and negative at the other. The arrows show the path of electron flow during this half cycle. The electron current passes through diode D_1, but it cannot go through D_4 because its anode is connected directly to the negative side of the source during this half cycle.

Continuing with the current path, it flows through the load resistor and through diode D_2. Current cannot flow through D_4 for the reason mentioned above. Furthermore, diode D_3 has its cathode connected directly to the positive side of the ac source during this half cycle, and therefore, it cannot conduct. To summarize, then, D_1 and D_2 conduct, while at the same time D_3 and D_4 are reverse-biased and cannot conduct.

On the next half cycle, as shown in Fig. 17-8*c*, the input power has reversed. The terminal that was positive during the last half cycle is now negative, and the terminal that was negative is now positive. Again, the arrows show

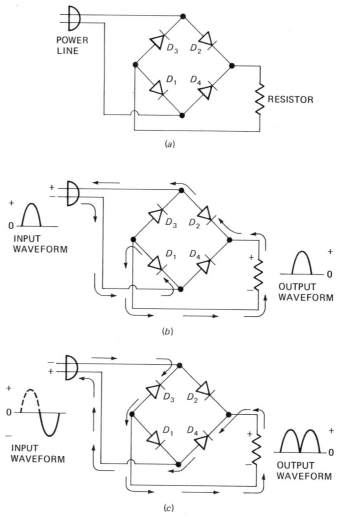

Figure 17-8.   The bridge rectifier circuit. (a) *This is a bridge rectifier circuit.* (b) *During one-half cycle of input power, current flows through* D_1 *and* D_2. (c) *On the next half cycle current flows through* D_3 *and* D_4.

the direction of electron flow. The electrons cannot flow through diode D_2 during this half cycle because its anode is connected to the negative terminal. In other words, it is reverse-biased. Likewise, diode D_1 has its positive cathode connected to the positive terminal, and being reverse-biased, it cannot conduct. However, diodes D_3 and D_4 are forward-biased, and the electron current flow is along the path shown by the arrows.

The output waveform of the full-wave rectifier shows that current flows through the load resistance for both half cycles of input power. Therefore, full-wave rectification is obtained. The bridge rectifier circuit requires the use of four diodes, but since diodes are much cheaper than transformers, this is not a serious disadvantage. In an ac meter, the ac voltage being measured can be applied to the input of the bridge rectifier and the rectified output delivered to the meter movement. Since current flows through the meter movement for both half cycles of input power, the average value of the bridge rectifier is greater than for a simple half-wave rectifier. This increases the deflection for a greater amount of ac input.

HOW IS A METER CALIBRATED TO READ RMS VOLTAGE?

You have learned that the amount of needle deflection of a meter movement is directly related to the average value of the current flow. The average value of voltage or current is of little value to a technician making measurements in a circuit. He is most often concerned with the rms, or effective, value. For this reason the manufacturer marks the scale reading on the meter to indicate rms values and ignores the fact that the deflection is proportional to the average value. The procedure for marking a meter scale for current indication of values is called *calibration*.

Suppose, for example, that the manufacturer wants the maximum reading on the meter scale to be 10 volts rms. The circuit of Fig. 17-9 can be used to calibrate the meter. Remember that the rms value of voltage is that value of dc that has the same effect as the ac voltage. In other words, if an ac voltage has an rms value of 10 volts, it will produce the same effect in a circuit as a dc value of 10 volts. A 10-volt battery is used to produce the 10 volts. A meter known to be accurate is used to measure the voltage to assure that it is correct. The meter being calibrated is connected across the meter of known accuracy. The voltage across all parts of a parallel circuit is the same, so the two voltmeters should read the same value. A mark can be made on the meter being calibrated to indicate the 10-volt mark. The process is repeated for different values of voltage until the scale of the meter has been marked. The procedure just described is called *calibrating* the meter.

It may be necessary to check the accuracy of a voltmeter. The setup shown in Fig. 17-9 can also be used for this purpose. A voltmeter of known accuracy is placed in parallel with the meter being checked. If the meter being checked does not give a correct reading, it is sometimes possible to make an adjustment in order to get it to read correctly. This is also known as calibration.

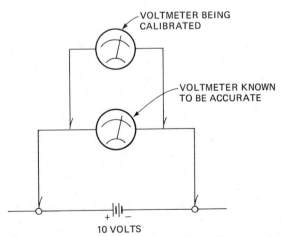

Figure 17-9. This setup can be used to calibrate a voltmeter.

HOW IS A MULTIMETER CONSTRUCTED?

A multimeter is a combination voltmeter, ohmmeter, and milliammeter constructed into a single package. If may also include an ammeter for measuring larger currents.

There are many versions of multimeters, and Fig. 17-10 shows a typical example. In the top row of resistors there is a meter jack marked "common." This jack is connected to the negative side of the meter movement, and it is used for one terminal of the voltmeter or milliammeter. In other words, one of the meter leads is connected to this common jack.

Resistors R_1, R_2, R_3, R_4, R_5, and R_6 are shunts for the meter movement when it is being used for measuring current. Suppose, for example, that the meter is being used to measure a current and the meter probes are connected to the "common" and "10-milliampere" jacks. For measuring current, these leads must be placed in series with the current being measured. If you redraw the circuit, you will see that it looks like the one shown in Fig. 17-11. Only the common and 10-milliampere jacks are shown in this illustration. Note in this equivalent diagram that resistors R_4, R_5, and R_6 form a shunt around the meter movement as required in current measurement. Resistors R_1, R_2, and R_3 are in series with the meter movement. The values of resistors in the series-parallel combination are selected so that a current of 10 milliamperes in the meter leads causes the needle of the meter movement to deflect to the maximum value on the scale.

The *maximum* current that can be measured with the meter probes in the position shown in Fig. 17-11 is 10 milliamperes. Figure 17-12 shows the scale for the multimeter. If a current larger than 10 milliamperes is measured, the needle will try to turn past the maximum point, and it will be damaged. However, a current less than 10 milliamperes can be measured. In order to use the

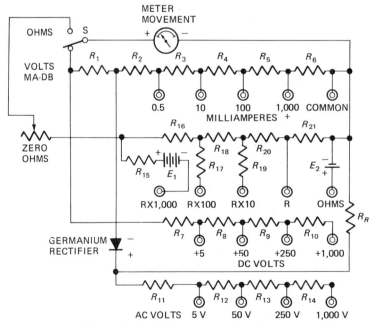

Figure 17-10. Basic circuit for a multimeter.

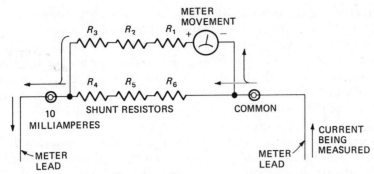

Figure 17-11. This is the circuit of the multimeter when it is being used to measure current. The meter is being used in the 10-milliampere range.

meter of Fig. 17-10 as a milliammeter it is necessary for the switch (S) to be in the position shown.

When a meter is used for measuring a dc voltage, one of the meter leads is connected to the common jack, and the other meter lead is connected to the appropriate dc voltage jack. Resistors R_7, R_8, R_9, and R_{10} are in series with the meter movement and the common terminal. The switch (S) must be in the position shown in Fig. 17-10.

In order to measure an ac voltage, the common jack at the negative side of the meter movement is used for one meter probe, and the other meter probe is connected into the appropriate jack in the ac voltage range. In this case resistors R_{11}, R_{12}, R_{13}, and R_{14} are used for meter multipliers, and a germanium rectifier serves as a half-wave rectifier for converting the ac to dc. The switch (S) must be in the position shown in Fig. 17-10.

To use the meter of Fig. 17-10 as an ohmmeter, the switch (S) must be turned into the *ohms* position. The jack marked "ohms" serves as one lead of

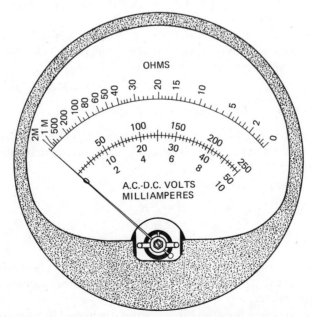

Figure 17-12. Meter scale for the multimeter.

the ohmmeter, and the other lead is connected to the appropriate jack—that is, to the R, R×10, R×100, or R×1,000 jack. Batteries E_1 and E_2 are used for supplying voltage to the unknown resistance being measured. The object is to cause a current to flow through the unknown resistance and then measure the voltage drop across it with the meter movement. However, the meter movement is calibrated to read ohms rather than volts. A "0 ohm" variable resistor is adjusted so that the meter will read zero when the ohmmeter leads are touched together. (This indicates a resistance of 0 ohm.)

Suppose, for example, that one ohmmeter lead is connected into the "ohms" jack and the other lead is connected into the terminal marked R. When the two leads are shorted together, the current will flow through R_{21}. The resulting voltage drop is measured by the meter movement in series with resistors R_{16}, R_{18}, and R_{20} and the "ohms adjust" variable resistor. The resistors in series serve as multipliers for the meter movement, permitting it to read the voltage across R_{21}. The 0-ohm resistor is first adjusted to read zero on the meter scale with the two terminals shorted.

Now suppose the leads are connected across an unknown resistance. This will decrease the current flowing through R_{21}, and therefore, decrease the voltage across it. The amount of voltage decrease depends upon the size of resistance placed between the terminals. The meter movement is calibrated to read the exact value of resistance placed across the terminal. You can verify for yourself that the R×10 and the R×100 positions of the ohmmeter work in a similar manner except that the number of resistors used for a multiplier has reduced.

When the lead is connected in the R×1,000 jack, an additional voltage source (E_1) is added in series with E_2 and resistors R_{16}, R_{18}, R_{20}, and R_{21}. The overall result is that a larger current is available for making resistance measurements.

An important thing to note about the simple meter movement of Fig. 17-10 is that one meter movement is used for measuring milliamperes, dc volts, ac volts, and resistance.

Since only one rectifier is used, the ac voltages being measured are half-wave rectified. In more expensive and elaborate meters, bridge rectifiers are used instead of single germanium rectifiers.

HOW CAN POWER BE MEASURED?

Power in an electric circuit is a measure of how fast energy is being expended or how fast work is being done. There are three mathematical equations for calculating power:

$$P = E \times I$$
$$P = I^2 R$$
$$P = \frac{E^2}{R}$$

The importance of these equations is in the fact that they show that you can determine the power if you know any two of the three measurements of voltage, current, and resistance.

Figure 17-13 shows how the power being dissipated by a resistor can be determined by voltage and current measurement. The voltage drop (V) is being

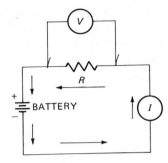

Figure 17-13. The power dissipated by the resistor can be determined by measuring the current (I) through the resistor and the voltage (V) across it. Then
$$P = E \times I \ or \ P = V \times I.$$

measured by the voltmeter, and the current through the resistor (I) by the ammeter. As shown in the list of power equations, multiplying the voltage times the current gives the power being dissipated. It would also be possible to find the power by measuring the current and then measuring the resistance and using the equation $P = I^2R$. Finally, the voltage across the resistor and the resistance value could be measured independently, and the equation $P = E^2/R$ could be used to determine the power dissipated.

Instead of using the technique shown in Fig. 17-13, it is also possible to measure the power with an instrument called the *wattmeter*. The wattmeter is actually a combination of two meters, a voltmeter and an ammeter. The needle in the meter is caused to deflect to a value which is proportional to the product of the voltage and the current, and the wattmeter scale is calibrated to read power in watts directly.

The power dissipated in a resistor is directly related to the amount of heat that the resistor generates when current flows through it. Another method of measuring the power is to measure the heat and then relate it to the actual power dissipated. This is called the *calorimetric method* of measuring power. It is used for measuring power in radio stations and other high-power applications.

The instrument that the power company puts in your house to determine the amount of your electric bill is *not* a wattmeter. Actually, it is a *watt-hour meter*. Its purpose is to measure the total amount of energy used from month to month. A wattmeter could tell only how much power is being dissipated at any instant of time, and this would not be useful to the power company. In other words, the power company wants to know how much power is used over a period of time.

SUMMARY

1. A bridge rectifier is a full-wave rectifier circuit that does not require a transformer for its operation.
2. Bridge rectifiers are ideal for use in ac measuring instruments because their output waveform has a high average value.
3. The procedure for marking a meter scale so that it properly shows the values being measured is called *calibration*.
4. A multimeter is a combination voltmeter, ohmmeter, and milliammeter. Some multimeters are also capable of measuring current in amperes.
5. A wattmeter measures the power dissipated in a circuit at any instant.
6. Some wattmeters are a combination of a voltmeter and an ammeter. The reading on the wattmeter gives the product of volts times amperes, which is the power in watts.

7. An indirect method of measuring power consists of measuring the heat that results from current flowing in a resistance. The heat is directly related to the power. This is called the *calorimetric method* of measuring power.
8. The meter that the power company installs in a house for determining the electric bill is *not* a wattmeter. It is a watt-hour meter.

Programmed Review Questions

(Instructions for using this programmed section are given in Chap. 1.)

We will review the important concepts of this chapter. If you have understood the material, you will progress easily through this section. Do not skip this material, because some additional theory is presented.

1. Which of the current waveforms shown in Fig. 17-14 is for a full-wave rectifier circuit?
   A. The waveform of Fig. 17-14*a*. (Proceed to block 7.)
   B. The waveform of Fig. 17-14*b*. (Proceed to block 17.)

2. *The correct answer to the question in block 10 is* **B**. *Since the diode is reverse-biased, no current will flow through it.* Here is your next question.
   A semiconductor diode can rectify an ac voltage. Another component that can rectify a voltage is
   A. a resistor. (Proceed to block 12.)
   B. a vacuum-tube diode. (Proceed to block 15.)

3. *The correct answer to the question in block 15 is* **B**. *A bridge rectifier circuit converts an ac voltage to a dc voltage. It is sometimes used in ac instruments. Its purpose is to convert the ac to dc so that it can be measured by a dc meter movement.* Here is your next question.
   If an ac sine-wave voltage is applied directly to a dc meter movement,
   A. the needle will not deflect, since the average value of a sine-wave voltage or current is zero. (Proceed to block 21.)
   B. the needle will deflect to the rms value. (Proceed to block 9.)

4. *The correct answer to the question in block 21 is* **A**. *The average value is 0.636 times the maximum value:*

$$AVERAGE\ VALUE = 0.636 \times MAXIMUM\ VALUE$$
$$= 0.636 \times 10\ volts$$
$$= 6.36\ volts$$

(a)   (b)

Figure 17-14. Output waveforms of two different kinds of rectifier circuits.

Here is your next question.

If a voltage is applied to a dc meter movement in such a way that it causes current to flow through the movement in the wrong direction,

    A. nothing will happen. It is a common practice to connect meters this way. (Proceed to block 20.)

    B. the meter movement may be damaged. (Proceed to block 8.)

5. *Your answer to the question in block 13 is **B**. This answer is wrong. While it is true that the amount of deflection of the needle is proportional to the average value of the waveform, the meter is normally calibrated to read the rms value.* Proceed to block 24.

6. *The correct answer to the question in block 16 is **B**. Although a square wave will cause the needle of an ac voltmeter to deflect, the reading of the meter will **not** be the rms value of that square wave. By using a complicated mathematical procedure, it is possible to determine the rms value of certain non-sinusoidal waves by the amount of deflection they cause on an ac meter. Since the sine wave is the most commonly encountered waveform in basic electricity, it is not necessary to learn the math procedure at this time.* Here is your next question.

Will current flow in the circuit of Fig. 17-15?

    A. Yes (Proceed to block 13.)

    B. No (Proceed to block 18.)

7. *The correct answer to the question in block 1 is **A**. The waveform shows that there is current flow for both half cycles of input power.* Here is your next question.

A watt-hour meter is used by the electric company to determine your electric bill. A watt-hour meter measures

    A. power. (Proceed to block 11.)

    B. energy. (Proceed to block 16.)

8. *The correct answer to the question in block 4 is **B**. If the wrong polarity of voltage is placed across a dc meter movement, the needle will slam against the needle stop and will surely be damaged!* Here is your next question.

A circuit that changes ac to dc is called a _____. (Proceed to block 26.)

9. *Your answer to the question in block 3 is **B**. This answer is wrong. In order to*

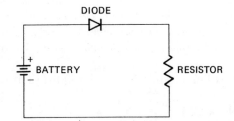

Figure 17-15. *Will current flow in this circuit?*

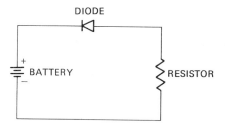

Figure 17-16.   Will current flow in this circuit?

use a dc meter movement to measure an ac voltage, it is necessary to use a rectifier. Proceed to block 21.

10.   *The correct answer to the question in block 24 is **A**. A diode is forward-biased whenever its anode is positive with respect to its cathode. This is always true, regardless of whether the diode is a vacuum tube or a semiconductor.* Here is your next question.
Will current flow in the circuit of Fig. 17-16?
A.   Yes (Proceed to block 25.)
B.   No (Proceed to block 2.)

11.   *Your answer to the question in block 7 is **A**. This answer is wrong. A watt-meter is used for measuring power, but a watt-hour meter is used for measuring energy.* Proceed to block 16.

12.   *Your answer to the question in block 2 is **A**. This answer is wrong. A resistor is a "bilateral component"—that is, it conducts current equally well in either direction. In order for a component to rectify, it must be able to pass current in one direction, but not in the reverse direction.* Proceed to block 15.

13.   *The correct answer to the question in block 6 is **A**. The only difference between the circuit of Fig. 17-15 and the circuit of Fig. 17-3b is that a resistor has been added in series with the diode. The resistor limits the amount of current flow, and therefore protects the diode from excessive current.* Here is your next question.
An ac voltmeter is normally calibrated to read
A.   the rms voltage value. (Proceed to block 24.)
B.   the average voltage value. (Proceed to block 5.)

14.   *Your answer to the question in block 21 is **B**. This answer is wrong. The average value of a sine wave is 0.636 times the maximum value. Reconsider your answer.* Then proceed to block 4.

15.   *The correct answer to the question in block 2 is **B**. A vacuum-tube diode does the same job as a semiconductor diode—that is, it passes current in one direction but not in the reverse direction. As with semiconductor diodes, current flows through a vacuum-tube diode from cathode to anode.* Here is your next question.
Which type of full-wave rectifier circuit does not require a transformer for its operation?
A.   Inverter (Proceed to block 22.)
B.   Bridge (Proceed to block 3.)

16. *The correct answer for the question in block 7 is* **B**. *Energy is defined as the capacity to do work, while power is defined as the rate of doing work or expending energy. The power company is not so much concerned with how fast you use power (as far as figuring your bill is concerned). Instead, they are interested in the total amount of energy delivered to your house over a period of time.* Here is your next question.

    Is this statement true or false? "An ac voltmeter can be used to measure the rms value of a square wave."
    A. True (Proceed to block 23.)
    B. False (Proceed to block 6.)

17. *Your answer to the question in block 1 is* **B**. *This answer is wrong. The waveform shown in Fig. 17-14b is identical to the one shown in Fig. 17-4c. Refer to this illustration and note that it shows the waveform for a half-wave rectifier.* Then proceed to block 7.

18. *Your answer to the question in block 6 is* **B**. *This answer is wrong. Compare the circuit of Fig. 17-3b. Note that the positive terminal of the battery is connected to the anode of the diode in each case.* Proceed to block 13.

19. *Your answer to the question in block 24 is* **B**. *This answer is wrong. When a diode is reverse-biased, as shown in Fig. 17-3c, no current will flow through it.* Proceed to block 10.

20. *Your answer to the question in block 4 is* **A**. *This answer is wrong. A voltage must never be placed across a meter movement that will cause the needle to deflect in the wrong direction. The meter movement could be damaged permanently as a result.* Proceed to block 8.

21. *The correct answer to the question in block 3 is* **A**. *The needle of the dc meter movement deflects to a value that is proportional to the average value of the waveform. The average value of a sine wave taken over a complete cycle is zero. Therefore, the needle will not deflect when a sine-wave voltage is applied.* Here is your next question.

    If the maximum value of a sine-wave voltage is 10 volts, the average value for a half cycle is
    A. 6.36 volts. (Proceed to block 4.)
    B. 5 volts. (Proceed to block 14.)

22. *Your answer to the question in block 15 is* **A**. *This answer is wrong. An inverter is actually a circuit that converts a dc voltage to an ac voltage. However, we have not discussed the use of an inverter in this chapter, and it is not the answer to the question in block 15.* Proceed to block 3.

23. *Your answer to the question in block 16 is* **A**. *This answer is wrong. An ac voltmeter will measure only the rms value of a sine wave.* Proceed to block 6.

24. *The correct answer to the question in block 13 is* **A**. *The rms value is also known as the "effective value." It is useful for determining power in ac circuits, and for other applications.* Here is your next question.

When a voltage is placed across a diode in such a way that current will flow through it, the diode is said to be
A. forward-biased. (Proceed to block 10.)
B. reverse-biased. (Proceed to block 19.)

25. *Your answer to the question in block 10 is **A**. This answer is wrong. You will note that the cathode of the diode is connected to the positive terminal of the battery. This means that it is reverse-biased and current cannot flow through it.* Proceed to block 2.

26. *The correct answer to the question in block 8 is **rectifier**. A **diode** is a component that allows current to flow through it in one direction only. It is a **rectifier**. The **circuit** that the diode is connected into may also be called a rectifier.* You have now completed the programmed questions. The next step is to put some of these ideas to work in laboratory experiments. Proceed to the "Experiment" section that follows.

Experiment

(The experiment described in this section may be performed on the circuit board described in Appendix C or on a similar laboratory setup.)

EXPERIMENT 1

Purpose—The purpose of this experiment is to demonstrate that the value of the voltage and current from a full-wave rectifier is greater than the value of voltage and current from a half-wave rectifier.

Theory—It has been shown in this chapter that the average value of a full-wave rectifier output is twice the average value of a half-wave rectifier output. In order to demonstrate this, the dc motor on the circuit board will be connected to the rectifier circuit through two light bulbs (which serve as current-limiting resistors). (See Fig. 17-17.) The circuit will first be connected to a half-wave rectifier and its speed noted. Then the full-wave rectifier will be added. The motor speed should be observed to increase with full-wave rectification due to the increase in average current through it.

Test Setup—Wire the circuit board as shown in the pictorial drawing of Fig. 17-17. The schematic representation of this circuit is also shown in Fig. 17-17. Note that diode D_2 is connected through a switch. When the switch is in the *off* position, the circuit is a half-wave rectifier. When the diode is switched into the circuit, the circuit becomes a full-wave rectifier.

Procedure—
Step 1—Wire the circuit as shown in Fig. 17-17.
Step 2—Insert the plug into the power outlet. With the switch in the *off*

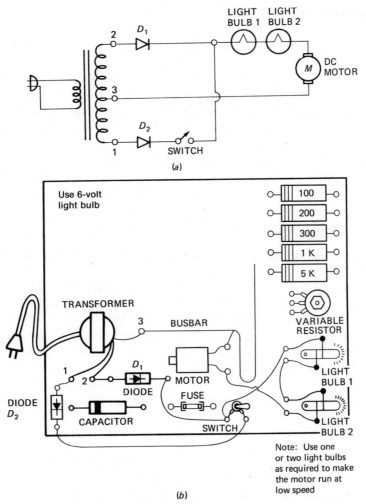

Figure 17-17. Circuit for comparing half-wave and full-wave rectification. (a) Circuit for the experiment. (b) This is the way your circuit board will look if you have wired the circuit correctly.

position note the speed at which the motor runs. With an ammeter, measure the motor current with half-wave rectification.

Step 3 — Close the switch so that both diodes are connected into the circuit as a full-wave rectifier. The speed of the motor should increase due to an increase in the average value of the current — that is, an increase in the average value of the full-wave rectifier current over a half-wave rectifier current.

Conclusion — Since the diodes do not generate a voltage, but rather, simply rectify the current, it can be concluded that the use of two diodes in the full-wave rectifier circuit has resulted in the increase in the average current. This is the cause of an increase in motor speed when the full-wave rectifier is used.

Self-Test with Answers

(Answers with discussions are given in the next section.)

1. In order to use a dc meter movement to measure an alternating current, you need a (*a*) fixed resistor; (*b*) rheostat; (*c*) rectifier; (*d*) switch.
2. When a current that is varying rapidly between two dc values is applied to a dc meter movement, the needle will point to (*a*) the rms value; (*b*) the average value; (*c*) the maximum value; (*d*) the minimum value.
3. Which of the following equations is not correct?
   (*a*) $I = ER$
   (*b*) $P = EI$
   (*c*) $R = \dfrac{E}{I}$
   (*d*) $P = I^2R$
4. Current will not flow in a diode if it is (*a*) forward-biased; (*b*) reverse-biased.
5. To forward-bias a diode, its (*a*) anode is made positive with respect to its cathode; (*b*) cathode is made positive with respect to its anode.
6. When an electron current flows through a resistor, there is *always* a voltage drop across the resistor. The negative side of the voltage drop is always at the side of the resistor where the electrons are (*a*) leaving the resistor; (*b*) entering the resistor.
7. A half-wave rectifier converts an ac voltage to a pulsating dc voltage. A disadvantage of the half-wave rectifier is that it (*a*) is too expensive; (*b*) is too large; (*c*) is only useful when the ac power is in the form of square-wave voltages and currents; (*d*) produces a pulsating output that has a low average value.
8. For use as an ac meter rectifier circuit, which of the following is not a disadvantage of a full-wave rectifier circuit that uses a transformer with a center-tapped secondary and two diodes? (*a*) It is expensive; (*b*) It is bulky; (*c*) It may distort the wave; (*d*) It produces an output waveform that has a higher average value than a half-wave rectifier.
9. Over a complete cycle of sine-wave voltage, the average value of the voltage is zero. Over a half cycle the average value is (*a*) 0.391 times the maximum value; (*b*) 0.518 times the maximum value; (*c*) 0.636 times the maximum value; (*d*) twice the maximum value.
10. A wattmeter is actually a combination of two instruments. They are (*a*) ammeter and ohmmeter; (*b*) ammeter and voltmeter; (*c*) voltmeter and ohmmeter; (*d*) ohmmeter and watt-hour meter.

Answers to Self-Test

1.	(*c*)	6.	(*b*)
2.	(*b*)	7.	(*d*)
3.	(*a*)	8.	(*d*)
4.	(*b*)	9.	(*c*)
5.	(*a*)	10.	(*b*)

18.
What Is Inductance and Capacitance?

Introduction

In order to understand how electric circuits work, it is necessary to have a clear idea of what the components in the circuits are used for. You have already studied one important component — the resistor. Resistors are used for a number of purposes in electric circuits:

To limit the amount of current flow in the circuit
To produce a voltage drop whenever current flows through the resistor
To radiate heat whenever current flows through the resistor

Resistors are used in both ac and dc circuits for the purposes listed above.
   In this chapter you will study two additional components — the *inductor* and the *capacitor*. They are used primarily in ac circuits. We will start the study of inductors and capacitors by giving basic definitions of each component.

An inductor is a component that opposes any change in the flow of current through it.
An inductor is a component that stores energy in the form of an electromagnetic field.
A capacitor is a component that opposes any change in voltage across its terminals.
A capacitor is a component that stores energy in the form of an electrostatic field.

The definitions for inductor and capacitor are important because they explain what the circuit component does in an electric circuit. Note that both components can store energy. In this chapter you will study how the components are able to perform their jobs.

It is not only important to know what resistors, inductors, and capacitors do in an electric circuit, and how they accomplish their job, but also necessary to know how these components are connected in combinations to perform certain jobs in electric circuits. After you have studied how inductors and capacitors work, you will then study a few examples of how these components are connected together to perform jobs such as tuning and filtering.

When you have a circuit made of resistors, inductors, and capacitors of the type that you buy in a parts supply store, then the circuit is said to be composed of *lumped components*. However, you should understand that it is possible to have circuits with resistance, inductance, and capacitance in which these properties are part of the circuit, but do not exist in separately mounted components. When resistance, inductance, and capacitance are dispersed throughout the circuit, rather than being mounted as individual components, the circuit is said to be composed of *distributed components*.

An example of a distributed component occurs in wire. You generally think of wire as being a conductor, but you must understand that all conductors have some electrical resistance. In short pieces of wire the resistance can be ignored, but in long stretches of wire the resistance becomes an important factor. The resistance of the wire is distributed throughout the length of the wire, and does not exist as a single component called a *resistor*. In other words, the resistance of a long piece of wire is a form of *distributed resistance*.

Figure 18-1 shows the schematic symbols for two types of inductors and two types of capacitors. These are symbols used to represent lumped components in a circuit. For distributed components the symbols are similar except they are made with dotted lines.

You will be able to answer these questions after studying this chapter:

What is inductance?
What is a henry?
What determines the inductance of a coil?
What is capacitance?
What is a farad?
What determines the capacitance of a capacitor?
How are resistors, inductors, and capacitors used in circuits?

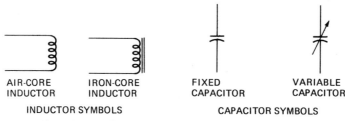

| AIR-CORE INDUCTOR | IRON-CORE INDUCTOR | FIXED CAPACITOR | VARIABLE CAPACITOR |

INDUCTOR SYMBOLS   CAPACITOR SYMBOLS

Figure 18-1. Schematic symbols for inductors and capacitors.

Instruction

WHAT IS INDUCTANCE?

Inductance may be defined simply as the property of a coil that enables it to oppose any change in current through it. To understand how inductance occurs, we will first review some very basic properties of electric circuits.

Figure 18-2 shows the magnetic field around the conductor which is carrying current. The direction of the magnetic field around the wire is easily determined by the left-hand rule, which states that if you grasp the wire (in your imagination) so that your thumb points in the direction of electron current flow, your fingers will circle the wire in the direction of the magnetic field. Applying the left-hand rule to the current-carrying wire of Fig. 18-2, it is seen that the direction of the magnetic field for the current in Fig. 18-2*a* is opposite to the direction of the current in Fig. 18-2*b*.

The strength of a magnetic field around a conductor is directly related to the amount of current through it. This is illustrated in Fig. 18-3. In Fig. 18-3*a* a small current produces a relatively small magnetic field around the wire. In Fig. 18-3*b* the strength of the current has been increased, and this increase in current strength is accompanied by an increase in the number of flux lines around the conductor.

Figure 18-4 shows an important relationship between voltages which are combined by addition or subtraction. In Fig. 18-4*a* the two batteries are connected in *series-aiding*. Note that the positive voltage of the 8-volt battery is directly connected to the negative terminal of the 6-volt battery. The two batteries combine their efforts to produce a total output voltage of 14 volts. For all practical purposes, then, the two batteries of Fig. 18-4*a* could be replaced by a single 14-volt battery as shown in the illustration.

In Fig. 18-4*b* the two batteries are connected *series-opposing*. In this case the negative terminal of the 8-volt battery is connected to the negative terminal of the 6-volt battery. The two voltages are connected in opposite directions, and their combined voltage is obtained by subtracting the two. Thus, the output

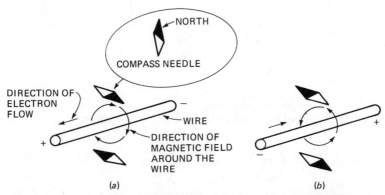

Figure 18-2. The direction of the magnetic field around a current-carrying wire depends upon the direction of electron flow. (a) *The direction of electron flow determines the direction of the magnetic field around the conductor.* (b) *Reversing the direction of current flow reverses the direction of the magnetic field.*

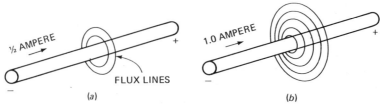

Figure 18-3. The strength of the magnetic field around a conductor depends upon the amount of electron current flow. (a) *With a small current the magnetic field around the wire is weak.* (b) *When the current is increased, the strength of the magnetic field around the wire is also increased.*

voltage is 2 volts, and the output voltage takes the polarity of the larger of the two batteries. For practical purposes the two batteries of Fig. 18-4*b* could be replaced by a single 2-volt battery as shown in the illustration.

Although the method of combining voltages shown in Fig. 18-4 is described in relation to dc voltages of batteries, it is also true for ac voltages at any single instant. For example, at some instant in an ac circuit two voltages may be applied simultaneously. These voltages might be 6 and 8 volts just as the two batteries in Fig. 18-4. If this should happen, then at the instant the voltages are being considered, the overall voltage is obtained by either adding or subtracting—depending upon whether the voltages at that instant are series-aiding or series-opposing.

One additional basic theory must be taken into consideration before continuing the study of inductance. Whenever there is relative motion between a conductor and a magnetic field, a voltage is always induced. This principle was introduced when the different methods of generating a voltage were discussed. It does not make any difference if the conductor is held at a fixed position and the magnetic field is moved across it, or if the magnetic field is at a fixed posi-

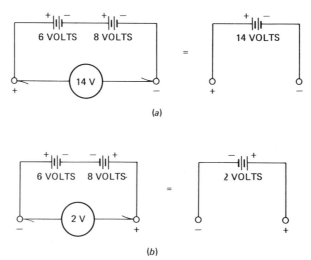

Figure 18-4. Two ways of connecting batteries in series. (a) *Batteries connected in series-aiding. The equivalent battery is shown on the right.* (b) *Batteries connected in series-opposing. The equivalent battery is shown on the right.*

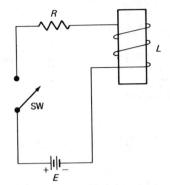

Figure 18-5. A simple R-L circuit.

tion and the conductor is moved through it. In either case, a voltage will always be induced in the conductor. The principle of inducing a voltage with a moving magnetic field is called *Faraday's law of magnetic induction.*

Having reviewed some basic properties of electric circuits, we can now apply them to the study of inductance.

The circuit of Fig. 18-5 will be used to describe the property of inductance. This circuit consists of a switch (*SW*), a current-limiting resistor (*R*), an inductor (*L*), and a dc source (*E*). Figure 18-6 shows what happens when the switch in the basic circuit is closed. The moment the switch is closed (Fig. 18-6*a*), the current through the circuit begins to increase. This causes the magnetic field around winding *a* to increase and cut across winding *b*. Likewise, the increase in current in winding *b* causes an expanding magnetic field which cuts across winding *a*. According to Faraday's law, *any* motion of a magnetic field across a conductor will produce a voltage in that conductor, and so a voltage will be produced in the windings. The voltage induced in the wires of the coil is called a *countervoltage.*

At any instant the countervoltage can be represented as a separate voltage in the circuit. This is done in Fig. 18-6*b*. It is important to note that the countervoltage is opposite in polarity to the applied voltage *E*. This means that the countervoltage is opposing the increase in current that occurs when the switch

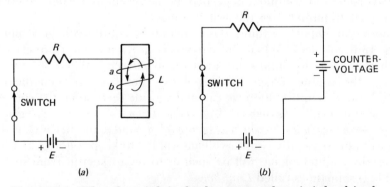

Figure 18-6. When the switch is closed, a countervoltage is induced in the windings. This countervoltage prevents the current from rising instantly to a maximum value. (a) *When the switch is closed, the current increases. The expanding magnetic field of each turn of wire cuts across the adjacent turns, introducing a countervoltage.* (b) *The countervoltage opposes the current.*

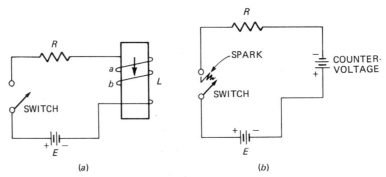

Figure 18-7.   Opening the switch reverses the polarity of the countervoltage. (a) *When the switch is opened, the magnetic field collapses. The collapsing field due to decreasing current in one turn cuts across each adjacent turn, introducing a countervoltage.* (b) *In this case, the countervoltage tries to keep the current flowing. This causes a spark across the switch contacts.*

is closed. If it was not for this countervoltage, the current in the circuit would rise immediately to its maximum value. The countervoltage is slightly less than the applied voltage, and the current will gradually increase in the circuit until it reaches its maximum value.

As soon as the current through the coil reaches its maximum value, the magnetic field is no longer increasing and the countervoltage drops to zero. Then the only limit to the current flow is resistor R, which is connected in series with the coil.

Figure 18-7 shows what happens when the switch is opened. This illustration assumes that the maximum current is already flowing in the coil at the moment the switch is opened. Opening the switch causes the current in the circuit to decrease rapidly, and the magnetic field around each turn of the coil collapses. The arrow on the coil of Fig. 18-7a indicates that the field is collapsing back to conductor b. The current in a is also decreasing, so its magnetic field is collapsing and cutting across conductor b. The overall result is that a countervoltage is induced in the windings. The difference in this case is that the polarity of the voltage is now in series-aiding with E as shown in Fig. 18-7b. The applied voltage and countervoltage may be large enough when combined to cause a spark to jump across the switch contacts.

There is an important thing to note about the countervoltage as illustrated in Figs. 18-6 and 18-7. When the current is *increasing*, the countervoltage is such that it tends to oppose an increase in current. When the current is *decreasing*, the countervoltage has a polarity that tries to keep the current flowing. From this information we can make a rule then about the countervoltage induced in a coil: *The countervoltage induced in a coil is always such that it tends to oppose any change in the current flowing through it.* Stated another way, if the current tries to decrease, the countervoltage will try to keep it going. If it tries to increase, the countervoltage will be such as to try to keep it from increasing. This rule is sometimes called *Lenz' law.*

From the description of the behavior of the coil, it is now possible to define inductance as *the property of a coil that enables it to store energy in the form of an electromagnetic field.* The field surrounds the coil when there is current flowing

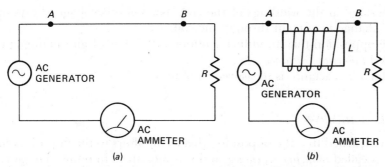

Figure 18-8. The effect of a coil on the flow of alternating current in a circuit. (a) *In this circuit, the only opposition to the flow of alternating current is* **R.** (b) *When the wire is coiled, it becomes an inductor. This causes a greater opposition to current flow, which is indicated by a decrease in current.*

through it. When the current stops, the field collapses and returns the energy.

Although the operation of inductors has been described by using circuits with batteries, it should be thoroughly understood that inductors are primarily ac components. The effect of inductance in an ac circuit is shown in Fig. 18-8. In Fig. 18-8a an ac generator is connected into a circuit containing a resistor and an ac ammeter. Current flowing through this circuit is limited only by the resistance of the resistor. In Fig. 18-8b the wire between points A and B has been wound into a coil. Nothing else in the circuit has been changed, but the current has been reduced by the addition of the coil. This illustrates that a coil will oppose the flow of alternating current. The opposition that the coil offers to the flow of alternating current is called *inductive reactance.*

It is easy to understand why an inductor opposes the flow of alternating current. Remember that alternating current is a current that is continually changing, and an inductor is a component that opposes any change in current through it. Inductive reactance is measured in ohms. However, you cannot add the inductive reactance and resistance values of a circuit to find the combined opposition to the flow of current. The combined opposition, which is called *impedance,* will be discussed in a later chapter.

SUMMARY

1. A resistor is a component that limits current flow, produces a voltage drop, or generates heat.
2. An inductor is a component that stores energy in the form of an electromagnetic field. An inductor opposes any change in current through it.
3. A capacitor is a component that stores energy in the form of an electric field. A capacitor opposes any change in voltage across it.
4. The direction of the magnetic field around a wire depends upon the direction of current flow through the wire.
5. The strength of the magnetic field surrounding a current-carrying wire is directly related to the amount of current.
6. Two voltages may be connected *series-aiding,* in which case their values are added. They may also be connected *series-opposing,* in which case their values are subtracted.
7. An increasing or decreasing current in a coil causes a countervoltage to be

generated in the windings of the coil. The countervoltage always opposes any change in current through the coil.

8.   The opposition that an inductor offers to the flow of alternating current is called *inductive reactance.*

9.   Inductive reactance is measured in ohms.

WHAT IS A HENRY?

You have learned that the opposition that a coil offers to the flow of alternating current is called *inductive reactance,* and it is measured in ohms. The amount of inductive reactance of a circuit is dependent upon the frequency of the alternating current and also upon the inductance of the coil.

The inductance of a coil, which is sometimes defined as a measure of the ability of the coil to oppose any change in current, is measured in units called *henrys.* A henry is defined as follows: *When a current through a coil is changing at a rate of one ampere per second, and this change in current produces a countervoltage of one volt, then the inductance of the coil is one henry.*

The henry unit is too large for most practical applications, so millihenrys and microhenrys are used instead. A millihenry is one-thousandth of a henry, and a microhenry is one-millionth of a henry.

WHAT DETERMINES THE INDUCTANCE OF A COIL?

In the circuit of Fig. 18-8 a piece of straight wire between points *A* and *B* was coiled to make an inductor. It was noted in the explanation of this circuit that coiling the wire to make an inductor causes the alternating current in the circuit to decrease. A logical question at this time would be: does it make any difference how many turns of wire you use for the inductor? The answer to the question would be *yes,* there is a direct relationship between the amount of inductance and the number of turns of a coil. What we are saying is that a coil of 10 turns has a greater inductance than a coil of 5 turns.

The inductive reactance of a coil is directly dependent upon its inductance. Therefore, increasing the number of turns of a coil increases the inductance, and also increases the inductive reactance.

The shape of the coil is another factor that determines its inductance. For example, a long thin coil has less inductance than a short fat one with the windings close together.

The inductance of a coil depends upon the magnetic field of each turn cutting across adjacent turns. By winding the coil on a piece of iron, the magnetic flux will flow through the center of the coil more easily. This means that the flux around each turn will cut across a greater number of adjacent turns than it would with an air core. With a greater number of turns being cut by the flux of each turn, the inductance of the coil is greater.

To summarize, then, the inductance of a coil depends upon the type of material used in the center of the coil. Coils with iron cores have more inductance than coils with air cores. It might seem from this that all coils would be wound on iron. However, as you will learn when you study transformers, some additional factors related to the core material must also be taken into consideration.

SUMMARY

1. The amount of inductive reactance in an ac circuit depends upon two things: the frequency of the alternating current and the inductance in the circuit.
2. The inductance of a coil is measured in henrys.
3. When the current through a coil is changing at the rate of 1 ampere per second, and this changing current causes a countervoltage of 1 volt, then the coil has an inductance of 1 henry.
4. The inductance of a coil depends upon the number of turns of wire that it has, the type of core material in the center of the coil, and the shape of the coil.

WHAT IS CAPACITANCE?

In the early days of experimenting it was believed that electricity was a form of fluid. This impression of electricity is still evident today from some of the terms used to describe it. For example, we talk about an electric "current," "current flow," and electrical "pressure" (voltage).

The idea that electricity is a fluid may have been responsible for the construction of the first capacitor. The story goes that the inventors were trying to store electricity in jars (the same way as water can be stored). You may be surprised to learn that the experiments were successful. A specially constructed jar—called a *Leyden jar*—was able to store a considerable amount of electrical charge.

Figure 18-9 shows the cross-sectional view of the Leyden jar. It gets its name from the fact that it was invented at the University of Leyden. You will see from Fig. 18-9 that it is a glass jar with metal foil on the inside and outside. A metal rod passing through an insulating cork makes connection with the inside foil. The Leyden jar is actually a form of capacitor. When a voltage is applied across the metal rod and the outside metal foil, as shown by dotted lines, a charge of electricity is actually stored in the Leyden jar.

Leyden jars are not used today in practical electric circuits, but capacitors

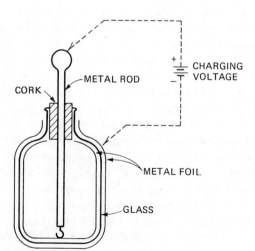

Figure 18-9. Cross section of a Leyden jar.

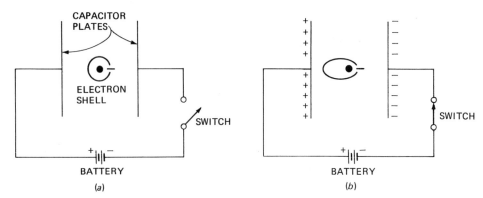

Figure 18-10.   How charging a capacitor affects the electrons in the dielectric. (a) *The electron shell in the dielectric is circular when the capacitor is not charged.* (b) *Charging the capacitor distorts the electron path.*

are used extensively. A capacitor, like the Leyden jar, consists of two conductors separated by an insulator. The conductors are called *plates*, and the insulator between the plates is called the *dielectric*. Electrical energy is stored in capacitors by placing a voltage across the plates. This is called *charging the capacitor.*

To understand how a capacitor becomes charged, consider the cross-sectional area of a capacitor as shown in Fig. 18-10. This capacitor consists of two parallel metal plates separated by a dielectric. As shown in Fig. 18-10*a*, an electron moving in its shell around the nucleus of an atom in the dielectric has a circular orbit.

When the switch is closed, as shown in Fig. 18-10*b*, electrons at the negative pole of the battery move to one plate of the capacitor. Electrons on the other plate of the capacitor are attracted away from the plate by the positive terminal of the battery. This leaves a surplus of electrons on one plate and a deficiency of electrons on the other plate, as indicated by the minus and plus signs. The surplus of electrons on the negative plate repel the electron that is moving around its nucleus in the dielectric. When this happens, the electron in the shell is pulled out of its circular orbit, making an elliptical orbit as shown.

If you open the switch, the surplus of electrons on one plate and deficiency of electrons on the other plate become trapped, and the capacitor is said to be charged. The process of charging the capacitor, then, is simply the process of making one of the plates negative with respect to the other. The electrons in the dielectric material retain their elliptical orbits like the one shown. The fact that electrons are pulled out of their circular orbits in the dielectric represents energy stored in the capacitor. An ellipse is an unnatural orbit for an electron, and it will try to return to its circular orbit if given a chance. However, the electrons are held in elliptical orbit as long as there is a difference in charge in the plates.

The fact that there is a charge on the plates of the capacitor indicates that the capacitor is capable of delivering current to a circuit under the right conditions. Figure 18-11 shows the condition for removing the charge from the capacitor. The process is called *discharging the capacitor.* In Fig. 18-11*a* the capacitor has been charged, and the charge is trapped on the capacitor. The electrons in the dielectric material are in an elliptical orbit.

When the switch of the circuit is closed, as shown in Fig. 18-11*b*, there is a path of current flow from the negative plate of the capacitor back to the posi-

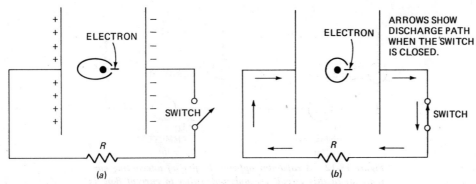

Figure 18-11. *Discharging a capacitor is the process of removing the stored energy.* (a) *A charged capacitor.* (b) *When the switch is closed, the capacitor discharges through the resistance.*

tive plate. The arrows in the illustration show the path of electron flow when the switch is closed. When enough electrons have moved from the negative plate to the positive plate to neutralize the charge—that is, when there is the same number of electrons on both plates—the discharge current will stop. The very fact that electron current will flow through the circuit of Fig. 18-11 indicates that the capacitor has stored energy and is returning the energy to the circuit when the switch is closed.

It is important to make a comparison between the inductor and capacitor. In the inductor the energy is stored by virtue of a magnetic field surrounding the inductor when current is flowing through it. In the capacitor the energy is stored in the dielectric when there is a voltage across it. To be scientifically precise, the energy in a capacitor is stored in the dielectric in the form of an electric field between the plates.

We can now define the word "capacitance." Capacitance is *a measure of the ability of a capacitor to store energy in the form of an electric field.* Capacitance is measured in *farads*, but this unit is too large for most practical applications, so microfarads (millionths of a farad) and picofarads (millionths of a millionth of a farad) are more commonly used. A 10-microfarad capacitor can store more energy than a 5-microfarad capacitor.

Earlier we defined capacitance as the property of a capacitor that makes it oppose any change in voltage on its terminals. To understand this property, consider again the simple discharging circuit of Fig. 18-11. When the switch is closed, the electrons begin to discharge through the resistance R. The resistor serves to limit the discharge current. Even without the resistor, the electrons would take a certain amount of time to get from one plate to the other. Therefore, when the switch is closed the voltage across the capacitor is not reduced to zero instantly, but decreases at a given rate. We can say that the voltage across the capacitor does not change instantly because the capacitor opposes any change in voltage on its terminals.

The capacitor also opposes an increase in voltage across its terminals. This is understood by referring to the circuit of Fig. 18-10. Closing the switch does not cause the voltage across the capacitor to change instantly. Instead, the voltage must build up from zero as the charges move into the plates. Thus, the capacitor opposes the change in voltage from zero to the applied value of voltage. Capacitors, like inductors, are primarily ac components.

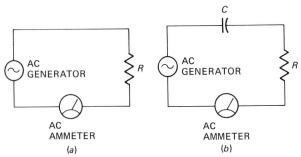

Figure 18-12.   A capacitor opposes the flow of alternating current. (a) *In this circuit the only opposition to current flow is provided by resistor* **R.** (b) *Adding a capacitor to the circuit decreases the alternating current. This indicates that the capacitor opposes the flow of alternating current.*

We have discussed the charge and discharge of a capacitor in terms of dc voltages, but in actual practice the capacitor is more often used in ac circuits.

The effect of a capacitor on current flow is illustrated in Fig. 18-12. In Fig. 18-12a a resistor is connected across a generator, and an ammeter measures the amount of current flow. When a capacitor is added to the circuit, as shown in Fig. 18-12b, the amount of current in the circuit decreases. This illustrates that the capacitor opposes the flow of alternating current.

The opposition that the capacitor offers to the flow of alternating current is called *capacitive reactance.* Capacitive reactance, like inductive reactance and resistance, is measured in ohms. However, you cannot add the capacitive reactance and resistance in ohms to find the total opposition of the circuit to the flow of current. Furthermore, you cannot add inductive reactance and resistance to find the total opposition to current flow. The combined opposition to current flow offered by capacitance, inductance, and resistance in an ac circuit is called *impedance.*

It is important to understand the difference between inductive reactance and capacitive reactance. The inductive reactance of an inductor is directly related to the amount of inductance it has. For example, a coil with a large inductance offers a large inductive reactance, and a coil with a small inductance offers a small inductive reactance. For capacitive reactance the opposite condition exists. Capacitive reactance is inversely related to capacitance. This means that a large capacitor offers a low reactance (opposition) to alternating-current flow, and a small capacitor offers a high reactance (opposition) to alternating-current flow.

SUMMARY

1. The earliest form of capacitors was the Leyden jar.
2. In its simplest form, a capacitor consists of two conductors separated by an insulator. The conductors are called the *plates* and the insulation is called the *dielectric.*
3. Applying a voltage to a capacitor in order to store electrical energy in it is called *charging the capacitor.*
4. Removing the charge from a charged capacitor is called *discharging the capacitor.*

5. Energy is stored in the dielectric of a capacitor.
6. A capacitor opposes any change in voltage across its terminals.
7. Capacitance is a measure of the ability of a capacitor to store energy.
8. Capacitors, like inductors, are used in ac circuits.
9. The opposition that a capacitor offers to the flow of alternating current is called *capacitive reactance*.
10. Capacitive reactance, like inductive reactance and resistance, is measured in ohms.

WHAT IS A FARAD?

You will remember that a coulomb is a unit of electric charge which is equal to the combined charge of 6.24×10^{18} electrons. The amount of charge on a capacitor can be measured in coulombs. This is simply a measure of the number of electrons on the negative plate compared with the number on the positive plate. For example, if the negatively-charged plate has 6.24×10^{18} more electrons than the positively-charged plate, then the capacitor is said to have a charge of 1 coulomb.

If the charge of a capacitor is 1 coulomb, and at the same time there is a voltage difference across the terminals of that capacitor of 1 volt, then the capacitor is said to have a capacitance of 1 farad.

As mentioned before, microfarads and picofards are more useful units, because it would take a very large capacitor to produce a capacitance of 1 farad.

WHAT DETERMINES THE CAPACITANCE OF A CAPACITOR?

Any time you have two conductors separated by an insulator or a dielectric, you have a capacitor. The amount of capacitance of a given capacitor depends upon three things: the area of the capacitor plates facing each other, the distance between the plates, and the type of material used for a dielectric.

The capacitance is directly related to the area of the plates facing each other. In other words, if you increase the area of the plates, you increase the capacitance directly.

Variable capacitors are sometimes made by moving one plate (or set of plates) so that the amount of area facing is varied. Also, manufacturers of capacitors get a larger amount of capacitance in a given area by connecting a number of plates in combination so as to get a larger plate area.

The capacitance of a capacitor is indirectly related to the distance between the plates. Moving the plates of a capacitor closer together causes the capacitance to increase, and moving the plates further apart causes the capacitance to decrease. This relationship between capacitance and distance between the plates is called an *inverse relationship* because an increase in one causes a decrease in the other.

The capacitance of a capacitor is directly dependent upon the type of material used for a dielectric. For example, if you have a capacitor that has air between the plates, the air serves as a dielectric. You can increase the capacitance of this capacitor simply by putting paper between the plates. (This assumes that the distance between the plates and the area of the plates remains unchanged.)

SUMMARY

1. The charge on a capacitor may be measured in coulombs. If the negatively-charged plate has 6.24×10^{18} more electrons than the positively-charged plate, then the charge on the capacitor is 1 coulomb.
2. When 1 volt across a capacitor causes it to charge to 1 coulomb, then the capacitor has a capacitance of 1 farad.
3. The capacitance of any capacitor depends upon the area of its plates facing each other, the distance between its plates, and the type of dielectric material between its plates.
4. Increasing the area of the capacitor plates increases its capacitance.
5. Decreasing the distance between the capacitor plates increases its capacitance.
6. Changing the dielectric material between the plates of a capacitor changes its capacitance.

HOW ARE RESISTORS, INDUCTORS, AND CAPACITORS USED IN CIRCUITS?

Inductors and capacitors are frequency-sensitive components, and therefore, they can be used for selecting one frequency or range of frequencies and rejecting all others. The relationship between inductors and capacitors and frequency is shown in Fig. 18-13. Figure 18-13*a* shows that an inductor will pass low frequencies but reject high frequencies. You should remember that low frequencies include dc, which is actually 0 cycles per second, or 0 hertz. That is the lowest possible frequency.

Figure 18-13*b* shows that capacitors will pass high frequencies but reject low frequencies. Again, the low frequency may be dc. When capacitors are used for the purpose of passing ac and rejecting the dc, they are called *blocking capacitors*.

Capacitors and inductors are sometimes used in combinations like the one shown in Fig. 18-14. This series circuit, comprised of *L* and *C* (Fig. 18-14*a*), is called a *series-tuned circuit*. It will pass one frequency or narrow range of frequencies, and reject all others. A graph of the frequency against current for this circuit is shown in Fig. 18-14*b*. Note that the current is relatively high at f_r (the *resonant frequency* of the tuned circuit). The resonant frequency is defined as the frequency that will pass through the series circuit with the least opposition. All frequencies other than f_r are opposed. This kind of circuit is used in your radios and television sets to enable you to select one station and reject all

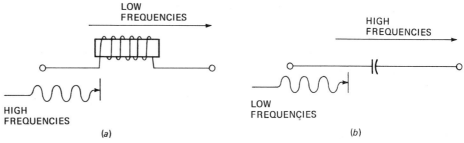

Figure 18-13. Inductors and capacitors are frequency-sensitive components. (a) *Inductors will pass low frequencies more readily than they will pass high frequencies.* (b) *Capacitors will pass high frequencies more readily than they will pass low frequencies.*

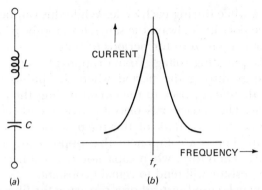

(a)   (b)

Figure 18-14. *One application of capacitors and inductors.* (a) *A series-tuned circuit.* (b) *This graph shows that the current flow is maximum at* f_r.

others. If it were not for the use of such tuned circuits, you would receive all the stations at one time.

The ability of capacitors and inductors to store energy makes them useful as *filtering* components. This application is illustrated in Fig. 18-15. The half-wave rectifier circuit shown in Fig. 18-15a has been used in a number of experiments described in some of the chapters. This simple circuit converts the ac voltage out of the transformer to a pulsating dc voltage, which is shown in Fig. 18-15b. This waveform shows that voltage is present for a while, and then

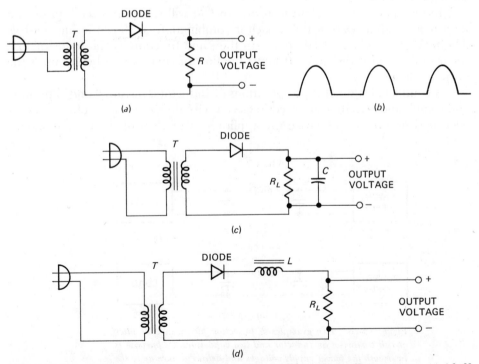

(a)   (b)

(c)

(d)

Figure 18-15. *A capacitor can be used to filter a voltage, and an inductor filters a current.* (a) *A half-wave rectifier.* (b) *Voltage of a half-wave rectifier.* (c) *The capacitor filters the output voltage.* (d) *The inductor filters the current.*

drops to zero for a while during each cycle. While this can be classified as a dc voltage because the polarity is always the same, it would be a better dc voltage if the voltage did not drop to zero for periods of time.

To prevent the pulsating voltage from dropping to zero, a capacitor can be used to store energy during the period when the pulses are applied. The capacitor releases the energy back to the circuit during the period when there is no pulse applied. The capacitor is placed across a load resistor as shown in Fig. 18-15c. An easy way to think of the purpose of this capacitor is to remember that a capacitor is a component that opposes any change in voltage across its terminals. Therefore, when capacitor C is connected across R_L, the voltage across this resistor will tend to remain constant.

An inductor may be used instead of a capacitor for filtering. This application is shown in Fig. 18-15d. Here the inductor is placed in series with the resistor. It tends to prevent the pulsations of current that flow through the load resistor as a result of the pulsating voltage applied. Remember that an inductor opposes changes in current, and therefore, it will try to make the current through the resistor a constant value. Thus the output voltage across R_L will be more nearly like a pure dc.

When used separately, the capacitor can filter the output, but it is a more common practice to use them in combinations, as shown in Fig. 18-16. Figure 18-16a shows an *LC* filter which changes the pulsating output from the power supply to a dc supplied to the load. Since this circuit looks like the Greek letter π, it is called a *pi-section filter*. The capacitors oppose a change in voltage across the supply and across the load. The inductor opposes changes in current through the load.

Inductors cost more than resistors, so you will sometimes see a pi-section filter made with a resistor and capacitor combination, as shown in Fig. 18-16b. The resistor does not exactly perform the same function as the inductor, but the overall result is the same. It limits the flow of current and hence limits the output voltage.

A complete list of all the applications of inductors, resistors, and capacitors and combinations as they are used in electric circuits would take volumes. However, it is useful to summarize the applications of each of these components.

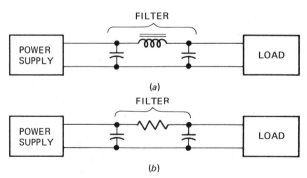

Figure 18-16.   Two examples of pi-section filters. (a) *The filter circuit comprises an inductor and two capacitors. Its purpose is to smooth the power supply voltages and deliver a pure dc to the load.* (b) *The filter may use a resistor and two capacitors as shown in this illustration. The purpose of the filter is the same as in Fig. 18-16a.*

Resistors are used for limiting current flow, producing a voltage drop, and generating heat. Capacitors are used for storing energy in the form of an electric field, producing an ac voltage drop, and passing high frequencies while rejecting low frequencies. Inductors are used for storing energy in the form of an electromagnetic field, producing an ac voltage drop, and passing low frequencies while rejecting high frequencies.

SUMMARY

1. An inductor will pass low frequencies and reject high frequencies.
2. A capacitor will pass high frequencies and reject low frequencies.
3. The tuned circuit of a radio consists of an inductor and a capacitor in series. This circuit permits the receivers to select one station and reject all others.
4. Capacitors and inductors may be used for filtering.

Programmed Review Questions

(Instructions for using this programmed section are given in Chap. 1.)

We will review the important concepts of this chapter. If you have understood the material, you will progress easily through this section. Do not skip this material, because some additional theory is presented.

1. Is the capacitance of a capacitor related in any way to the area of the plates of the capacitor?
   A. Yes (Proceed to block 7.)
   B. No (Proceed to block 17.)

2. *The correct answer to the question in block 21 is* **B**. *The amount of* **charge** *on a capacitor is determined by the voltage across it. However, the capacitance is not dependent upon the voltage. Changing the dielectric will change the capacitance.* Here is your next question.
   Which of the following statements is true?
   A. The countervoltage induced in a coil always opposes the flow of current through the coil. (Proceed to block 25.)
   B. The countervoltage induced in a coil always opposes any change in current through the coil. (Proceed to block 20.)

3. *The correct answer to the question in block 22 is* **B**. *The shape of a coil is an important factor in determining its inductance.* Here is your next question.
   In Fig. 18-17 the voltage between terminals *A* and *B* is
   A. 6 volts. (Proceed to block 27.)
   B. 9 volts. (Proceed to block 23.)

4. *Your answer to the question in block 16 is* **A**. *This answer is wrong. Remember that inductors are dependent upon a changing current, not voltage.* Proceed to block 26.

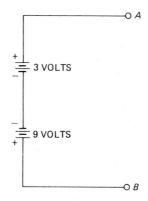

Figure 18-17.   How do the voltages combine?

5.   *The correct answer to the question in block 18 is **A**. The batteries are connected in series-aiding, so their voltages are added. The two batteries of Fig. 18-18 could be replaced with a single 10-volt battery with its positive lead connected to terminal* A. Here is your next question.

The tuned circuit in a radio permits you to select one station and reject all others. The components used in the tuned circuit are _____ and _____. (Proceed to block 30.)

6.   *Your answer to the question in block 20 is **B**. This answer is wrong. The direction of current determines the direction of the magnetic field that accompanies it.* Proceed to block 14.

7.   *The correct answer to the question in block 1 is **A**. The capacitance of a capacitor depends upon three things: the area of plates facing each other, the distance between the plates, and the type of material used for a dielectric.* Here is your next question.

The opposition that a capacitor offers to the flow of alternating current is called
A.   capacitive reactance. (Proceed to block 21.)
B.   capacitive resistance. (Proceed to block 15.)

8.   *The correct answer to the question in block 13 is **A**. Moving the plates of a capacitor closer together increases the capacitance of a capacitor. One way of making a variable capacitor is to use a movable plate that can be positioned with respect to another fixed plate. Moving the plate changes the capacitance.* Here is your next question.

A charge of 1 coulomb on a certain capacitor produces a voltage of 1 volt across its terminals. The capacitor has a capacitance of
A.   1 microfarad. (Proceed to block 24.)
B.   1 farad. (Proceed to block 18.)

9.   *Your answer to the question in block 21 is **A**. This answer is wrong. The capacitance of a capacitor does not depend upon the amount of voltage across it.* Proceed to block 2.

10.   *Your answer to the question in block 22 is **A**. This answer is wrong. The resistance of the wire will help to limit the current through the coil, but it does not affect the inductance.* Proceed to block 3.

11. *Your answer to the question in block 18 is **B**. This answer is wrong. Note that the positive terminal of the 6-volt battery is connected to the negative terminal of the 4-volt battery.* Proceed to block 5.

12. *Your answer to the question in block 27 is **A**. This answer is wrong. Inductance is not measured in volts.* Proceed to block 16.

13. *The correct answer to the question in block 26 is **B**. Inductive reactance and capacitive reactance are measured in ohms. Impedance, which is "the complete opposition to the flow of alternating current in a circuit having inductive reactance, capacitive reactance, and resistance," is also measured in ohms.* Here is your next question.
Moving the plates of a capacitor closer together will
A. increase its capacitance. (Proceed to block 8.)
B. decrease its capacitance. (Proceed to block 29.)

14. *The correct answer to the question in block 20 is **A**. The greater the amount of current flow, the greater the strength of the magnetic field around it.* Here is your next question.
Which of the following is *not* a typical use of resistors in a circuit?
A. Limit current (Proceed to block 28.)
B. Store energy (Proceed to block 22.)

15. *Your answer to the question in block 7 is **B**. This answer is wrong. Remember the term "reactance," because it is used for both inductance and capacitance.* Proceed to block 21.

16. *The correct answer to the question in block 27 is **B**. The conditions described for the coil are exactly right for a 1-henry inductance.* Here is your next question.
A component that opposes any change in voltage across its terminals is the
A. inductor. (Proceed to block 4.)
B. capacitor. (Proceed to block 26.)

17. *Your answer to the question in block 1 is **B**. This answer is wrong. The capacitance of a capacitor is directly dependent upon the area of the plates facing each other.* Proceed to block 7.

18. *The correct answer to the question in block 8 is **B**. A farad is much too large for a practical unit of measurement, so microfarads (millionths of a farad) and picofarads (millionths of a millionth of a farad) are more often used. To convert farads to microfarads, move the decimal place to the right six places. To convert farads to picofarads, move the decimal place to the right 12 places.*
Examples:

0.000 001 farad = 1.0 microfarad
0.000 000 000 100 farad = 100 picofarads

Here is your next question.

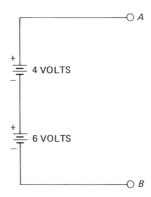

Figure 18-18.   Circuit for the question in block 18.

In Fig. 18-18 the voltage between terminals *A* and *B* is
A.   10 volts. (Proceed to block 5.)
B.   2 volts. (Proceed to block 11.)

19.   *Your answer to the question in block 26 is **A**. This answer is wrong. The opposition that coils and capacitors offer to the flow of alternating current is called **reactance**.* Proceed to block 13.

20.   *The correct answer to the question in block 2 is **B**. The countervoltage does not oppose the flow of current, but rather, it opposes **any** change in the current.* Here is your next question.
The strength of the magnetic field around a current depends upon
A.   the amount of current. (Proceed to block 14.)
B.   the direction of current. (Proceed to block 6.)

21.   *The correct answer to the question in block 7 is **A**. Capacitive reactance is the opposition that a capacitor offers to the flow of alternating current, and inductive reactance is the opposition that an inductor offers to the flow of alternating current.* Here is your next question.
Which of the following will change the capacitance of a capacitor?
A.   Increase the voltage across it (Proceed to block 9.)
B.   Change the dielectric material (Proceed to block 2.)

22.   *The correct answer to the question in block 14 is **B**. Storing energy is **not** a typical use of resistors in electric circuits. Inductors and capacitors are used for storing energy.* Here is your next question.
The inductance of a coil is dependent upon
A.   the resistance of the wire used to make the coil. (Proceed to block 10.)
B.   the shape of the coil. (Proceed to block 3.)

23.   *Your answer to the question in block 3 is **B**. This answer is wrong. The batteries are connected in opposition, so their voltages must be subtracted. Review the discussion related to Fig. 18-4b.* Then proceed to block 27.

24.   *Your answer to the question in block 8 is **A**. This answer is wrong. A microfarad is a convenient unit of measurement, but the description in the question actually identifies a farad.* Proceed to block 18.

25. *Your answer to the question in block 2 is **A**. This answer is wrong. The countervoltage will try to prevent an increase in current, and it will also try to prevent a decrease in current.* Proceed to block 20.

26. *The correct answer to the question in block 16 is **B**. A capacitor may be defined as a component that opposes any change in voltage across its terminals.* Here is your next question.
    The opposition that a coil offers to the flow of alternating current is called
    A. resistive inductance. (Proceed to block 19.)
    B. inductive reactance. (Proceed to block 13.)

27. *The correct answer to the question in block 3 is **A**. The voltages are in opposition and must be subtracted. The two batteries could be replaced with a single 6-volt battery having its positive terminal connected to terminal B.* Here is your next question.
    When the current through a certain coil is changing at a rate of 1 ampere per second, the countervoltage induced in the coil windings is 1 volt. What is the inductance of the coil?
    A. 1 volt (Proceed to block 12.)
    B. 1 henry (Proceed to block 16.)

28. *Your answer to the question in block 14 is **A**. This answer is wrong. An important use of resistors in electric circuits is to limit current.* Proceed to block 22.

29. *Your answer to the question in block 13 is **B**. This answer is wrong. The capacitance of a capacitor is inversely related to the distance between the plates.* Proceed to block 8.

30. *The tuned circuit is made by connecting an inductor and a capacitor in series. Either the inductor or the capacitor is made variable so that the resonant frequency can be changed. This makes it possible to select any station in the broadcast band.*
    You have now completed the programmed questions. The next step is to put some of these ideas to work in laboratory experiments. Proceed to the "Experiments" section that follows.

Experiments

(The experiments described in this section may be performed on the circuit board described in Appendix C or on a similar laboratory setup.)

EXPERIMENT 1

Purpose—The purpose of this experiment is to demonstrate the effect of inductance on current flow in alternating-current circuits.

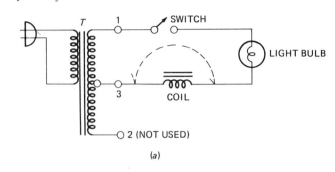

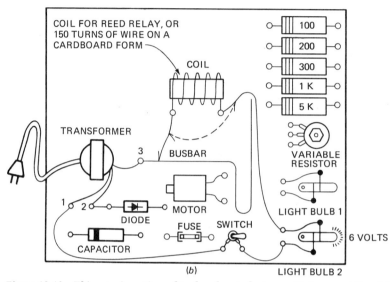

Figure 18-19. This experiment is used to show how inductance affects current flow. (a) *Test setup for showing the effect of inductance in a circuit.* (b) *This is the way your circuit board will look if you have wired the circuit correctly.*

Theory — An inductor opposes changes in current through it. Since alternating current is continually changing, an inductor will oppose it. The opposition is called *inductive reactance*. This experiment will show that inductance opposes the flow of alternating current.

Test Setup — Wire the circuit board as shown in Fig. 18-19. The schematic representation of this circuit is also shown in Fig. 18-19. You will need a piece of insulated wire to short across the inductor. The connection for the short is shown with a dotted line in the illustration. Do not connect this short until told to do so.

Procedure —

   Step 1 — Energize the circuit by closing the switch. Note the brightness of the light.

   Step 2 — Touch the short circuit across the coil momentarily. Note that the light is brighter when the coil is shorted.

Conclusion — When the inductor is in series with the light, the light does not glow at full brightness. This is because the inductor limits the current flow.

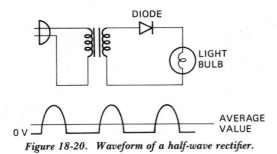

Figure 18-20. Waveform of a half-wave rectifier.

Shorting across the inductor has the same effect as removing the inductor from the circuit and replacing it with a piece of wire. Without the inductor, the light glows brighter. This indicates that the current is greater (and the opposition is lower) without the inductor.

EXPERIMENT 2

Purpose — To show how a capacitor is used to filter a pulsating voltage.

Theory — The output waveform of a half-wave rectifier is shown in Fig. 18-20. The average value of this wave is very low, as shown in the illustration.

The solid line of Fig. 18-21 shows the waveform after it has been filtered with a capacitor. (The pulsating wave is shown with a dotted line.) The capacitor is charged to the peak value at point *a*, and then holds its charge (with a slight loss) until the next pulse comes along at point *b*.

The important thing to note in the circuit of Fig. 18-21 is that the average value of voltage is higher. If you used the voltage waveform of Fig. 18-20 to light a light, the light would not glow as brightly as it would if you used the voltage waveform of Fig. 18-21.

Test Setup — Wire the circuit as shown in Fig. 18-22. Both the wiring diagram and the schematic are shown in the illustration. Do not connect the lead shown with a dotted line at this time.

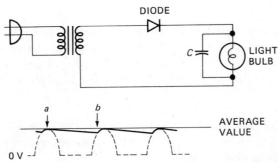

Figure 18-21. Waveform of a half-wave rectified voltage with capacitive filtering.

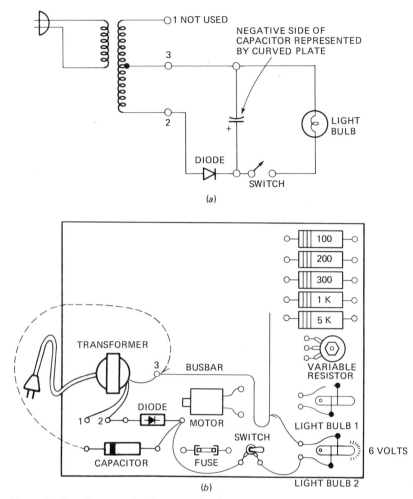

Figure 18-22. *Test setup for demonstrating how capacitors are used for filtering. The capacitor accomplishes its job by storing energy for part of the cycle and then releasing it back to the circuit.* (a) *Test setup for demonstrating that capacitors can store energy.* (b) *This is the way your circuit board will look if you have wired the circuit correctly.*

Procedure—

Step 1—Energize the circuit by closing the switch. Note the light brightness.

Step 2—Momentarily connect the lead that places the capacitor across the light, as shown by the dotted line in Fig. 18-22. This places the capacitor across the light bulb. Note that the light is glowing more brightly with the capacitor in place.

Conclusion—The capacitor stores energy during the peaks of voltage and then returns it to the circuit between pulses. This raises the average value of voltage and makes the light glow brighter.

Self-Test with Answers

(Answers with discussions are given in the next section.)

1. A certain ac circuit has an inductor, a capacitor, and a resistor. The combined opposition that the three components offer to the flow of alternating current is called (*a*) resistance; (*b*) reactance; (*c*) impedance.
2. Inductive reactance is measured in (*a*) ohms; (*b*) henrys.
3. Capacitive reactance is measured in (*a*) ohms; (*b*) farads.
4. Which of the following could store more energy? (*a*) A 10-microhenry inductor; (*b*) A 10-megohm resistor.
5. Increasing the number of turns of wire on a coil will (*a*) increase its inductance; (*b*) decrease its inductance.
6. Moving the plates of a capacitor closer together will (*a*) increase its capacitance; (*b*) decrease its capacitance.
7. Which of the following would offer greater opposition to the flow of an alternating frequency? (*a*) A 1.0-microfarad capacitor; (*b*) A 1.0-picofarad capacitor.
8. In a series-tuned circuit, what is the name given to the frequency that will pass through the circuit? _____.
9. List three uses of capacitors in electric circuits. _____, _____, and _____.
10. List three uses of inductors in electric circuits. _____, _____, and _____.

Answers to Self-Test

1. (*c*)
2. (*a*)—Inductance is measured in henrys. Inductive reactance is measured in ohms.
3. (*a*)—Capacitance is measured in farads. Capacitive reactance is measured in ohms.
4. (*a*)—A resistor does not store energy.
5. (*a*)
6. (*a*)
7. (*b*)—The smaller the capacitance, the greater its reactance for any given frequency.
8. Resonant frequency.
9. Store energy; reject low frequencies and pass high frequencies; produce an ac voltage drop. (Other answers are possible.)
10. Reject high frequencies and pass low frequencies; limit alternating-current flow; store energy. (Other answers are possible.)

19.
What Is a Transformer?

Introduction

You have learned in an earlier chapter that all conductors of electricity, such as copper and aluminum, have some electrical resistance. In many cases the resistance of a piece of wire is so small it can be neglected. However, consider the problem of generating the electricity in Hoover Dam in the state of Nevada and transmitting it over 275 miles to Los Angeles, California. The resistance of the very long transmission line cannot be neglected in such a case.

Whenever a current flows through an electrical resistance, there is *always* a voltage drop. This means that some of the voltage generated at the power station is lost in the form of the voltage drop when the power is conducted from the generator (in the power station) to the load (at the consumer's location). If the voltage generated at Hoover Dam were dc, then the voltage drop throughout the transmission line could not be recovered. Houses and industries along the path of the transmission line would receive less and less voltage as the distance from the generator became greater and greater. It is not possible to adjust the voltage at the power station so that a consumer at the end of the line is receiving the correct amount of voltage—that is, the amount of voltage needed for proper operation of his electrical equipment. To do this would mean that the consumer at the beginning of the line (nearest the power station) would receive a voltage that is too high.

Some early power stations did generate and deliver a dc voltage for consumer use. The problem of power line voltage drop has been reduced in modern power plants by distributing ac, rather than dc, to the consumer. An ac component called a *transformer* makes it possible to increase the power line voltage at points along the transmission line. The transformer has no moving parts, so it does not require continuous maintenance. Very little power is lost within the transformer, and it can be used for increasing the voltage to compensate for any losses due to voltage drop along the line. This chapter explains how a transformer can be used to change an ac voltage.

In this chapter some of the uses of transformers will be discussed, and the theory of transformer operation will also be covered.

You will be able to answer these questions after studying this chapter:

What are transformers used for?
What is mutual inductance?
What is the coefficient of coupling?
What are step-up and step-down transformers?
What are losses in transformers?
What is reflected impedance?
What is meant by "impedance matching"?

Instruction

WHAT ARE TRANSFORMERS USED FOR?

A good way to start the study of transformers is to get an idea of what they are used for and what different types of transformers are available. In its simplest form the transformer consists of two coils located near to each other. Power is applied to one of the coils, called the *primary*, and power is taken away from the other coil, called the *secondary*.

Figure 19-1 shows the symbols for some typical transformers. The primary and secondary windings are identified on these symbols. Loops in the symbols

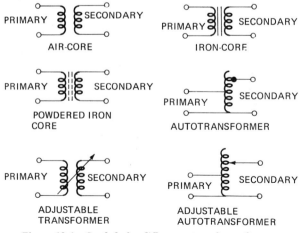

Figure 19-1. Symbols for different types of transformers.

represent turns of wire in the transformer, but these loops are not necessarily representative of the exact number of turns in the transformer coils. For example, in the transformers represented in Fig. 19-1 there may be hundreds of turns of wire in both the primary and secondary windings, but the symbols would be the same.

If the primary coil and the secondary coil are wound on a piece of plastic, wood, or other nonmagnetic material, it is called an *air-core transformer*. The reason for the name "air core" is the fact that plastic, wood, and many other materials have the same magnetic properties as air. The magnetic property of the material upon which the coils are wound has an important bearing on the operation of the transformer.

The iron-core transformer has the primary and secondary windings wound on a magnetic material. The material usually used is called *soft iron*. In most applications it is undesirable to use a material that can retain permanent magnetism. "Soft iron" is the name given to magnetic materials which can be easily magnetized but which cannot retain the magnetism when the magnetizing force is removed. The two lines between the primary and secondary coils of the iron-core transformer represent the iron core.

Transformers designed for special applications such as high-frequency work may be wound on powdered iron cores. The symbol for this type of transformer is shown in Fig. 19-1. The dotted lines between the primary and secondary windings indicate that the material upon which the coil is wound is made of powdered iron.

In an *autotransformer* a single coil serves as the primary and secondary windings. As indicated by the symbol in Fig. 19-1, part of the coil is used for the primary winding and part of the coil is used for the secondary. *Adjustable transformers* and *adjustable autotransformers* make it possible to vary the voltage across the secondary winding of the transformer. With these components it is possible to select the desired output voltage for a given input voltage. The operation of a transformer is based on electrical fundamentals which were discussed earlier in this book. These fundamentals (relating to the operation of a transformer) will be reviewed at this time.

Whenever an electric current flows in a coil, a magnetic field is produced. The strength of the magnetic field depends directly upon the amount of cur-

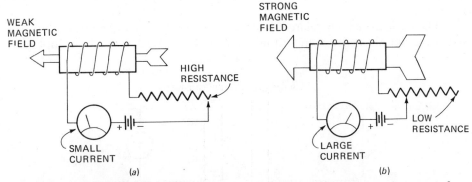

Figure 19-2. *The strength of the magnetic field around a coil depends directly upon the amount of current through the coil.* (a) *A small current produces a weak magnetic field.* (b) *A large current produces a strong magnetic field.*

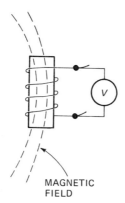

MAGNETIC
FIELD

*Figure 19-3. A moving magnetic field
generates a voltage across the coil.*

rent in the circuit. This is shown in Fig. 19-2. In Fig. 19-2*a* the arm of the variable resistor is positioned so that there is a large amount of resistance in the circuit. This causes a small amount of current to flow, and a weak magnetic field is produced. In Fig. 19-2*b* the variable resistor arm is positioned so that there is very little resistance in the circuit. In this case a large current flows, and the strength of the magnetic field in the coil is much greater.

The direction of the magnetic field is determined directly by the direction of the current through the coil. If the current is reversed in the circuit of Fig. 19-2 (by reversing the connection of the battery), then the direction of the magnetic field as indicated by the dark arrow will also be reversed.

If an alternating current were used in the coil instead of the direct current supplied by the battery, the magnetic field would periodically reverse every time the current reversed. Furthermore, the strength of the magnetic field would vary from moment to moment as the strength of the alternating current varied from cycle to cycle.

It should be understood that the magnetic field at the coil in Fig. 19-2 exists not only through the center of the coil but also around it.

It has just been shown that an alternating current flowing through a coil would produce a varying magnetic field. Another important fundamental related to transformer action is illustrated in Fig. 19-3. In this case a moving magnetic field is cutting across the turns of the coil. Whenever this happens a voltage is *always* induced in the windings of the coil. The voltmeter shows that all the voltages of the windings are combined to produce a voltage across the coil.

The basic principles related to transformer operation are combined to construct the simple transformer shown in Fig. 19-4. An ac voltage is applied to

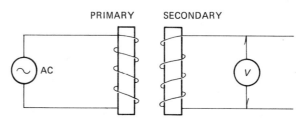

PRIMARY   SECONDARY

Figure 19-4. When an alternating current flows in the primary, an alternating voltage is induced across the secondary.

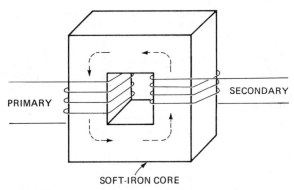

Figure 19-5. When the primary and secondary coils are wound on a core of soft-iron material, the power is coupled with less loss.

the primary winding of this simple transformer. Alternating current, flowing in the primary, produces an expanding and contracting magnetic field that cuts across the secondary winding. Since this is a moving magnetic field that varies with the alternating current from instant to instant, it follows that a voltage will be induced across the secondary winding. Transformer action is based on the fact that the moving flux from the primary cuts across the secondary turns and induces a voltage.

A disadvantage of the simple transformer of Fig. 19-4 is that many flux lines are lost between the primary and secondary. In other words, the amount of *coupling* between the windings is low. In an ideal transformer every magnetic flux line along the primary would be coupled, or *linked,* with the secondary turns. Of course this is not possible with the arrangement of Fig. 19-4.

Figure 19-5 shows a more desirable arrangement. Here the primary and secondary coils are wound on a piece of soft iron. When a current flows in the primary winding, a magnetic field is produced through the center of the coil, and this magnetic field temporarily magnetizes the soft-iron core. The broken arrows show the direction of the flux path during one-half cycle of primary alternating current. (On the next half cycle the direction of the arrows will reverse.) Since the soft iron is easily magnetized, almost all the magnetic flux of the primary coil flows through the center of the secondary coil, and the amount of coupling between the two coils is much greater than for the one shown in Fig. 19-4.

As shown in Fig. 19-6, a transformer is strictly an ac component. In Fig. 19-6*a* an ac voltage is applied to the primary winding. This causes an alternating current to flow in the primary, and an ac voltage to be induced in the secondary.

When a dc voltage is applied to the primary, as shown in Fig. 19-6*b*, there is no secondary voltage. (There will be a momentary surge of voltage across the secondary at the instant the dc voltage is connected to the primary. This is due to the increase in primary current from zero to the maximum value. However, once the direct current is established, there is no longer a changing flux, and therefore, there is no voltage induced in the secondary.)

An important use of transformers is to pass an alternating current and at the same time block the passage of a direct current. In other words, if the

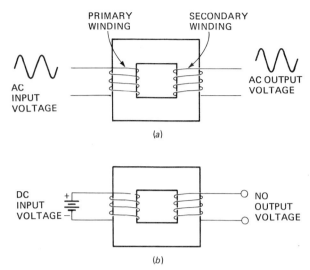

Figure 19-6. A transformer passes an ac voltage but will not pass a dc voltage. (a) When an ac voltage is applied to the primary winding, a voltage is generated in the secondary winding. (b) When a dc voltage is applied to the primary winding, there is no voltage generated in the secondary winding.

primary winding has an alternating and a direct current flowing, only the alternating current energy will be coupled to the secondary.

Nothing useful would be accomplished if the only purpose of the transformer was to couple energy from the primary to the secondary winding. If this was the only objective, then the primary and secondary circuits could be directly connected and the transformer would not be needed. However, the transformer accomplishes certain things in coupling the energy from one winding to another. One of these things is the elimination of all dc energy. As mentioned before, dc energy cannot be coupled from the primary to the secondary.

Another important application of transformers is to change the value of voltage across the primary to a larger or smaller value. This is accomplished by putting more or less turns of wire in the secondary winding. Voltage is induced in each turn, and the total secondary voltage is the sum of all the voltages of all the turns. More will be said about this application of transformers later in this chapter.

Since the primary and secondary windings of a transformer are actually inductors, they have all the properties of coils used in other circuits. One of these properties is the ability of *tuning*. Placing a capacitor across the primary or the secondary (or both) causes the transformer to be frequency-selective. In other words, by using capacitors across the windings, a transformer can be made to pass one frequency (or range of frequencies) and reject all others. This important application of transformers is used in electronic circuits.

SUMMARY

1. When transmitting power over long transmission lines, the resistance of the conductors in the transmission line causes a voltage drop. If the current flow is dc, the voltage loss is not easily compensated for.

2. Transformers can be used to step up or step down a voltage.

3. Transformers have no moving parts, and this makes them practically maintenance-free. An important feature of iron-core transformers is their very high efficiency. In other words, there is very little loss of power within the transformer.

4. Transformers are often identified by the type of core material used in their construction. If the coils are wound on nonmagnetic materials, they are called *air-core transformers.* If they are wound on magnetic materials, they are called *iron-core transformers.*

5. *Soft iron* is a magnetic material which can be readily magnetized but which cannot retain its magnetism.

6. Some of the uses of transformers are (*a*) stepping an ac voltage up or down; (*b*) passing an ac voltage while at the same time preventing a dc voltage from passing; and (*c*) when used with capacitors, selecting a frequency or range of frequencies to be passed from one circuit to another.

WHAT IS MUTUAL INDUCTANCE?

It has been shown that a varying current in the primary winding of a transformer produces a voltage across the secondary. This is called *transformer action.* The voltage across the secondary winding will produce a current flow if a resistor is placed across the winding terminals to provide a complete current path. This is illustrated in Fig. 19-7. Remember that the current flowing through *R* must be an alternating current, because only an alternating voltage can be produced across the secondary winding. Since any flow of electron current results in a magnetic field, it follows that current flowing in the secondary circuit of the transformer will produce flux lines around the secondary coil. These flux lines will be alternating because the current in the secondary circuit will be alternating.

A varying flux in the secondary will, of course, cut across the primary turns. This is also illustrated in Fig. 19-7. In this illustration the dotted lines represent the flux from the primary at some instant, either expanding or contracting, and the solid line represents the secondary flux—that is, the flux around the secondary winding which results in current flowing through the secondary circuit. Since both coils have varying magnetic fields which cut across the opposite coil, it follows that voltage is being induced in *both* the primary and secondary windings. It is said that the inductance of the two windings is "mu-

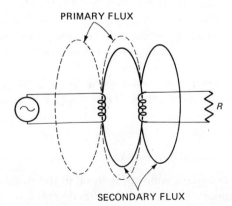

PRIMARY FLUX

SECONDARY FLUX

Figure 19-7. When alternating current flows in the primary and secondary windings, there is a mutual inductance.

tual," because the primary induces a voltage in the secondary and the secondary induces a voltage in the primary.

It is important to give consideration to all the voltages present in the simple transformer of Fig. 19-7. First, there is a voltage applied across the primary which causes an alternating current to flow through that winding. The alternating current through that winding causes a self-induced voltage or a countervoltage, which was described in a previous chapter. The expanding and contracting magnetic flux not only induces a countervoltage in the primary, but also induces a voltage in the secondary. If there is a complete circuit in the secondary, as shown in Fig. 19-7, a varying current will flow in the secondary winding. This causes an expanding and contracting magnetic field which induces a voltage back into the primary.

If we were to combine all these voltages (or more properly, the *effects* of these voltages) we could call it the effect of *mutual inductance*. If the secondary winding did not exist, then the self-inductance of the primary winding would limit the amount of current flow. This is due to the inductive reactance of that coil. The secondary winding causes an additional countervoltage to be induced in the primary, which also affects the amount of current flow. We are no longer simply interested in the inductance of the primary in terms of how it will limit the flow of alternating current. We are now also interested in the inductance of the secondary. This also will affect the amount of primary current flow. This is due, of course, to the voltage induced by the flux around the secondary. In other words, we no longer have simply the inductance of the primary but rather the mutual inductance between the primary and the secondary that affects the amount of current flowing in a primary circuit.

WHAT IS THE COEFFICIENT OF COUPLING?

It has been shown that a transformer is more efficient when the primary and secondary windings are wound on the same iron core. The efficiency is higher because of the greater number of flux linkages that occur. The *coefficient of coupling* is a measure of how well the primary flux couples with the secondary winding. Basically, if all the flux lines from a primary linked with all the turns of the secondary, then the coefficient of coupling would have a value of 1.0, or 100 percent. In actual practice the coefficient of coupling is always some value less than 1.

The term *tight coupling* is used to indicate a high coefficient of coupling—that is, a condition where the majority of the flux lines from the primary are linking with the secondary. The term *loose coupling* means that the coefficient of coupling is low. There is a point between tight coupling and loose coupling that is called *critical coupling*. This is simply the dividing line between loose coupling and tight coupling.

The coefficient of coupling in iron-core transformers is as high as 98 percent or better. With air-core transformers the coefficient of coupling does not exceed 65 percent.

WHAT ARE STEP-UP AND STEP-DOWN TRANSFORMERS?

If the number of turns of wire in the secondary winding is equal to the number of turns in the primary, then the secondary voltage will be exactly equal to the

Figure 19-8. When the number of turns of wire in the secondary equals the number of turns in the primary, the primary and secondary voltages are equal.

primary voltage. Figure 19-8 shows this type of transformer. (This statement ignores the fact that there are losses in the transformer. In an iron-core transformer the losses may be so low that for all practical purposes the voltages would be the same.)

It is important to understand that voltage is induced in each turn of the secondary winding in a transformer, and that the total secondary voltage is the sum of the voltages induced in each of the turns. It follows, then, that by increasing the number of turns of wire in the secondary, you can increase the output voltage. Therefore, whenever the secondary coil contains more windings than the primary, the secondary voltage will be greater than the primary voltage. This is called a *step-up transformer*. An example is shown in Fig. 19-9. If the secondary winding has less turns than the primary, the secondary voltage will be less than the primary. This is called a *step-down transformer*, and an example is shown in Fig. 19-10.

An important method of rating transformers is by the *turns ratio*. This is equal to the number of turns in the transformer secondary coil divided by the number of turns in the transformer primary coil. Suppose, for example, that the secondary winding has 200 turns and the primary winding has 100 turns. The turns ratio of this step-up transformer is calculated as follows:

$$TURNS\ RATIO = \frac{NUMBER\ OF\ TURNS\ IN\ THE\ SECONDARY\ COIL}{NUMBER\ OF\ TURNS\ IN\ THE\ PRIMARY\ COIL}$$

$$= \frac{200}{100}$$

$$= \frac{2}{1}$$

The turns ratio is $2:1$ (read "two to one").

The schematic symbols of Figs. 19-8, 19-9, and 19-10 are used to indicate the type of transformer used in a circuit. However, you cannot determine the turns ratio of a transformer by counting the number of turns shown in the coil symbols. You can determine the turns ratio by dividing the secondary voltage by the primary voltage, since the turns ratio and the voltage ratio in transformers are identical. For example, in Fig. 19-8 the secondary and primary voltages

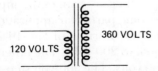

Figure 19-9. An example of a step-up transformer. There are more turns of wire in the secondary of this transformer than in the primary.

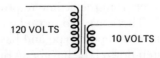

Figure 19-10. An example of a step-down transformer. This transformer has fewer turns of wire in the secondary than in the primary.

are shown to be 120 volts each. The turns ratio is

$$TURNS\ RATIO = \frac{SECONDARY\ VOLTAGE}{PRIMARY\ VOLTAGE}$$

$$= \frac{120}{120}$$

$$= \frac{1}{1}$$

The turns ratio is 1 : 1.

In Fig. 19-9 the secondary voltage is 360 volts, and the primary voltage is 120 volts. The turns ratio of this transformer is

$$TURNS\ RATIO = \frac{SECONDARY\ VOLTAGE}{PRIMARY\ VOLTAGE}$$

$$= \frac{360}{120}$$

$$= \frac{3}{1}$$

The turns ratio of this step-up transformer is 3 : 1.

Figure 19-10 shows a step-down transformer with a turns ratio of

$$TURNS\ RATIO = \frac{SECONDARY\ VOLTAGE}{PRIMARY\ VOLTAGE}$$

$$= \frac{10}{120}$$

$$= \frac{1}{12}$$

The turns ratio of this step-down transformer is 1 : 12.

If the primary winding of the transformer in Fig. 19-10 had 120 turns, then the secondary would have 10 turns. However, this is not necessarily the exact number of turns in the primary and secondary. It is just as possible that the primary has 240 turns and the secondary has 20 turns. In that case the turns ratio is still the same (20 divided by 240, or 1 : 12).

There may be more than one secondary winding in a given transformer, and they may be either step-up, step-down, or both. Figure 19-11 shows the schematic symbol of a transformer designed for operating from the power line. This type of transformer is often called a *power transformer*. However, the name is misleading because it does not supply power, but rather, it changes the input power from the ac line to power that is needed for some particular application.

There are two secondary windings in the power transformer of Fig. 19-11. One is a step-up type and has an output voltage equal to 200 volts. The other is a step-down type having an output voltage equal to 6 volts. The 200-volt step-up winding is divided into two sections, each containing 100 volts. This type of winding is said to be *center-tapped*. It permits two voltages to be obtained from a single winding.

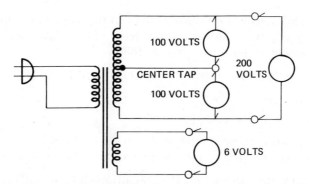

Figure 19-11. This power transformer has both a step-up and a step-down winding.

SUMMARY

1. The coefficient of coupling of a transformer is a measure of how well the secondary winding is coupled to the primary.
2. If all the flux lines of the primary are linked to the secondary, then maximum coupling occurs. Under this condition, the coefficient of coupling is 1.0, or 100 percent.
3. If the coefficient of coupling is high—that is, if it approaches a value of 1.0—the windings are said to be "tightly coupled." If the coefficient of coupling is low, the windings are said to be "loosely coupled."
4. Critical coupling is a value between loose coupling and tight coupling.
5. The secondary voltage of a transformer is related to the primary voltage by the transformer turns ratio. This is given by the equation

$$TURNS\ RATIO = \frac{NUMBER\ OF\ TURNS\ IN\ THE\ SECONDARY}{NUMBER\ OF\ TURNS\ IN\ THE\ PRIMARY}$$

6. If there are more turns of wire in the secondary than in the primary, it is called a *step-up transformer*.
7. If there are more turns of wire in the primary than in the secondary, it is called a *step-down transformer*.
8. The voltage ratio is the same as the turns ratio:

$$TURNS\ RATIO = VOLTAGE\ RATIO = \frac{SECONDARY\ WINDING\ VOLTAGE}{PRIMARY\ WINDING\ VOLTAGE}$$

WHAT ARE LOSSES IN TRANSFORMERS?

You might get the impression from the discussion on step-up and step-down transformers that it is possible to get more output power than input power. This is not possible in any mechanical device. The power in a purely resistive circuit is equal to the voltage times the current and is written mathematically as

$$P = E \times I$$

Suppose that the input power and the output power are equal in a transformer. (This is only possible in ideal transformers, because there is always some power loss within the transformer itself. However, we will just suppose that the input and output powers are equal.) At the same time suppose that the secondary voltage is twice the primary voltage. The relationship between the primary and secondary power is given as follows:

PRIMARY POWER = SECONDARY POWER

Therefore, since power $= E \times I$,

$E_1 I_1 = E_2 I_2$, where
E_1 and $I_1 =$ the voltage and current in the primary
E_2 and $I_2 =$ the voltage and current in the secondary

Now, if E_2 is twice the value of E_1, the equation will be unbalanced unless I_2 is ½ of the value of I_1. In other words, if you double the secondary voltage and cut the secondary current in half, the secondary power will still be equal to the primary power. We can make a general rule then: *When the voltage is stepped up in a transformer, the current at the same time must be stepped down.* Likewise, *if the voltage is stepped down, the amount of current flowing in the secondary winding is stepped up.* The secondary current, of course, always depends upon the amount of load connected to the secondary winding.

It has been mentioned that the output power of a transformer can never be greater than or even equal to the input power. The reason for this is that there is always some loss in the coupling circuit, which is, in this case, the transformer. There are three important places where power is lost in the transformer itself.

First, the coils and the connections to the coils are made from conductors (usually copper). All conductors have a certain amount of resistance. When current flows through this resistance, heat is generated, and this heat represents a power loss.

The *second* place where power is lost in a transformer is in the iron core itself. Here, again, the basic idea of Faraday's law is important. Everytime there is a moving magnetic field cutting across a conductor, a voltage is induced. The iron used for the core of a power transformer is a conductor. Therefore, an expanding and contracting flux in the primary and secondary cuts across this iron core and induces a voltage in it, and this causes the current to flow in a circular path within the iron core itself. The circulating currents are called *eddy currents.* Since iron is a conductor which has resistance, it follows that current flowing within the iron causes the conductor to heat. This heat represents a power loss.

One method of reducing the losses due to eddy-current flow is illustrated in Fig. 19-12. Figure 19-12*a* shows the eddy currents in the iron core. Figure 19-12*b* shows how they are reduced by slicing the core into thin sections called

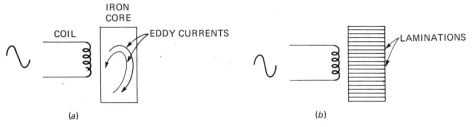

Figure 19-12. *One type of loss in transformers is the result of eddy currents flowing in the core.* (a) *A varying magnetic field from the coil causes eddy currents to flow in the iron.* (b) *When the core is laminated, the eddy currents are reduced.*

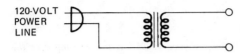

120-VOLT
POWER
LINE

laminations. The circular eddy currents cannot flow easily because of the insulating material between the laminations, but laminating the iron does not seriously affect its magnetic characteristics.

A *third* kind of loss in the transformer occurs as a result of the fact that the iron core becomes magnetized during each half cycle of current flow. When the current on each half cycle drops to zero, the iron core retains a small amount of the magnetism. This is true even when the iron core is made of soft iron. The soft iron becomes partially magnetized, and this magnetism must be removed before the flux can be reversed on each half cycle. It is the fact that the remaining flux on each half cycle must be removed by energy from the primary that causes a loss in transformers. This is called *hysteresis loss.*

WHAT IS REFLECTED IMPEDANCE?

If you connect a transformer to a power line with no load connected across the secondary, as shown in Fig. 19-13, the transformer takes virtually no power from the line. The current flow in the primary is limited by the inductive reactance of that coil, but no power is dissipated in a purely inductive reactance circuit. Of course, there is a small amount of resistance in the primary winding, and this resistance will result in heat dissipation when a small current flows through it. The amount of power dissipation is so small that it can usually be neglected.

When a load is connected across the secondary, it means that power is being taken from the transformer. This power must be supplied by the power source which is connected to the primary. In other words, you cannot draw power from the secondary of a transformer without supplying it to the primary. Thus, connecting a load across a secondary of a transformer causes a direct change in the amount of current flowing in the primary.

Figure 19-14 shows what happens when a high resistance is connected to the secondary winding. The arm of the variable resistor is adjusted so that the secondary current must flow through all the resistance. With a high resistance in the secondary, very little current is delivered to the secondary winding. This results in a very low primary current flow. Although it is low, it is still somewhat higher than it would be if no resistance were connected at all across the secondary—that is, if the secondary winding were open.

Figure 19-15 shows what happens when a low resistance value is connected to the secondary. In this case, the arm of the potentiometer is adjusted so that

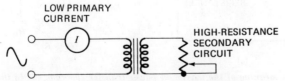

LOW PRIMARY
CURRENT

I

HIGH-RESISTANCE
SECONDARY
CIRCUIT

Figure 19-14. A high resistance in the secondary circuit causes a low current flow in the primary of the transformer.

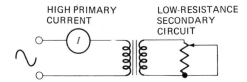

Figure 19-15. A low resistance in the secondary circuit causes a high current flow in the primary of the transformer.

there is actually very little resistance across the secondary. Note that there is a very large current flow in the secondary, and in order to supply this current, it is necessary for the primary current to be large. It can be concluded from the foregoing that the amount of current flowing in the primary is directly dependent upon the amount of current flowing in the secondary. Since the impedance of the secondary circuit directly affects the amount of current flowing in the primary circuit, the impedance is said to be "reflected" into the primary. This is what is meant by *reflected impedance*.

The transformer in Figs. 19-13, 19-14, and 19-15 is presumed to have a turns ratio of $1:1$. Under this condition any change in secondary current is directly reflected back to the primary. If the turns ratio is other than $1:1$, then there will still be a reflected impedance, but the amount will depend on the actual turns ratio. For example, if the turns ratio is $1:1$, then doubling the secondary current will also cause the primary current to double. If the turns ratio is not $1:1$, then doubling the secondary current will not cause the primary current to double. The actual change in the latter case depends upon the turns ratio.

WHAT IS MEANT BY "IMPEDANCE MATCHING"?

In order to understand the need for impedance matching it will be useful to review the meaning of the *maximum power transfer theorem*. This theory is explained in Fig. 19-16. Figure 19-16a shows a simple dc circuit with a battery connected across the load resistance (R_L). The battery has an internal resistance R_i and voltage E. (This is a practical circuit because *all* sources of power, whether ac or dc, have internal resistance.) If you measured the voltage across the terminals of the battery without a load connected to the terminals, the terminal voltage would be equal to E. However, when current is flowing in the circuit, there is a voltage drop across R_i, which subtracts from E. This makes the

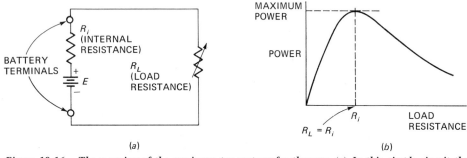

     (a)   (b)

Figure 19-16. The meaning of the maximum power transfer theorem. (a) *In this simple circuit the amount of power dissipated by the load resistance depends upon the value of internal resistance* (R_i) *and also upon the value of* R_L. (b) *This graph shows that the maximum power dissipated by the load resistance occurs when* $R_L = R_i$.

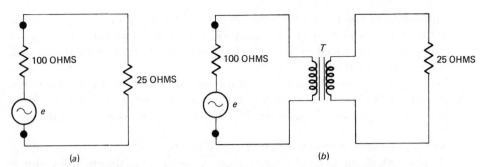

Figure 19-17. *Impedance matching is an important application of transformers.* (a) *There is an impedance match in this circuit.* (b) *The transformer serves as an impedance-matching device.*

terminal voltage less than the value of E. The greater the amount of current flow in the circuit, the greater the drop across R_i and the lower the terminal voltage.

If the value of load resistance in the circuit of Fig. 19-16a is changed, the value of current through the load resistance also changes. At the same time the terminal voltage of the battery changes.

The power dissipated by R_L depends on the resistance of R_L. The value of power and the value of load resistance are plotted on the graph shown in Fig. 19-16b. This curve indicates that the power rises very rapidly from the point where the load resistance is zero, up to a maximum value of power that occurs when the load resistance is equal to R_i ($R_L = R_i$).

If the load resistance is increased beyond the value of R_i, the power begins to decrease again. The curve of Fig. 19-16b is called the *maximum power transfer curve*. It shows that the maximum possible power that you can get from any power source occurs when an external resistance equals the internal resistance. A similar thing occurs in ac circuits, although it is somewhat more complicated when the load includes inductances and capacitances. However, the basic principle is the same.

Figure 19-17 shows how the maximum power transfer theorem applies to transformer circuitry. A simple case of an ac generator with a pure resistance load is shown in Fig. 19-17a. In this case, the internal resistance of the generator is a pure resistance. The load resistance has a value of 25 ohms, and the internal resistance of the generator is 100 ohms. This condition will not produce a maximum power transfer. However, a transformer can be inserted between the generator and the load to match the impedances, as shown in Fig. 19-17b.

If the turns ratio of the transformer is properly chosen, then the ac generator will look into an impedance of 100 ohms and the load will look into an impedance of 25 ohms. When the impedances are matched, the maximum possible power transfer can be obtained. This is an important application of transformers in electricity and electronics—that of matching impedances.

SUMMARY

1. There are three important kinds of losses in transformers: copper loss, eddy-current loss, and hysteresis loss.
2. Copper loss occurs because the primary and secondary windings are made with copper wire. Current flowing through the wire resistance causes a power loss.

3. Eddy-current loss occurs because the expanding and contracting magnetic fields of the primary and secondary cut across the iron core. This causes currents to be induced in the core. The currents, flowing in the iron, result in a power loss in the form of heat.
4. Hysteresis loss occurs because a small amount of permanent magnetism remains in the core after each half cycle. This magnetism must be removed before the core can be magnetized in the reverse direction during the next half cycle.
5. The impedance connected across the secondary of a transformer is reflected back to the primary. Any change in secondary current will result in a change in primary current.
6. A transformer can be used to match the internal impedance of an ac generator to a load having a different value.

Programmed Review Questions

(Instructions for using this programmed section are given in Chap. 1.)

We will now review the important concepts of this chapter. If you have understood the material, you will progress easily through this section. Do not skip this material, because some additional theory is presented.

1. In a transformer, power is applied to the
    A. primary winding. (Proceed to block 7.)
    B. secondary winding. (Proceed to block 17.)

2. *Your answer to the question in block 3 is **A**. This answer is wrong. The transformer described in the question has less turns of wire in the secondary than in the primary. This is a step-down transformer. Therefore, there must be less voltage in the secondary than in the primary. If there is 100 volts applied to the primary, there cannot be 500 volts across the secondary.* Proceed to block 13.

3. *The correct answer to the question in block 11 is **A**. Eddy currents get their name from the fact that they flow in circles like whirlpools, or eddies. Eddy-current losses can be reduced in power transformers by laminating the core. Transformers that operate at higher frequencies must use air cores to prevent excessive eddy-current losses.* Here is your next question.
    The primary of a certain power transformer has 250 turns and the secondary has 50 turns. If 100 volts is applied across the primary, the secondary voltage will be
    A. 500 volts. (Proceed to block 2.)
    B. 20 volts. (Proceed to block 13.)

4. *The correct answer to the question in block 12 is **B**. Although there are other advantages to using ac for power distribution, the one emphasized in this chapter is that of using the transformer to compensate for a drop in the line voltage.* Here is your next question.
    If a direct and an alternating current are flowing at the same time in the primary winding of a transformer,

A. the transformer will be destroyed. (Proceed to block 16.)
B. only the ac energy will be coupled to the secondary. (Proceed to block 9.)

5. *Your answer to the question in block 13 is* **B**. *This answer is wrong. Copper loss occurs in a transformer when current flows through the windings having copper wire. These windings have a small amount of resistance, and when current flows through the resistance, power is lost in the form of heat.* Proceed to block 10.

6. *Your answer to the question in block 14 is* **B**. *This answer is wrong. In a step-down transformer there are less turns of wire in the secondary than there are in the primary.* Proceed to block 11.

7. *The correct answer to the question in block 1 is* **A**. *Power is applied* **to** *the primary winding and is taken* **from** *the secondary winding.* Here is your next question.
   For the transformer of Fig. 19-18,
   A. there are three turns of wire in the primary and three turns of wire in the secondary. (Proceed to block 15.)
   B. the number of turns of wire in the primary and in the secondary are not known. (Proceed to block 12.)

8. *Your answer to the question in block 12 is* **A**. *This answer is wrong. Transformers do not convert ac to dc. (Rectifiers are components that convert ac to dc.)* Proceed to block 4.

9. *The correct answer to the question in block 4 is* **B**. *Dc energy cannot pass from the primary to the secondary in a transformer. Transformer action depends upon a* **varying** *magnetic field that is created around the primary coil as a result of a varying current in that coil.* Here is your next question.
   In an ideal transformer all the magnetic flux lines from the primary are linked with all the secondary turns. The coefficient of coupling in such an ideal transformer would be
   A. 1.0. (Proceed to block 14.)
   B. 0. (Proceed to block 19.)

10. *The correct answer to the question in block 13 is* **A**. *When two circuits are inductively coupled, a changing current in the first circuit induces a voltage in the second circuit. Likewise, a varying current in the second circuit produces a varying current in the first. This is called* **mutual inductance**. Here is your next question.
    In a certain transformer the voltage across the primary is 120 volts and the voltage across the secondary is 240 volts. Does this mean

Figure 19-18. Symbol for a transformer.

that there is more power in the secondary than in the primary? (Proceed to block 20.)

11. *The correct answer to the question in block 14 is **A**. There are more turns of wire on the secondary than there are on the primary, so this is a step-up transformer.* Here is your next question.
Moving magnetic fields cutting across the iron core of a transformer cause circulating currents to generate heat in the core. These currents are called
A. eddy currents. (Proceed to block 3.)
B. hysteresis currents. (Proceed to block 18.)

12. *The correct answer to the question in block 7 is **B**. It is not possible to tell the number of turns of wire from the symbol. The manufacturer may give this information in a catalog. However, the turns ratio is more important than the actual number of turns in each coil.* Here is your next question.
An advantage of using ac instead of dc for power distribution in cities is that
A. transformers can be used to convert the ac to dc whenever it is needed. (Proceed to block 8.)
B. transformers can be used to step the voltage up in order to compensate for a voltage drop in the line. (Proceed to block 4.)

13. *The correct answer to the question in block 3 is **B**. The turns ratio is*

$$\frac{50}{250} = \frac{1}{5}$$

There are 1/5 as many turns in the secondary as in the primary. Therefore, the secondary voltage will be 1/5 of the value of the primary voltage. 1/5 × 100 = 20. Here is your next question.
When a varying magnetic field in one circuit causes a voltage to be induced in another circuit, the effect is called
A. mutual inductance. (Proceed to block 10.)
B. copper loss. (Proceed to block 5.)

14. *The correct answer to the question in block 9 is **A**. In an ideal transformer, where all the flux lines of the primary are linked with the secondary, the coefficient of coupling would be 1.0. This value cannot be obtained in practice, but with iron-core transformers it can be closely approached.* Here is your next question.
A certain power transformer has three times as many turns of wire on the secondary as on the primary. This is a
A. step-up transformer. (Proceed to block 11.)
B. step-down transformer. (Proceed to block 6.)

15. *Your answer to the question in block 7 is **A**. This answer is wrong. Did you obtain your answer by counting the loops in the transformer? You cannot do this.* For further explanation, proceed to block 12.

16. *Your answer to the question in block 4 is **A**. This answer is wrong. The transformer will not be destroyed unless the input power applied to the primary*

winding is excessive, or unless the output power taken from the secondary winding is excessive. Proceed to block 9.

17. *Your answer to the question in block 1 is* **B**. *This answer is wrong. Power is always applied to the transformer primary. This must be an ac power if there is to be a voltage across the secondary.* Proceed to block 7.

18. *Your answer to the question in block 11 is* **B**. *This answer is wrong. There is no such thing as a hysteresis current in the transformer.* Proceed to block 3.

19. *Your answer to the question in block 9 is* **B**. *This answer is wrong. If the coefficient of coupling between two coils is 0, it means that none of the flux lines from one cuts across the turns of the other. This would not be an ideal transformer.* Proceed to block 14.

20. *The correct answer to the question in block 10 is* **no**. *Even though there is more voltage across the secondary winding, there will be less secondary current than in the primary. Thus, the secondary power* (E × I) *will never be greater than the primary power.*

    You have now completed the programmed questions. The next step is to put some of these ideas to work in laboratory experiments. Proceed to the "Experiment" section that follows.

Experiment

(The experiment described in this section may be performed on the circuit board described in Appendix C or on a similar laboratory setup.)

EXPERIMENT 1

Purpose — To demonstrate the relationships in transformers.

Theory — With relatively simple measurements it is possible to determine the turns ratio and the transformer losses in a transformer. The turns ratio (N_2/N_1) is simply obtained by dividing the secondary voltage (E_2) by the primary voltage (E_1):

$$\frac{E_2}{E_1} = \frac{N_2}{N_1}$$

The power loss in a transformer can be obtained by measuring the input power (P_1) and the output power (P_2). The loss in the transformer is obtained by the simple equation

TRANSFORMER POWER LOSS = INPUT POWER − OUTPUT POWER

Test Setup — Figure 19-19 shows the test setup for making the measurements required for this experiment.

If you have a volt-ohm-milliammeter, you can make each of the measurements shown in Fig. 19-19 one at a time.

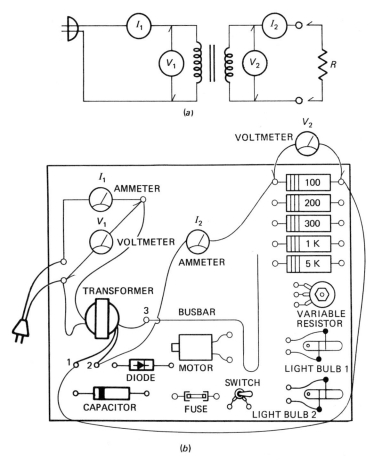

Figure 19-19.   Test setup for the experiment. (a) *Circuit for measuring input power* ($V_1 \times I_1$) *and output power* ($V_2 \times I_2$) *in a transformer circuit.* (b) *This is the way your circuit board will look if you have wired the circuit correctly.*

Procedure—

Step 1—Measure the resistance of R. Use an ohmmeter if you have one. If not, measure the current through R and the voltage across R when it is connected to the transformer secondary winding. Then calculate the resistance using Ohm's law:

$$R = \frac{E}{I} = \frac{\textit{THE VOLTAGE ACROSS THE RESISTOR}}{\textit{THE CURRENT THROUGH THE RESISTOR}}$$

If your meter measures the current in milliamperes, then you must divide the current by 1,000 to convert the value to amperes.

$$\frac{\textit{METER READING IN MILLIAMPERES}}{\textit{1,000}} = \frac{\textit{CURRENT READING}}{\textit{IN AMPERES}}$$

Remember this important fact: You can use Ohm's law only when the voltage is in volts and the current is in amperes.

The value of resistance is _____ (larger, equal to, lower than) the rated value. (The rated value is the value stated by the manufacturer. This is given by the color code, or by printing on the resistor.)

Step 2—Connect the resistor across the secondary winding. Measure V_2 and I_2:

$$V_2 = \underline{\hspace{2cm}} \text{ volts}$$
$$I_2 = \underline{\hspace{3cm}} \text{ milliamperes}$$

Step 3—Calculate the power in the secondary circuit by using the equation $P_2 = E_2 \times I_2$. Remember that if your meter gives the current in milliamperes, you must convert the current to amperes by moving the decimal place to the left three places before you use the power equation.

$$P_2 = \underline{\hspace{2cm}} \text{ watts}$$

WARNING: THE NEXT STEP COULD BE DANGEROUS. BE SURE THAT YOU DO NOT TOUCH ANY OF THE CIRCUITRY WITH YOUR BARE HANDS!

Step 4—Measure the voltage (V_1) and the current (I_1) in the transformer primary while the resistor is connected to the secondary.

$$V_1 = \underline{\hspace{2cm}} \text{ volts}$$
$$I_1 = \underline{\hspace{3cm}} \text{ milliamperes}$$

Step 5—Calculate the primary power using the equation $P_1 = V_1 \times I_1$. Again, convert I_1 from milliamperes to amperes.

$$P_1 = \underline{\hspace{2cm}} \text{ watts}$$

Step 6—Determine the power loss in the transformer by subtracting the secondary power from the primary power:

$$\text{Power loss} = P_2 - P_1 = \underline{\hspace{2cm}} \text{ watts}$$

Step 7—Calculate the turns ratio of the transformer by using the equation

$$TURNS\ RATIO = \frac{V_2}{V_1} = \underline{\hspace{3cm}}$$

Step 8—Calculate the turns ratio of the transformer by using the equation

$$TURNS\ RATIO = \frac{I_1}{I_2} = \underline{\hspace{3cm}}$$

Step 9—Do the values obtained in Steps 7 and 8 agree with reasonable closeness—that is, within 20 percent?

$$\underline{\hspace{4cm}}$$

Conclusion—The power loss in a power transformer is very small compared to the output power.

The turns ratio is directly related to the voltage ratio, but it is inversely related to the current ratio. Thus,

$$TURNS\ RATIO = \frac{N_2}{N_1} = \frac{V_2}{V_1} = \frac{I_1}{I_2}$$

(Note that the subscripts of the current ratio are opposite to the subscripts of the turns ratio or voltage ratio.)

Self-Test with Answers

(Answers with discussions are given in the next section.)

1. Which of the following would have a higher coefficient of coupling? (*a*) An iron-core transformer; (*b*) An air-core transformer.
2. Which of the following cannot be performed by a transformer? (*a*) Change an ac voltage from one value to a higher value; (*b*) Change an ac voltage from one value to a lower value; (*c*) Pass an ac voltage and reject a dc voltage; (*d*) Change an ac voltage to a dc voltage.
3. The type of transformer loss due to circular currents flowing in the iron core is (*a*) hysteresis loss; (*b*) eddy-current loss; (*c*) copper loss.
4. The iron core of a power transformer becomes partially magnetized on each half cycle, and this magnetism must be removed before the magnetism in the core can be reversed on the next half cycle. This produces a loss called (*a*) hysteresis loss; (*b*) eddy-current loss; (*c*) copper loss.
5. The windings of a power transformer are made of copper conductor (wires). Since all conductors have a certain amount of resistance, there will be a heat loss when current flows through the windings. This represents a power loss called (*a*) hysteresis loss; (*b*) eddy-current loss; (*c*) copper loss.
6. Which of the transformer losses listed below would occur in air-core as well as in iron-core transformers? (*a*) Hysteresis loss; (*b*) Eddy-current loss; (*c*) Copper loss.
7. A certain transformer has a turns ratio of 3 : 1. This is the ratio of secondary to primary turns. If the primary has a voltage of 50 volts across it, the secondary winding voltage will be (*a*) 3 volts; (*b*) 50 volts; (*c*) 150 volts; (*d*) 300 volts.
8. In a power transformer, the actual output power is (*a*) always slightly less than the input power due to losses within the transformer itself; (*b*) exactly equal to the input power; (*c*) always greater than the input power.
9. The type of transformer in which a single coil is used for both the primary and the secondary is the (*a*) air-core transformer; (*b*) autotransformer.
10. When there is no load connected to the secondary of a transformer, the power delivered to the primary is (*a*) maximum; (*b*) minimum.

Answers to Self-Test

1. (*a*)—An iron-core power transformer may have a coefficient of coupling value that is close to 1.0 (100 percent).
2. (*d*)—A *rectifier* converts ac to dc. A transformer cannot perform this.
3. (*b*)—Eddy currents cause the core of the transformer to become heated. In most cases, heat in electrical and electronic components means a loss of power. Eddy-current losses are reduced in power transformers by laminating the core material.
4. (*a*)—Hysteresis loss is reduced to a minimum by the proper choice of iron-core material.

5. (*c*)—Copper loss can be reduced by using larger-diameter conductors for the winding.

6. (*c*)—Eddy-current and hysteresis losses are due to the presence of iron in the transformer core. They could not occur in an air-core transformer. However, the windings of an air-core transformer, being made of copper, can produce copper loss.

7. (*c*)—There are three times as many turns in the secondary; therefore, the secondary voltage will be three times the primary voltage: $3 \times 50 = 150$ volts.

8. (*a*)—The actual relationship is given by the simple equation

INPUT POWER = OUTPUT POWER
+ POWER LOST IN THE TRANSFORMER

9. (*b*)

10. (*b*)—There is very little power delivered to the primary of a transformer when there is no load connected to the secondary.

20.
What Is
Impedance?

Introduction

When you are dealing with simple dc problems, your primary concern is with the values of voltage, current, resistance, and power. Dc circuit problems seldom become more complicated than calculating series and parallel resistance values. In ac circuits, inductance and capacitance, in addition to the circuit resistance, affect the amount of current flowing, the ac voltage drops, and the amount of power dissipated. Thus, calculations of current and voltage in ac circuits are usually a little more involved than for dc circuits.

An inductor in an ac circuit opposes the flow of alternating current. The opposition that it offers is called inductive reactance, and it is measured in ohms. A capacitor in an ac circuit opposes the flow of alternating current, and its opposition is called capacitive reactance. Capacitive reactance is also measured in ohms. An ac circuit may contain resistance, inductive reactance, and capacitive reactance. All these forms of opposition are combined to obtain the complete opposition of the circuit, called impedance.

It was mentioned in an earlier chapter that you cannot calculate the impedance simply by adding the resistance and reactance values. One of the things that you will learn in this chapter is how to determine circuit impedances. The mathematics involved in calculating ac circuit values can become quite involved. However, in this chapter we will ignore the complicated mathematics and concentrate on solutions by graphical methods. There is nothing undignified about a graphical solution. In fact, engineers use graphics in many ways when solving problems.

You will be able to answer these questions after studying this chapter:

How are voltage, current, and power related in a purely resistive circuit?
How are voltage, current, and power related in an inductive circuit?
How are voltage, current, and power related in a capacitive circuit?
How is inductive reactance determined?
How is capacitive reactance determined?
What is impedance?
What is the relationship between the true power and the apparent power in an ac circuit?

Instruction

HOW ARE VOLTAGE, CURRENT, AND POWER RELATED IN A PURELY RESISTIVE CIRCUIT?

A *purely resistive circuit* is one that does not contain inductance or capacitance. Such a circuit is shown and analyzed in Fig. 20-1. In Fig. 20-1a an ac voltage (*e*)

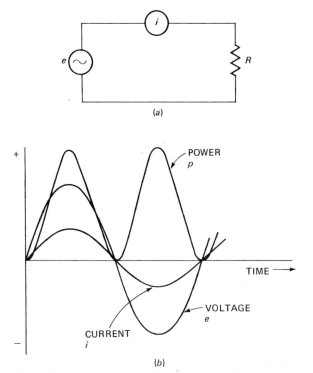

Figure 20-1. *Analysis of a simple resistive circuit.* (a) *A purely resistive circuit.* (b) *Relationship between voltage, current, and power in a purely resistive circuit.*

is applied across a resistor, and the resulting alternating current flow is measured by an ammeter (*i*). The voltage and current are represented by small letters in order to distinguish the ac circuit from the equivalent dc circuit. For this discussion we will presume that the ac generator produces a sine-wave voltage, and the current flowing through the circuit is a sine-wave current.

Figure 20-1*b* shows the relationship between the voltage, current, and power in the circuit. Notice that as the voltage rises to its maximum value, the current rises to its maximum value at the same instant. Also, at the instant the maximum positive or negative voltage and current occur, the power dissipated in the resistor is at a maximum. The power, of course, will rise and fall as the current rises and falls. In the graphical illustration a complete cycle of voltage and current are shown. It will be noted that the power goes through two complete cycles during this period.

The fact that the complete power wave is above the time axis in this graph is interpreted to mean that the power is always taken away from the circuit. In other words, we will define a positive value of power as being one in which power is taken from the circuit in the form of heat. (This will become important later. As you will learn, in other circuits it is not always true that the power is taken from the circuit.) Since the power is represented by a wave, you may get the impression that the resistor is heated and cooled rapidly as the current flows through it. Actually, the resistor heats to an average value between the minimum and maximum value of power. You know that when you turn a light off, the light bulb does not immediately get cool, but rather, it takes some time to cool off. The same is true of a resistor. If it becomes heated, it cannot follow the rapid power variations, so it stays at a temperature that is between the maximum and minimum value.

HOW ARE VOLTAGE, CURRENT, AND POWER RELATED IN AN INDUCTIVE CIRCUIT?

A *purely inductive circuit* is one that has only inductance, as shown in Fig. 20-2. If there are other components in the circuit, but the inductance is predominant, then it is called an *inductive circuit*. For example, there may be an inductor and a resistor in the circuit, but the inductor has the greater amount of influence on the flow of current. In such a case the circuit is said to be inductive because the inductor has the greater influence on the current flow.

Figure 20-2*a* shows a purely inductive circuit. The voltage, current, and power for this circuit are shown in the graph of Fig. 20-2*b*. Notice that the voltage and current do not go through their maximum values at the same instant of time. On a graph such as this the time becomes later as you move from left to right. Since the maximum value of the current is to the right of the maximum value of the voltage, you can say that the current lags the voltage. In a purely inductive circuit, the amount of lag is 90°, or one-fourth cycle. This simply means that when the voltage is maximum, the current is zero, and when the voltage is zero, the current is maximum. The relationship is shown in Fig. 20-2*b*. The power for a purely inductive circuit is also shown on this graph. It is important to note that the power has both positive and negative values. Earlier we defined positive power as being power that is dissipated in the form of heat. It must follow, then, that negative power is power that is delivered to the circuit rather than being taken away from it. Notice then in Fig. 20-2*b* that in one-half

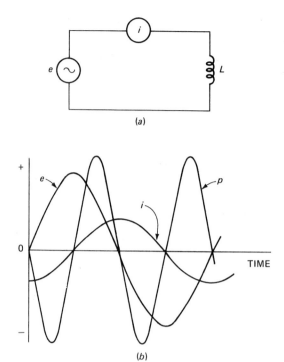

Figure 20-2. Analysis of a simple inductive circuit. (a)
A purely inductive circuit. (b) *Relationship between voltage, current, and power in a purely inductive circuit.*

cycle, the power is being delivered to the circuit, while in the next half cycle,
the power is being taken away. The average value between the positive peaks
and the negative peaks of the power wave is zero. This means that *there is no
power dissipated in a purely inductive circuit.*

The negative peaks of power occur in the circuit when the magnetic field
collapses and induces a voltage in the coil. This produces a voltage in series
with the applied voltage. The two voltages are additive and can produce quite a
large current. Since no power is dissipated in the inductor, it is necessary for all
the power to be dissipated within the internal resistance of the generator. The
large current flowing through the internal resistance can actually cause the
generator to burn out. For this reason, we can make an important rule: *Never
connect an inductor directly across an ac generator as shown in Fig.* 20-2a.

Figure 20-3 shows the relationship between voltage, current, and power in
an *R-L* (resistor-inductor) circuit. The circuit is shown in Fig. 20-3*a*, and the
graph is shown in Fig. 20-3*b*. Note that in the graphical representation the current is still lagging behind the voltage, but not by as much as it was in a purely
inductive circuit. In other words, the effect of adding resistance is to bring the
current and voltage in step, or *in phase*. The greater the amount of resistance
added, the more closely the current and voltage will be in phase.

It is important to note that in the circuit of Fig. 20-3 there is a small
amount of negative power, indicating that some power is being fed back into
the circuit. However, most of the power wave is positive, indicating that power
is being taken away in the form of heat.

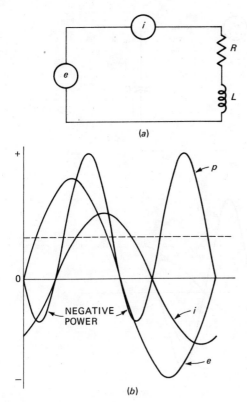

Figure 20-3. Analysis of an R-L circuit.
(a) *An* **R-L** *circuit.* (b) *Relationship between voltage, current, and power in an* R-L *circuit.*

SUMMARY

1. In an ac circuit the current flow is affected by inductance, capacitance, and resistance.
2. The opposition that an inductor offers to the flow of alternating current is called inductive reactance. It is measured in ohms.
3. The opposition that a capacitor offers to the flow of alternating current is called capacitive reactance. It is measured in ohms.
4. Positive power is defined as power taken from the circuit.
5. In a purely resistive circuit the power is positive. The current and voltage are in series in a purely resistive circuit.
6. In a purely inductive circuit, the current lags behind the voltage by 90°. No power is dissipated in a purely inductive circuit.

HOW ARE VOLTAGE, CURRENT, AND POWER RELATED IN A CAPACITIVE CIRCUIT?

Figure 20-4 shows the relationship between voltage, power, and current in a capacitive circuit. The circuit is shown in Fig. 20-4*a*. Figure 20-4*b* shows the relationship between voltage, current, and power. You will note that the current is at its maximum value first, and then the voltage follows. In other words, the current is leading the voltage.

In a purely capacitive circuit the current leads the voltage by one-fourth

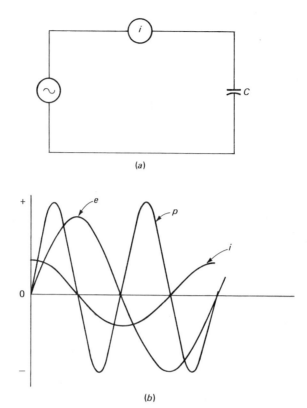

Figure 20-4. *Analysis of a simple capacitive circuit.* (a) *A purely capacitive circuit.* (b) *Relationship between voltage, current, and power in a purely capacitive circuit.*

cycle, or 90°. Saying that the current leads the voltage is the same as saying that the voltage lags behind the current in a capacitive circuit.

The graph of power in Fig. 20-4 is similar to the one shown for the purely inductive circuit. Note that the power wave is equally distributed in the positive and negative areas. This means that power is first taken away from the circuit and then returned to the circuit.

There are two cycles of power for every one cycle of voltage or one cycle of current. The fact that power is being dissipated only on one half of the power cycle, and returned on the next, means that a capacitor should never be connected directly across the generator terminals. The power returned to the circuit must be dissipated someplace, and in this circuit it is dissipated inside the generator. The result is an overheated generator.

Figure 20-5 shows the relationship between the voltage, current, and power in an *R-C* circuit. The actual circuit is shown in Fig. 20-5a. As in the case of an inductive circuit, the effect of the resistance is to pull the voltage and current together. The greater the amount of resistance, the more closely the voltage and current will be in phase.

In Fig. 20-5b you will note that the power is mostly in the positive area, indicating that power is being taken away from the circuit in the form of heat. A small amount of the power goes below the time axis in this circuit, showing that some power is returned to the circuit.

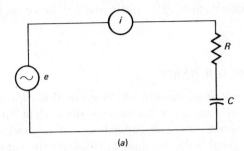

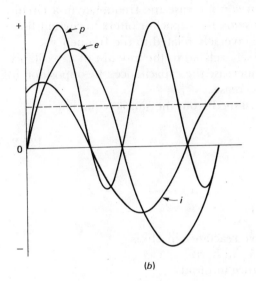

Figure 20-5. Analysis of an R-C circuit. (a) An R-C circuit. (b) Relationship between voltage, current, and power in an R-C circuit.

HOW IS INDUCTIVE REACTANCE DETERMINED?

You have read that an inductor placed in an alternating-current circuit opposes the flow of alternating current, and the opposition is called inductive reactance. Inductive reactance, like resistance, is measured in ohms. There are two things that affect the amount of inductive reactance that a coil has. First, the amount of inductance is definitely related to the amount of inductive reactance. In other words, in comparing two coils in a given circuit, the one with the larger inductance will have the greater inductive reactance. The second thing that determines the amount of inductive reactance is the frequency of the circuit. The higher the frequency, the greater the opposition that a coil will offer to the flow of alternating current.

The relationship between inductive reactance, frequency, and inductance is given by the simple equation

$$X_L = 2\pi fL \tag{20-1}$$

or

$X_L = 6.28fL$, where
X_L = the inductive reactance in ohms
f = the frequency in hertz
L = the inductance in henrys

The important thing about this equation is that it shows that increasing either the frequency or the inductance causes the inductive reactance to increase.

HOW IS CAPACITIVE REACTANCE DETERMINED?

Capacitive reactance is the opposition that a capacitor offers to the flow of alternating current. It is measured in ohms. Capacitive reactance is dependent upon two things: the frequency of the circuit, and the capacitance of the circuit. Unlike the inductive reactance, the greater the frequency, the *lower* the capacitive reactance. In other words, when you increase the frequency in a circuit, you will decrease the amount of opposition the capacitor offers to current flow. The capacitive reactance is said to be inversely related to the frequency.

Capacitive reactance is also inversely related to the amount of capacitance in the circuit. This means that if you increase the capacitance, the opposition to the flow of alternating current will decrease.

Capacitive reactance can be determined from the simple equation

$$X_C = \frac{1}{2\pi f C} \qquad\qquad (20\text{-}2)$$

or

$$X_C = \frac{0.159}{fC}, \text{ where}$$

X_C = the capacitive reactance in ohms
 f = the frequency in hertz
 C = the capacitance in farads

The importance of this equation is that it shows you that the frequency and the capacitance are inversely related to the capacitive reactance. If you increase either f or C, the value of the denominator increases in the equation, and if you increase the size of the denominator, you decrease the size of the fraction. Thus, a greater f or C means a smaller X_C.

SUMMARY

1. In a purely capacitive circuit, the current leads the voltage by 90°. No power is dissipated in a purely capacitive circuit.
2. The relationship between inductive reactance, frequency, and inductance is given by the equation

$$X_L = 2\pi f L$$

Inductive reactance is measured in ohms.

3. The relationship between capacitive reactance, frequency, and capacitance is given by the equation

$$X_C = \frac{1}{2\pi f C}$$

Capacitive reactance is measured in ohms.

WHAT IS IMPEDANCE?

Resistance, inductive reactance, and capacitive reactance are all forms of opposition to the flow of alternating current. When more than one of these is in the circuit at the same time, the combined opposition to the flow of current is called *impedance*. Notice that we do not say that impedance is the "total opposition," because this would imply that you could find it by adding the values. The actual process is just a little bit more involved than simple addition.

To understand how the impedance in a circuit can be determined, we must first consider some standard conventions. Figure 20-6 shows three vectors used to represent the resistance, inductive reactance, and capacitive reactance. These vectors are said to be in their *standard positions*. In other words, the inductive reactance vector is always drawn upward at an angle of 90° with the resistance vector, and the capacitive reactance vector is always drawn downward at an angle of 90° with the resistance vector. Note that the inductive reactance and capacitive reactance are in a straight line. In a circuit that contains both X_C and X_L, the effective reactance is obtained simply by subtracting them. In other words,

$$X = X_L - X_C$$

or

$$X = X_C - X_L$$

To find the reactance of the circuit simply subtract the smaller reactance from the larger one.

> *Example 20-1* — Suppose a certain circuit has an inductive reactance of 10 ohms and a capacitive reactance of 4 ohms. What is the reactance (X) of the circuit?
> *Solution* —
>
> $$X = X_L - X_C$$
> $$= 10 - 4$$
> $$= 6 \text{ ohms} \qquad\qquad Answer$$

Since the X_L is larger than the X_C, the circuit is said to be inductive.

Figure 20-7 shows X_L, R, and X_C in their standard positions on an impedance triangle. This triangle clearly illustrates the relationship between the

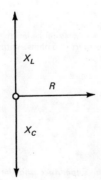

Figure 20-6. Standard positions for the vectors. To be scientifically correct, these vectors should be called phasors.

R
L
THE CIRCUIT
THE IMPEDANCE TRIANGLE
Z
θ
X_L
R
(a)
R
C
THE CIRCUIT
THE IMPEDANCE TRIANGLE
R
θ
Z
X_C
(b)

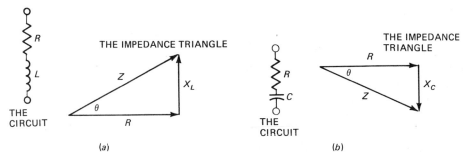

Figure 20-7. Standard impedance triangles for finding impedance and phase angle. (a) ***An impedance triangle for an* R-L *circuit.*** (b) ***An impedance triangle for an* R-C *circuit.***

impedance and the circuit values. Note that the impedance is the hypotenuse of the right triangle.

In Fig. 20-7*a* the triangle is obtained by adding the R and X_L phase *vectorially*. The hypotenuse of the triangle and the phase angle between the voltage and current is shown as θ. We will show you how to find the impedance of an *R-L* circuit, but first consider the impedance triangle in Fig. 20-7*b*. In this case the circuit involved contains a resistor and a capacitor in series. The impedance of the combined opposition to current flow is the hypotenuse of a right triangle formed by combining the R and X_C vectors at right angles. Notice that the X_C vector is pointing downward in this case instead of upward as shown in Fig. 20-7*a*.

The angle θ is the phase angle between the voltage and current in the circuit. This is interpreted to mean that the voltage lags behind. Figure 20-8 shows a

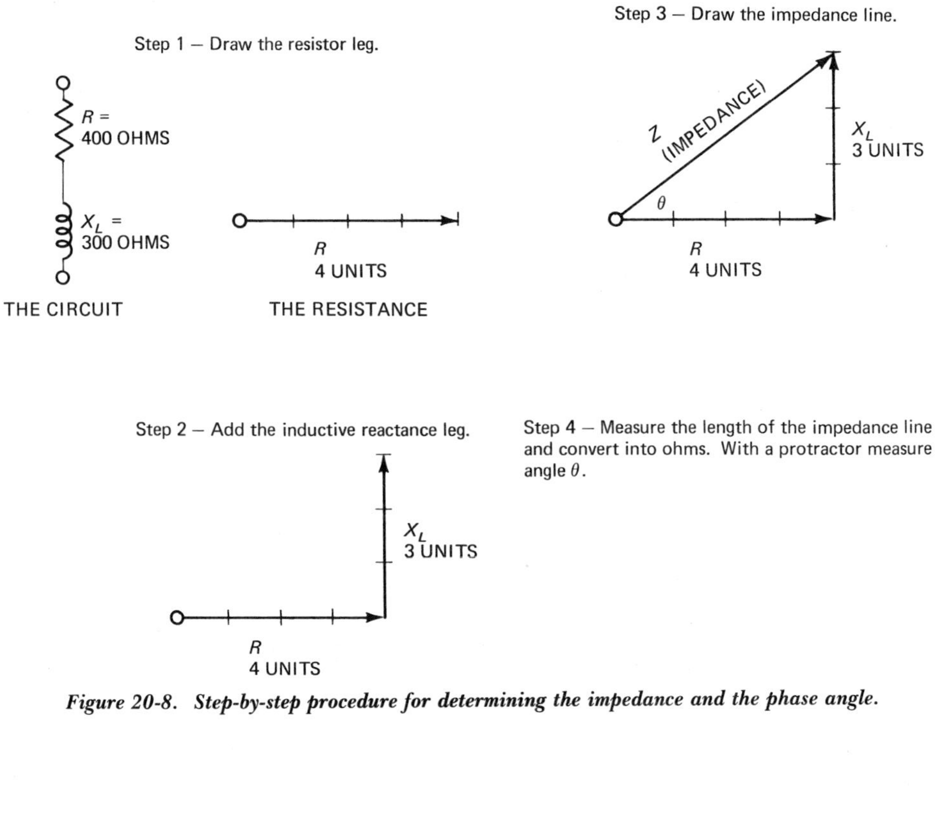

Step 1 – Draw the resistor leg.
R = 400 OHMS
X_L = 300 OHMS
THE CIRCUIT
R 4 UNITS
THE RESISTANCE
Step 3 – Draw the impedance line.
Z (IMPEDANCE)
θ
X_L 3 UNITS
R 4 UNITS
Step 2 – Add the inductive reactance leg.
X_L 3 UNITS
R 4 UNITS
Step 4 – Measure the length of the impedance line and convert into ohms. With a protractor measure angle θ.

Figure 20-8. Step-by-step procedure for determining the impedance and the phase angle.

step-by-step procedure for determining the impedance of an *R-L* circuit. The circuit in question is shown in Step 1. It consists of a 400-ohm resistor in series with a 300-ohm inductive reactance. The problem is to find out how much combined opposition (or impedance) this circuit offers to the flow of alternating current.

The first step is to draw a vector line representing the 400 ohms. In this case, the line can be 4 units long, which might be 4 inches or 4 centimeters, whichever is convenient. In Step 2 a vector line representing inductive reactance is drawn at right angles to the line representing resistance. The inductive reactance line must be exactly the same type of unit that was used in Step 1. In other words, if 4 inches represents 400 ohms in Step 1, then 3 inches will represent 300 ohms in Step 2. Step 3 shows how the impedance line is drawn. Actually, it is the hypotenuse of the triangle drawn from the start of the resistance vector to the end of the inductive reactance vector. The length of this impedance line (in this case 5 units long) represents the impedance of the circuit. Thus, an inductive reactance of 400 ohms in series with an inductive reactance of 300 ohms gives an impedance of 500 ohms.

The angle between the impedance line and the resistance line is called *theta* (θ). This angle can be measured with a protractor after the triangle has been constructed (Step 4). It represents the phase angle between the voltage and current in the circuit.

WHAT IS THE RELATIONSHIP BETWEEN THE TRUE POWER AND THE APPARENT POWER IN AN AC CIRCUIT?

In a purely resistive circuit, if you multiply the rms value of voltage by the rms value of current, you will obtain the power dissipated in watts. This is the average value of power, and it is given by the equation

$$P = E \times I, \text{ where} \tag{20-3}$$

$P =$ the average power of the circuit in watts
$E =$ the rms voltage in volts
$I =$ rms current in amperes

You will remember that in a purely resistive circuit the power is always positive, and the peak value of power occurs when the voltage and current are at either their positive or negative maximum values.

In a circuit that contains reactance, the voltage and the current do not reach their peak values at the same instant. If you multiply the rms voltage by the rms current in a reactive circuit, you will *not* get the actual power dissipated, but rather, you will get the *apparent power*. The apparent power is exactly what the term implies. It is the amount of power that appears to be dissipated, and it is determined by multiplying the rms value of voltage by the rms value of current. The true power is the power that is actually dissipated in the circuit as heat. It is a value somewhat less than the apparent power.

The relationship between true power and apparent power is shown in Fig. 20-9. Note that the apparent power is the hypotenuse of the triangle, and the true power is one of the legs. This is why the true power is always less than the apparent power. (In a purely resistive circuit the true power and the apparent power are exactly equal.)

The angle theta in the triangle has the same meaning as for the impedance

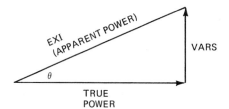

Figure 20-9.   A power triangle.

triangle. It represents the phase angle between the voltage and the current measured in degrees.

Vars is a term which stands for *reactive volt amperes.* The vars in an impedance triangle is equal to the voltage across a reactive component times the current through it. Obviously, this product will not give the value of power actually taken from the circuit. Remember that reactive components do not take any power from the circuit. Instead, they "borrow" power from the circuit during one-half cycle and return it during the next half cycle.

Appliances are sometimes rated according to their apparent power rather than their true power. Another name for apparent power is *volt amperes.*

SUMMARY

1. The combined opposition that a resistor, capacitor, and inductor offer to the flow of alternating current is called impedance. Impedance is measured in ohms.
2. Vectors can be used to represent resistance, capacitance, and inductance.
3. In an impedance triangle the resistance and reactance are at right angles and the impedance is represented by the hypotenuse.
4. The angle between the impedance vector and the resistor vector in an impedance triangle is the phase angle between the voltage and the current in the circuit represented by the impedance triangle.
5. There is no power dissipated by a reactive component in an ac circuit.
6. The apparent power in a reactive circuit is obtained by multiplying the rms voltage by the rms current. In a purely resistive circuit this product gives the real power.
7. The voltage across a reactive component (inductance or capacitance) multiplied by the current through it gives the vars.
8. The relationship between the apparent power, true power, and vars is shown in the power triangle.

Programmed Review Questions

(Instructions for using this programmed section are given in Chap. 1.)

We will review the important concepts of this chapter. If you have understood the material, you will progress easily through this section. Do not skip this material, because some additional theory is presented.

1. In a purely capacitive circuit
    A.   the current leads the voltage by 90°. (Proceed to block 7.)
    B.   the current and voltage are in phase. (Proceed to block 17.)

2. ***Impedance.*** *The impedance is obtained by combining the effects of resistance, inductive reactance, and capacitive reactance in a circuit. Impedance is measured in ohms.* Here is your next question.
   Which of the following equations is correct?
   A. $X_L = 2\pi fL$ (Proceed to block 3.)
   B. $X_C = 2\pi fC$ (Proceed to block 16.)

3. *The correct answer to the question in block 2 is **A**. The inductive reactance will increase if either the frequency or the inductance is increased.* Here is your next question.
   There is no negative power—that is, power returned to the circuit—in a purely
   A. resistive circuit. (Proceed to block 11.)
   B. inductive circuit. (Proceed to block 8.)

4. *Your answer to the question in block 13 is **A**. This answer is wrong. There is no such thing as capacitive resistance.* Proceed to block 18.

5. *Your answer to the question in block 11 is **A**. This answer is wrong. The equation for inductive reactance is* $X_L = 2\pi fL$. *From this equation you can see that decreasing the inductance and keeping the frequency the same will cause a decrease in inductive reactance.* Proceed to block 19.

6. *Your answer to the question in block 19 is **B**. This answer is wrong. A current leads the voltage in a purely capacitive circuit by 90°. However, in an inductive circuit the current lags the voltage by 90°.* Proceed to block 9.

7. *The correct answer to the question in block 1 is **A**. The current leads the voltage by 90°, or one-fourth cycle. This is another way of saying that the voltage lags the current by 90°.* Here is your next question.
   The combined opposition, including the resistance and the reactance, in an ac circuit is called the _____. (Proceed to block 2.)

8. *Your answer to the question in block 3 is **B**. This answer is wrong. In a purely inductive circuit the average power is zero. This means that power is taken away from the circuit on the positive half cycle and returned to the circuit on the negative half cycle.* Proceed to block 11.

9. *The correct answer to the question in block 19 is **A**. The current lags the voltage in an inductive circuit, and if it is a purely inductive circuit, the amount of lag is 90°, or one-fourth cycle. Instead of saying that the current lags the voltage, you could also say that the voltage leads the current. This is a restatement of the same thing.* Here is your next question.
   In a purely resistive circuit
   A. the voltage and current are in phase. (Proceed to block 13.)
   B. the voltage and current are 90° out of phase. (Proceed to block 10.)

10. *Your answer to the question in block 9 is **B**. This answer is wrong. In a purely resistive circuit the voltage and current are exactly in step, or "in*

phase." *This means that when the voltage is maximum, the current is also maximum.* Proceed to block 13.

11.   *The correct answer to the question in block 3 is **A**. In a purely resistive circuit power is always being taken away from the circuit in the form of heat. This is what we call "positive power." There is no negative power — that is, there is no power returned to the circuit.* Here is your next question.
If you decrease the inductance but do not change the frequency, the inductive reactance will be
A.   increased. (Proceed to block 5.)
B.   decreased. (Proceed to block 19.)

12.   *The correct answer to the question in block 18 is **B**. In a purely resistive circuit the true power and the apparent power are the same thing. It is obtained by multiplying the rms voltage by the rms current.* Here is your next question.
In the power circuit of Fig. 20-10, line *AB* represents
A.   the true power. (Proceed to block 15.)
B.   the apparent power. (Proceed to block 20.)

13.   *The correct answer to the question in block 9 is **A**. In an inductive or capacitive circuit the voltage and current are out of phase, but in a purely resistive circuit they are exactly in phase. Adding resistance to a capacitive or inductive circuit tends to reduce the amount of phase shift between the voltage and current. If there is no resistance in the circuit, the phase shift is exactly 90° in purely capacitive and inductive circuits.* Here is your next question.
The opposition that a capacitor offers to the flow of alternating current is called
A.   capacitive resistance. (Proceed to block 4.)
B.   capacitive reactance. (Proceed to block 18.)

14.   *Your answer to the question in block 18 is **A**. This answer is wrong. The word "vars" means reactive volt amperes. It is the volts times amperes (voltage times current) in a reactive circuit. In a purely resistive circuit there is no reactance. Therefore, you cannot get a value of vars in a purely resistive circuit.* Proceed to block 12.

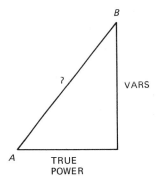

Figure 20-10.   Power triangle for the question in block 12.

15. *Your answer to the question in block 12 is **A**. This answer is wrong. Line AB in this illustration is the hypotenuse of a right triangle. The hypotenuse is longer than either of the sides. Remember that the apparent power in a reactive circuit is always greater than the true power. Therefore, line AB represents the apparent power and not the true power.* Proceed to block 20.

16. *Your answer to the question in block 2 is **B**. This answer is wrong. The frequency (f) and the capacitance (C) are inversely related to the capacitive reactance. In other words, the larger the frequency or capacitance, the lower the capacitive reactance.* Proceed to block 3.

17. *Your answer to the question in block 1 is **B**. This answer is wrong. In a purely capacitive circuit the voltage and the current are out of phase by 90° or by one-fourth cycle.* Proceed to block 7.

18. *The correct answer to the question in block 13 is **B**. Capacitive reactance is the opposition that a capacitor offers to the flow of alternating current.* Here is your next question.
    If you multiply the rms voltage (in volts) by the rms current (in amperes), and if the circuit is purely resistive, you will get
    A. vars. (Proceed to block 14.)
    B. true power. (Proceed to block 12.)

19. *The correct answer to the question in block 11 is **B**. The inductance and the inductive reactance are directly related. If you keep the frequency the same, increasing the inductance will cause the inductive reactance to increase. Decreasing the inductance will cause the inductive reactance to decrease.* Here is your next question.
    In a purely inductive circuit the current
    A. lags the voltage by 90°. (Proceed to block 9.)
    B. leads the voltage by 90°. (Proceed to block 6.)

20. *The correct answer to the question in block 12 is **B**. The apparent power is obtained by simply multiplying the rms voltage by the rms current in a reactive circuit.*
    You have now completed the programmed questions. The next step is to put some of these ideas to work in laboratory experiments. Proceed to the "Experiment" section that follows.

Experiment

(The experiment described in this section may be performed on the circuit board described in Appendix C or on a similar laboratory setup.)

EXPERIMENT 1

Purpose—To show that the voltage relationships in an ac circuit are not the same as for a dc circuit.

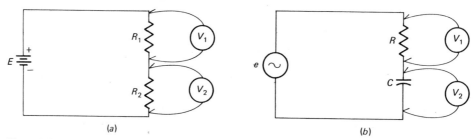

Figure 20-11. Comparison of dc and ac circuits and their voltage relationships. (a) *In the simple circuit the sum of the voltage drops* ($V_1 + V_2$) *equals the applied voltage* (E). (b) *In this simple circuit the sum of the voltage drops* ($V_1 + V_2$) *equals the applied voltage only if the voltage drops are added as vectors.*

Theory—You have learned in this chapter that the inpedance of a circuit cannot be determined by simply adding the resistance values to the reactance values. They must be combined vectorially. This can be done by simply drawing the resistance and reactance to scale to form a right triangle. The hypotenuse of the right triangle is equal to the impedance on the same scale, and the angle between the resistance and impedance lines is equal to the phase angle between the voltage and the current in the circuit.

Since the resistance and reactance must be combined vectorially to obtain the impedance, it is reasonable to expect that the voltage drops across these components must also be combined vectorially. Figure 20-11 illustrates this. In Fig. 20-11a the sum of the voltage drops across the resistors equals the applied voltage. This is known as Kirchhoff's voltage law.

In Fig. 20-11b the voltage across the resistor and the voltage across the capacitor *cannot be added to obtain the generator voltage!* You must combine the voltages vectorially. This is shown in Fig. 20-12. The voltage across the resistor is drawn horizontally, and the voltage across the capacitor is drawn vertically. Note that the voltage vectors (which are technically known as "phasors") are drawn in the same relative positions as the vectors representing the resistance and reactance in the impedance triangle.

Procedure—

Step 1—Connect the circuit as shown in Fig. 20-13.

Step 2—Measure the ac voltage drop across each resistor and record the values.

Voltage drop across $R_1 = V_1 =$ _____ volts
Voltage drop across $R_2 = V_2 =$ _____ volts
Voltage drop across $R_3 = V_3 =$ _____ volts

Step 3—Add the voltage drops obtained in Step 2.

$$V_1 + V_2 + V_3 = _____$$

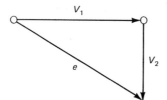

Figure 20-12. The voltages in Fig. 20-11b must be combined vectorially as shown here.

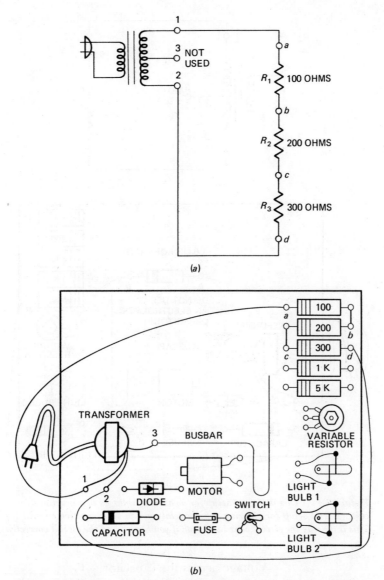

Figure 20-13. *This is the setup for measuring voltages in a resistive ac circuit. (a) Test setup for showing voltage relationships in a purely resistive circuit. (b) This is the way your circuit board will look if you have wired the circuit correctly.*

Step 4 — Measure the voltage across the transformer secondary — that is, the voltage between terminals 1 and 2 of the transformer. Record the value here.

$$e = \underline{\hspace{3cm}}$$

If you have performed the experiment properly, the total voltage obtained in Step 3 should equal the applied voltage in Step 4.

Step 5 — Wire the circuit as shown in Fig. 20-14.

Step 6 — Measure the voltage drop across the capacitor and the resistor and record the values here.

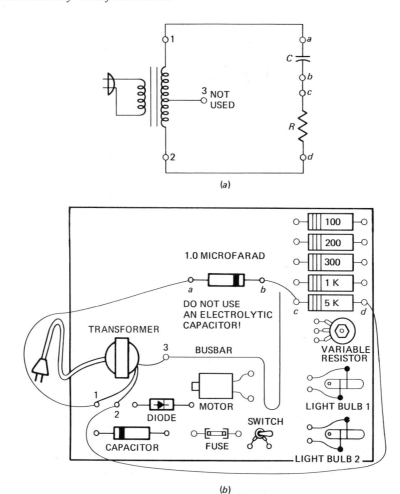

(a)

(b)

Figure 20-14.   This is the test setup for measuring voltages in a reactive ac circuit. (a) *Test setup for showing voltage relationships in a reactive circuit.* (b) *This is the way the circuit board will look if you have wired the circuit correctly.*

Voltage across the capacitor $= V_C =$ _____ volts
Voltage across the resistor $= V_R =$ _____ volts

Step 7—Combine the voltages vectorially as shown in Fig. 20-12. Find the applied voltage (*e*) and record the value here.

_____

Step 8—Measure the voltage across the transformer secondary and record the value here.

_____

If you have performed the measurements carefully, the voltage values in Steps 7 and 8 should be approximately the same.

Conclusion—The voltage drops across the components in a reactive circuit must be combined vectorially to equal the applied voltage.

Self-Test With Answers

(Answers with discussions are given in the next section.)

1. The opposition that an *R-L* or *R-C* circuit offers to the flow of alternating current is called (*a*) resistance; (*b*) reactance; (*c*) susceptance; (*d*) impedance.
2. Impedance is measured in (*a*) vars; (*b*) ohms; (*c*) mhos; (*d*) rels.
3. In a purely resistive ac circuit, the true power dissipated can be determined by multiplying (*a*) maximum voltage by maximum current; (*b*) rms voltage by rms current; (*c*) average voltage by average current; (*d*) current by resistance.
4. Inductive reactance is measured in (*a*) ohms; (*b*) henrys; (*c*) vars; (*d*) mhos.
5. Which of the following equations is correct?

$$(a)\ X_C = 2\pi fC \qquad (b)\ X_C = \frac{1}{2\pi fC}$$

6. The voltage drops across the resistor and inductor in a series *R-L* circuit (*a*) combine vectorially to equal the applied voltage; (*b*) can be added to determine the applied voltage.
7. Which of the following statements is true? (*a*) A large capacitor across the terminals of an ac generator will prolong its life; (*b*) A large capacitor across the terminals of an ac generator may cause the generator to burn out.
8. The term "vars" stands for (*a*) volts at resonance; (*b*) reactive volt amperes.
9. Which of the following series ac circuits would not dissipate any power? (*a*) *R-L;* (*b*) *R-C;* (*c*) pure capacitive; (*d*) *R-L-C.*
10. The voltage leads the current in (*a*) an inductive circuit; (*b*) a capacitive circuit.

Answers to Self-Test

1. (*d*) — If a circuit has both resistance and reactance, the combined opposition is called impedance.
2. (*b*)
3. (*b*)
4. (*a*)
5. (*b*)
6. (*a*) — In the experiment it was shown that the voltages across a resistor and across a capacitor combine "vectorially" to equal the applied voltage. The same is true of the voltages across a series *R-L* circuit.
7. (*b*) — A large capacitor would offer practically no opposition to current flow, so it would act like a short circuit across the generator terminals.
8. (*b*)
9. (*c*) — Power is always dissipated when current flows through resistance. However, no power is dissipated in a purely inductive or purely capacitive circuit.
10. (*a*)

21.
How Does an Automobile Electrical System Work?

Introduction

The basic principles of electricity have been discussed in this book. In this chapter you will study a practical circuit in which some of these fundamentals are put to use.

The power supply for an automobile is basically a dc system. The reason is that it is necessary to start the car in remote places where power is not available, and the only economical portable voltage source capable of starting an automobile engine is the lead-acid secondary battery. It is made in 6- and 12-volt sizes.

Since the starter system involves a dc source, it is convenient to use the same battery for the rest of the automobile electrical networks. This presents a special problem in the ignition system, where it is necessary to have a voltage of about 20,000 to 35,000 volts to operate the spark plugs. Figure 21-1 shows what the spark of the spark plug is used for. When the piston moves toward the spark plug, the gasoline and air mixture is compressed. This occurs just before a high voltage is delivered to the spark plug. When voltage arrives at the plug, a spark jumps across the spark-plug gap. This ignites the gasoline and air mixture, causing it to explode. The force of the explosion causes the piston to move downward, and the downward motion forces the crankshaft to turn.

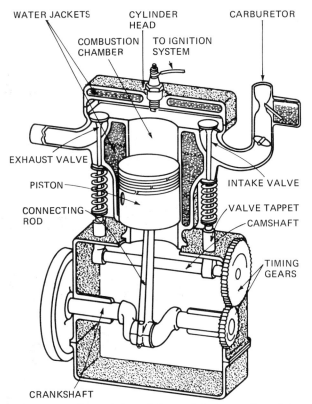

Figure 21-1. The spark is needed to explode the gasoline mixture.

If you connect the battery directly to the spark plug, it will not operate because the 12 volts from the battery is too low. One of the things you will learn in this chapter is how the 12 volts of the battery is converted to the high voltage required by the ignition system.

You will be able to answer these questions after studying this chapter:

What are the characteristics of the car battery?
What are the breaker points used for?
What is a condenser used for?
What is the ignition coil used for?
What is the distributor used for?
How are spark plugs constructed?
Why is electrical timing needed?

Instruction

WHAT ARE THE CHARACTERISTICS OF THE CAR BATTERY?

Figure 21-2 (see also Fig. 8-13) shows a cutaway view of a typical lead-acid battery. This is a form of *secondary battery*, which means that it can be recharged. A *primary cell* or *primary battery* cannot be recharged. In this definition of primary and secondary cells, the term "recharged" means to reverse the chemical action

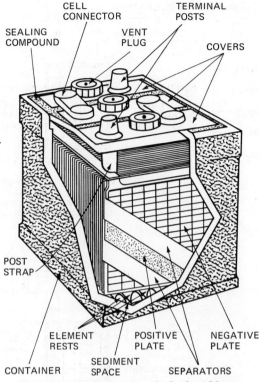

CELL
CONNECTOR

SEALING
COMPOUND

VENT
PLUG

TERMINAL
POSTS

COVERS

POST
STRAP

ELEMENT
RESTS

POSITIVE
PLATE

NEGATIVE
PLATE

CONTAINER

SEDIMENT
SPACE

SEPARATORS

Figure 21-2. Cutaway view of a lead-acid battery.

within the battery. The so-called charging circuits that are sold for charging dry cells are, in reality, *rejuvenators.* They increase the electrical action, but do not reverse the chemical action within the cell.

Figure 21-3 shows the difference in action between a discharging and a charging battery. During discharge, the battery is delivering power to the various parts of the car such as the starter, ignition, horn, lights, and radio. During the charging period, electricity is forced to flow through the battery in an opposite direction to the direction of flow during discharge. This is also shown in Fig. 21-3. Note that there is actually a chemical change that occurs in the electrolyte during discharge. The battery supplies power to the starter motor when the car is started. When the engine is running, electricity is obtained from a generator or from an alternator. The generator, or alternator, produces a dc voltage that serves two purposes: to help operate the electrical system of the car, and to charge the battery.

The solution of sulfuric acid (H_2SO_4) and distilled water in a battery is called the *electrolyte.* Actually, an electrolyte is any fluid which will conduct electricity. The condition of a lead-acid battery in an automobile can be determined by measuring the specific gravity of the electrolyte. Specific gravity compares the weight of a solution with the weight of an equal amount of distilled water at 4 degrees Celsius. When a lead-acid battery is fully charged, the specific gravity of its electrolyte is between 1.270 and 1.285. This means that it is heavier than an equal volume of water at the same temperature. Specific gravity is measured by a hydrometer as shown in Fig. 21-4. The procedure is to draw the electrolyte into the glass case of the hydrometer. This causes a float inside the glass

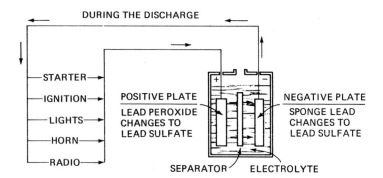

DURING THE DISCHARGE

STARTER
IGNITION
LIGHTS
HORN
RADIO

POSITIVE PLATE
LEAD PEROXIDE
CHANGES TO
LEAD SULFATE

NEGATIVE PLATE
SPONGE LEAD
CHANGES TO
LEAD SULFATE

SEPARATOR   ELECTROLYTE

Sulfate ion of sulfuric acid unites with active
materials on plates, leaving weaker acid solution.
Hydrogen of acid and oxygen of lead peroxide
combine to form water, diluting the solution.

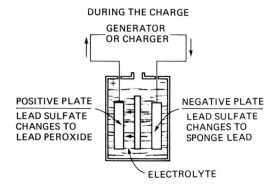

DURING THE CHARGE
GENERATOR
OR CHARGER

POSITIVE PLATE
LEAD SULFATE
CHANGES TO
LEAD PEROXIDE

NEGATIVE PLATE
LEAD SULFATE
CHANGES TO
SPONGE LEAD

ELECTROLYTE

Water is broken down by electrolysis.
Hydrogen combines with sulfate ion from plates
to form acid which makes electrolyte stronger.
Oxygen combines with lead of positive plate to
form lead peroxide.

**Figure 21-3.   *Current flow during charge and discharge of an automobile
battery.***

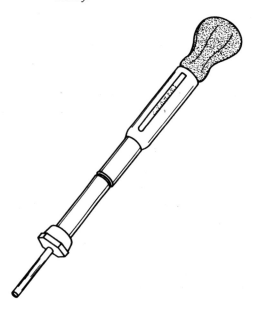

**Figure 21-4.   *A hydrometer can be used
to check the condition of a battery.***

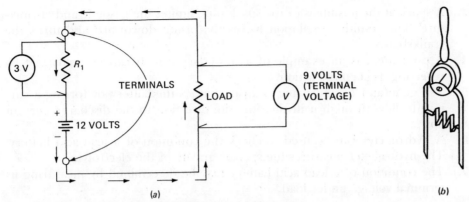

Figure 21-5. A battery can be checked by measuring its voltage under load. (a) The load current flowing through the internal resistance causes the terminal voltage to be lower than 12 volts. (b) This instrument checks the battery under load.

case to move up and down, and the specific gravity is read directly from the floating indicator. A specific gravity of 1.200 to 1.215 indicates that the battery is in a discharged condition, but that it is not completely discharged. Readings between 1.125 and 1.140 indicate that the battery is almost completely discharged and must be recharged before it can be used again.

You cannot determine the condition of a battery by simply measuring the voltage across its terminals unless you connect a load to the battery terminals first. Figure 21-5 shows how a battery is connected for measuring the terminal voltage. In Fig. 21-5a a 12-volt battery with its internal resistance is connected to a load. There is a complete current path here, so current will flow through the internal resistance R_i. The voltage drop across this internal resistance is shown to be 3 volts, and therefore, the terminal voltage is

$$\frac{TERMINAL}{VOLTAGE} = \frac{BATTERY}{VOLTAGE} - \frac{DROP\ ACROSS}{INTERNAL\ RESISTANCE}$$

$$= 12 - 3$$
$$= 9 \text{ volts}$$

When the load is removed, the circuit path is opened and there is no longer a current path through R_i. The drop across R_i is then 0 volt, and the terminal voltage is 12 volts.

Figure 21-5b shows an instrument that is designed for use in measuring the terminal voltage under load. It consists simply of two sharp probes which are made so they can be stuck into the lead terminals. A resistor is connected between the probes to complete the circuit. A voltmeter measures the voltage across the load, which is equal to the voltage across the terminals. This type of instrument is usually designed with a green-red scale. If the needle deflects to the green section, it indicates that the battery is good. If the needle deflects to the red section, it indicates that the battery is discharged.

SUMMARY

1. The power supply for an automobile is basically a dc system.
2. The ignition system delivers a high voltage to a spark plug.

3. A spark at the terminals of the spark plug ignites the gasoline and air mixture. The resulting explosion forces the piston downward and turns the crankshaft.
4. A car battery is an example of a secondary cell. It can be recharged by reversing its chemical action.
5. The generator or alternator of a car charges the battery by forcing a current to flow through it in the opposite direction to the discharge current flow.
6. A hydrometer can be used to check the condition of a lead-acid battery. The hydrometer measures the specific gravity of the electrolyte.
7. The condition of a lead-acid battery may be determined by measuring its terminal voltage under load.

WHAT ARE BREAKER POINTS USED FOR?

If you connect a dc voltage source to the primary of a transformer, there is no voltage generated across the secondary. The reason for this is that transformer action depends upon an expanding and contracting magnetic field from the primary winding. This magnetic field cuts across the secondary and induces a voltage in it. The principle is shown in Fig. 21-6.

Figure 21-6*a* shows that the output of the transformer is 0 volt when a battery is connected across the primary winding. In this case a step-up transformer is used, but the result would be the same with a step-down type.

Figure 21-6*b* shows one method of using a dc voltage to generate a high voltage needed for operating spark plugs. Note that the "coil" is in reality a step-up transformer. In this circuit the battery is connected across the primary with a switch connected in series. At the moment the switch is closed, a magnetic field around the primary expands and induces a voltage in the secondary. When the switch is opened, the magnetic field around the primary collapses and induces a voltage in the secondary. If you rapidly open and close the switch, a voltage will be constantly generated across the secondary.

In the automobile ignition system, the *breaker points* serve the purpose of the switch. The breaker points are automatically opened and closed by a rotating cam in the distributor housing. Figure 21-7 shows the breaker switch. When the cam turns, it causes the breaker arm to move back and forth, thus opening and closing the contacts. This opens and closes the circuit in the

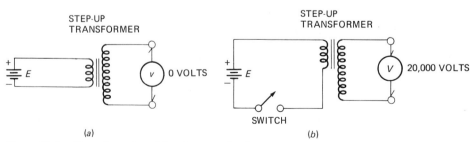

Figure 21-6.   How a low value of dc voltage can be changed into a high voltage. (a) *There is no output voltage when a dc voltage is applied to the primary of a transformer.* (b) *Opening and closing the switch rapidly causes a high voltage to be generated across the secondary.*

primary of the transformer and causes a high voltage to be generated across the secondary.

WHAT IS A CONDENSER USED FOR?

You will notice in Fig. 21-7 that there is a capacitor connected into the same circuit with the breaker points. Although it may not be clear from the drawing, the capacitor is actually connected across the breaker points. In an automobile ignition system this capacitor is often referred to as the *condenser*.

You will remember that a capacitor is a component that stores electrical energy. To understand the need for the capacitor, look again at Fig. 21-6*b*. Here the switch (actually the breaker points) opens and closes the primary circuit. When the switch is opened, the magnetic field around the primary winding collapses. This generates a very large voltage in the secondary, and at the same time, induces a large countervoltage in the primary. The counter-

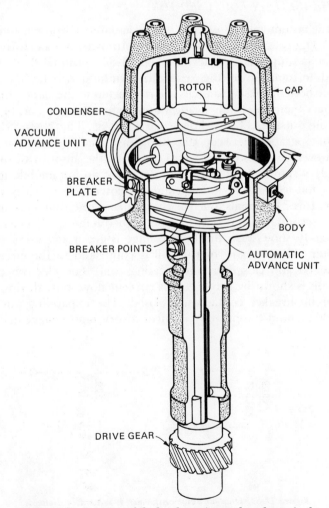

Figure 21-7. Location of the breaker points and condenser in the distributor housing.

voltage tries to keep the current going even though the switch contact is open. The countervoltage is large enough to cause an arc to jump across the switch contacts. (An interesting thing about an electric circuit is that it is almost impossible to open a switch in an inductive circuit without having arcing at the contacts.)

If you place a capacitor across the switch contacts, the countervoltage charges the capacitor when the switch is opened. The capacitor stores the energy that would cause a spark across the switch contacts. By using the capacitor, then, there is no arcing at the contacts. This is important because continual arcing at the breaker contacts will eventually cause them to burn out.

You should remember this important point: *If the breaker points in an ignition system are burned up, it may be an indication that the capacitor is no longer functioning.* A good mechanic will always replace the capacitor when he replaces the breaker points. In actual use the capacitor deteriorates somewhat over a period of time and may not be offering good protection for the breaker points.

WHAT IS THE IGNITION COIL USED FOR?

The coil of the automobile ignition system is actually an iron-core step-up autotransformer. Figure 21-8 shows how an autotransformer performs the same function as a step-up transformer. In Fig. 21-8a, terminal 2 of the primary winding and terminal 4 of the secondary winding are both connected to ground. Since both of these terminals are connected to the same point, it is possible to connect them internally inside the transformer case and bring a single terminal to the outside. This is shown in Fig. 21-8b. There is no difference in the theory of operation between the autotransformer and the step-up transformer. However, it is less expensive to construct the autotransformer.

Figure 21-9 shows the construction details of an automobile ignition coil. The primary and secondary windings and the iron core are clearly visible. Note that there are three terminals: one for the primary, one that is common for the primary and secondary, and one for the high-voltage or *high-tension* output.

Figure 21-10 shows how the high voltage is generated using the coil and circuit breaker. The primary of the coil is connected to the circuit breaker, which is opened and closed by the rotating cam. The electron path in the primary circuit is shown by arrows. This current flows only during the period when the circuit breaker contacts are closed. The expanding and collapsing magnetic field around the primary induces a very high voltage across the sec-

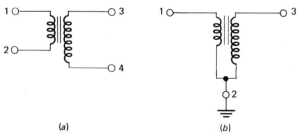

Figure 21-8. How an autotransformer is related to a simple transformer. (a) *A simple step-up transformer.* (b) *An autotransformer.*

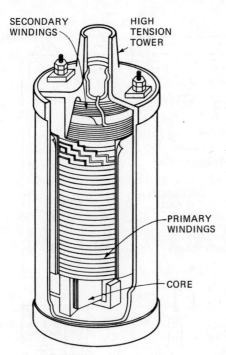

SECONDARY
WINDINGS

HIGH
TENSION
TOWER

PRIMARY
WINDINGS

CORE

Figure 21-9. Construction of an ignition coil.

ondary. Note that the capacitor (or condenser) *C* is connected directly across the terminals of the circuit breaker.

WHAT IS THE DISTRIBUTOR USED FOR?

Figure 21-11 shows the complete ignition system. The battery is the primary source of power, but after the car has been started, the generator or alternator (not shown in this illustration) supplements the battery power. The ammeter tells whether the battery is being charged or discharged. The ignition switch is the master *on/off* switch for the complete ignition system. When this switch is off, no power is delivered to the primary of the coil.

Assuming that the ignition switch is on, the primary of the coil is connected from the battery to ground through the breaker points located in the distributor housing. As the distributor cam turns, the breaker points open and

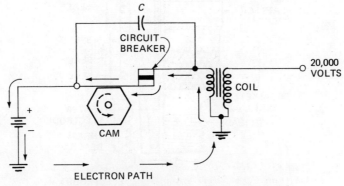

C

CIRCUIT
BREAKER

20,000
VOLTS

COIL

CAM

+

−

ELECTRON PATH

Figure 21-10. How the high voltage is generated.

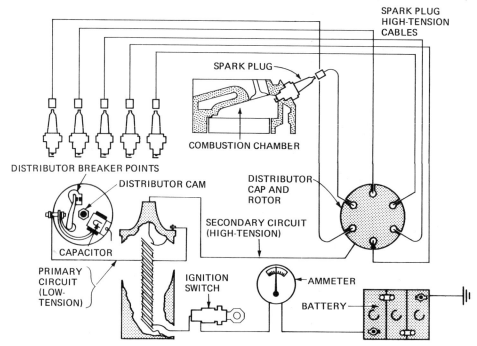

Figure 21-11.   The complete ignition system.

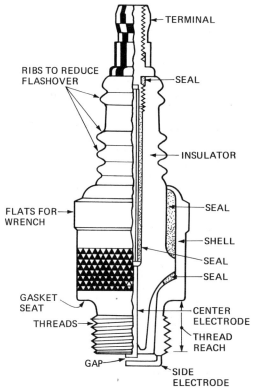

Figure 21-12.   Construction of a spark plug.

close. This produces pulsations of current through the primary. The varying magnetic field around the primary cuts across the secondary and induces a very high output voltage. The capacitor (or condenser) stores energy during the period when the breaker points would normally be arcing due to the counter-voltage induced in the coil.

The secondary voltage of the coil is delivered to the rotor in the distributor cam. This rotor turns with the distributor, and as it turns, it connects the high voltage to each individual spark plug, one at a time. When the rotor connects to a spark plug, that plug fires and ignites the gas in the gas chamber of the piston. The distributor, then, distributes the high tension or high voltage to each spark plug at just the right instant for firing the plug and igniting the gas when the piston is at the top of its travel.

HOW ARE SPARK PLUGS CONSTRUCTED?

A spark plug produces the arc discharge necessary for igniting the gasoline in the piston chamber. Figure 21-12 shows the construction of a spark plug. The voltage is delivered to the top of the plug, which in turn is delivered through the inside of the plug to one of the spark-plug electrodes. The other electrode is grounded through the thread to the plug. When the high voltage is delivered to the plug from the distributor, it causes a spark to arc across the contact points.

WHY IS ELECTRICAL TIMING NEEDED?

Earlier we pointed out that the spark plug fired at the moment the piston compressed the gas at the top of the cylinder. This is a theoretical condition. Actually, the plug may be caused to fire just before the piston has completed its upward travel in order to give the gas time to completely ignite and explode as the piston starts in a downward direction. One of the features of the automobile ignition system is that it must be properly timed so that the plug fires at exactly the right moment to produce the maximum efficiency of operation. If the plug fires too soon or too late, the result will be incomplete or inefficient burning of the gas and loss of power. The procedure for adjusting the ignition system so that the spark plugs fire at the proper time is called *electrical timing*.

SUMMARY

1. The breaker points are located in the distributor housing. They open and close the circuit between the battery and the primary of the coil.
2. By pulsing the current to the primary of the coil, the magnetic field around the primary expands and contracts. This moving field cuts across the secondary and produces a very high voltage.
3. The coil in the ignition system is in reality a step-up transformer.
4. The condenser is connected across the breaker points. It stores the energy that would normally produce a spark across the points. This prolongs the life of the points.
5. The distributor delivers the high voltage to the spark plugs one at a time. The voltage fires the plug. The arc across the plug contacts explodes the gasoline and air mixture in the cylinder.

6. Electrical timing assures that the plugs fire at exactly the right moment to explode the gasoline-air mixture.

Programmed Review Questions

(Instructions for using this programmed section are given in Chap. 1.)

We will review the important concepts of this chapter. If you have understood the material, you will progress easily through this section. Do not skip this material, because some additional theory is presented.

1. The primary source of power for the automobile ignition system is the
   A. ignition coil. (Proceed to block 7.)
   B. battery. (Proceed to block 17.)

2. *The correct answer to the question in block 9 is **A**. The electrical-timing adjustment determines the exact instant that the plugs fire.* Here is your next question.
   A certain transformer has a turns ratio (N_s/N_p) of $2:1$. If a 12-volt battery is connected to the primary, what will the secondary voltage be?
   A. 0 volt (Proceed to block 12.)
   B. 24 volts (Proceed to block 5.)

3. *The correct answer to the question in block 16 is **B**. Figure 21-13 shows a comparison of six 2-volt cells in parallel and in series. Note that all the positive terminals and all the negative terminals are connected together for parallel operation. (See Fig. 21-13a.) Note also that the output voltage of the parallel combination is the same as for one cell. When the cells are connected in series as shown in Fig. 21-13b, the positive terminal of one cell is connected to the negative terminal of the next. The output voltage is the sum of all the cell voltages.*
   The terminal voltage of a lead-acid cell is slightly over 2 volts. For a 6-volt battery, three series-connected cells are needed. For a 12-volt battery, six series-connected cells are used. Here is your next question.
   The purpose of the condenser across the breaker points is
   A. to prevent arcing at the breaker points. (Proceed to block 13.)
   B. to convert the battery dc to an ac voltage for the primary of the transformer. (Proceed to block 6.)

4. *Your answer to the question in block 18 is **A**. This answer is wrong. There is no instrument by the name of "electronometer" used for testing car batteries. A company could adopt this as a trade name, but there is no instrument called by this name.* Proceed to block 16.

5. *Your answer to the question in block 2 is **B**. This answer is wrong. There is no transformer action when there is a direct current flowing in the primary winding.* Proceed to block 12.

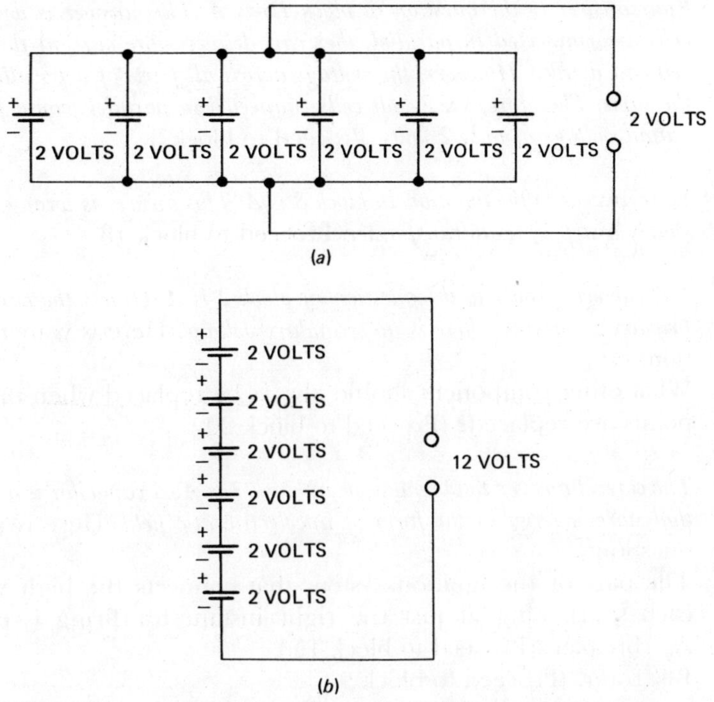

(a)

(b)

Figure 21-13. A comparison of parallel and series connections of cells. (a) *Parallel connection of six 2-volt cells. The output voltage is 2 volts, but the combination can deliver more current than a single cell.* (b) *Series connection of six 2-volt cells. The output voltage is 12 volts.*

6. *Your answer to the question in block 3 is* **B**. *This answer is wrong. A condenser, or capacitor, cannot convert dc to ac.* Proceed to block 13.

7. *Your answer to the question in block 1 is* **A**. *This answer is wrong. The ignition coil is actually a step-up transformer. It does not generate power. Instead, it converts the pulsating 12-volt dc voltage to a high voltage for operating the spark plugs.* Proceed to block 17.

8. *The correct answer to the question in block 17 is* **A**. *The generator or alternator is used for recharging the battery in normal use. The "voltage regulator" protects the battery from overcharging.* Here is your next question.
The breaker points are located in
A.   the coil housing. (Proceed to block 11.)
B.   the distributor housing. (Proceed to block 18.)

9. *The correct answer to the question in block 13 is* **B**. *The rotor turns inside the distributor shaft. Each spark plug is connected to a terminal in the cap. As the rotor turns, voltage is delivered to the spark plugs—one at a time.* Here is your next question.
If the ignition system is improperly adjusted so that the spark plugs fire too late, the condition would be corrected by
A.   electrical timing. (Proceed to block 2.)
B.   installing new breaker points. (Proceed to block 19.)

10. *Your answer to the question in block 16 is **A**. This answer is wrong. When cells are connected in parallel, they can deliver more current than a single cell can deliver. However, the voltage across all parts of a parallel circuit is the same. Therefore, six 2-volt cells connected in parallel would produce an output voltage of only 2 volts.* Proceed to block 3.

11. *Your answer to the question in block 8 is **A**. This answer is wrong. The coil is simply a step-up autotransformer.* Proceed to block 18.

12. *The correct answer to the question in block 2 is **A**. Unless the current in the primary is varying, there is no secondary voltage. Here is your next question.*
What other component should always be replaced when the breaker points are replaced? (Proceed to block 20.)

13. *The correct answer to the question in block 3 is **A**. A capacitor is a component that stores energy in the form of an electrostatic field. Here is your next question.*
The part of the ignition system that connects the high voltage to each spark plug at just the right instant for firing is called the
A. breaker. (Proceed to block 15.)
B. rotor. (Proceed to block 9.)

14. *Your answer to the question in block 17 is **B**. This answer is wrong. A primary cell cannot be recharged in the strictest interpretation of the word "recharge." However, a dry cell can be rejuvenated. This involves simply heating it (or pulsing current through it) to increase the chemical action.* Proceed to block 8.

15. *Your answer to the question in block 13 is **A**. This answer is wrong. The breaker converts the dc of the battery into a pulsating current so that the transformer — that is, the coil — can be operated. This is necessary because a transformer cannot operate with a steady direct current flowing in its primary.* Proceed to block 9.

16. *The correct answer to the question in block 18 is **B**. A hydrometer has a floating element. The electrolyte is drawn into the chamber by suction, and the floating element indicates the specific gravity. Here is your next question.*
A 12-volt lead-acid battery has
A. six cells connected in parallel. (Proceed to block 10.)
B. six cells connected in series. (Proceed to block 3.)

17. *The correct answer to the question in block 1 is **B**. The lead-acid battery provides the primary power for operating the ignition system. When the motor is running, the generator or alternator provides supplementary power. Here is your next question.*
The type of battery that can be recharged by forcing a current to flow through it in the opposite direction to the direction of flow during discharge is a

A. secondary cell. (Proceed to block 8.)

B. primary cell. (Proceed to block 14.)

18. *The correct answer to the question in block 8 is* **B**. *The breaker points are located on a metal platform—called the* **breaker plate**—*within the distributor case.* Here is your next question.

The condition of a lead-acid battery can be determined by measuring the specific gravity of its electrolyte. This measurement is made with

A. an electronometer. (Proceed to block 4.)

B. a hydrometer. (Proceed to block 16.)

19. *Your answer to the question in block 9 is* **B**. *This answer is wrong. Installing new breaker points may be necessary after an extended period of use, but this will not affect the timing of the explosion in the cylinder.* Proceed to block 2.

20. *The correct answer to the question in block 12 is* **condenser.** *Although a condenser is simply two conductors separated by an insulation, they do deteriorate with age. To be safe, and to assure a long life for the breaker points, the condenser should be replaced.*

You have now completed the programmed questions. There is no "Experiments" section for this chapter. Proceed to the "Self-Test" that follows.

Self-Test with Answers

(Answers with discussions are given in the next section.)

1. Which of the following statements is true? (*a*) A primary cell of the type used in flashlights can be recharged; (*b*) A primary cell of the type used in flashlights cannot be recharged.

2. The coil used in the automobile ignition system is (*a*) a choke coil for smoothing the current pulsations in the circuit; (*b*) a step-down air-core transformer; (*c*) used for converting dc to ac; (*d*) a step-up autotransformer.

3. A hydrometer measures (*a*) the water level in the battery; (*b*) the specific gravity of the electrolyte; (*c*) the water pressure in the radiator; (*d*) the temperature of the battery water.

4. The popular name for the capacitor used in the automobile ignition system is (*a*) condenser; (*b*) storer; (*c*) breaker; (*d*) timer.

5. In order to operate the coil, the direct current from the battery is converted to a pulsating current. This is accomplished by using the (*a*) ignition switch; (*b*) breaker points; (*c*) coil; (*d*) battery cables.

6. The breaker points are in the (*a*) primary winding circuit of the coil; (*b*) secondary winding circuit of the coil.

7. Each cell of a lead-acid battery has a voltage of about (*a*) ½ volt; (*b*) 2 volts; (*c*) 6 volts; (*d*) 12 volts.

8. To charge a secondary cell, (*a*) decrease the load resistance across it; (*b*) force a current to flow through it in the opposite direction to the normal current during discharge.

9. The spark plugs (*a*) are used to ignite the gasoline and air mixture in the cylinder; (*b*) are in the primary circuit of the coil; (*c*) do not conduct electricity; (*d*) are not part of the automobile ignition system.

10. Which of the following is not located in the distributor housing? (*a*) the breaker plate; (*b*) the breaker points; (*c*) the rotor; (*d*) the coil.

Answers to Self-Test

1. (*b*) – Primary cells can be rejuvenated but not recharged.
2. (*d*)
3. (*b*)
4. (*a*) – The term "condenser" does not describe the use of the component as well as the term "capacitor." The component does not condense anything. It stores energy. When you want to store something, you get a container with the correct capacity. To store electricity you get a capacitor.
5. (*b*)
6. (*a*) – See Fig. 21-8.
7. (*b*)
8. (*b*)
9. (*a*)
10. (*d*)

22.
How Are Houses Wired?

Introduction

The purpose of this chapter is to give you a knowledge of some of the requirements and basic circuits related to house wiring. It is not intended to be a complete course on wiring houses. House wiring is something that should be attempted only by trained and skilled electricians.

There are three basic things an electrician must be familiar with in order to accomplish his job. First, he must know which components are available for making the safest and most efficient installation. Of course, busy electricians do not have time to test all the components to see if they are safe. Fortunately, there is a nonprofit organization called the *Underwriters' Laboratories* (UL) that does this job for him. Manufacturers submit their components to the Underwriters' Laboratories, where the parts are tested to determine if they are safe for public use. If the UL finds a component to be acceptable, they will list it as acceptable. (They do not approve components; that is not their job.) Manufacturers desire to have their components listed by UL, and therefore, they try to design their components so that they will pass the UL tests. If an electrician selects only the components that are listed by the Underwriter's Laboratories, he will be assured that the installation contains the best possible parts.

The second thing the electrician must be familiar with is the National Electrical Code. It gives the requirements for wiring a home with a minimum of fire and electrical hazard.

The third thing that the electrician must be familiar with is the local code. Local codes do not contradict the national code, but they may extend it. In general, the local code will be more rigid than the national code. Local codes vary from place to place, and there is no general written code that covers all of them.

In many localities—in fact, in most—it is necessary for electricians to pass a test to show that they are qualified to wire homes safely. Usually, however, a person may work on his own home provided he gets the proper local permit and gets his wiring approved by an inspector after he has completed it.

You will be able to answer these questions after studying this chapter:

How is ac power distributed?
What is demand factor?
How are loads on power lines balanced?
How is wire size related to current capacity?
What is the minimum power requirement?
How many receptacles are needed?
How does a three-way switch work?

Instruction

HOW IS AC POWER DISTRIBUTED?

There are two ways of delivering ac power to your home. They are shown in Fig. 22-1. A simple two-wire system is shown in Fig. 22-1a. With this system, only 115 volts is available across the line. The white wire is grounded.

Figure 22-1b shows how power is delivered to the home using a three-wire system. Again, one of the wires, the white wire, is grounded. The ungrounded wires are called *hot lines*. There is 115 volts between the ground wire and either

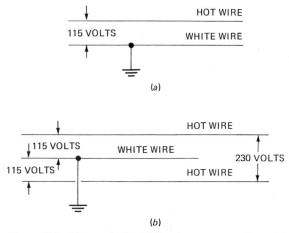

Figure 22-1. Two methods of delivering power to a home. (a) *A two-wire line delivers 115 volts.* (b) *A three-wire line delivers either 115 volts or 230 volts (or both).*

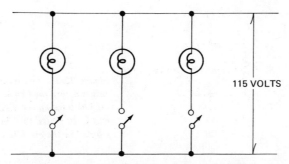

115 VOLTS

Figure 22-2. The light circuits are in parallel.

hot wire. The voltage between the two hot wires is 230 volts. This voltage is used for operating certain appliances which present a heavy demand on the power system. Examples are stoves and electric dryers.

At a point somewhere near where the power enters the house, there is a kilowatt-hour meter. This meter is used by the electric company to determine how much your electric bill will be. From the name "kilowatt-hour meter" you might think that this meter measures power in kilowatts. Actually, however, it measures energy.

Following the kilowatt-hour meter there is circuit breaker or fuse panel. Its purpose is to protect the house in the case of an overload in the wiring circuit or from a short circuit that may cause a fire. If there is an excessive demand from one of the circuits in the house, the overload will trip the overload relay, or burn out the fuse. It is important to remember that the circuit breakers or the fuse box is installed to protect you. If the circuit breaker should trip, or the fuse should burn out, it is necessary to determine what the cause of the excessive load is before returning the circuit to service.

The lighting circuits in the home are in parallel, for obvious reasons. This is shown in Fig. 22-2. Remember that the voltage across all parts of a parallel circuit is the same. Also, by placing the lights in parallel you can turn any one of the lights on, or all of them on, if you desire. When lights are connected in series, however, they must all be on or all off at the same time. It would be inconvenient to have a home wired in such a way that it was necessary to either turn all the lights on or turn all of them off. That is the reason for the parallel connection.

SUMMARY

1. National and local codes are used to ensure that the wiring in the house is safe.
2. The Underwriters' Laboratories is a nonprofit organization which assures that components are designed for maximum safety.
3. Power may be delivered to the home in either a two-wire or a three-wire system. In a two-wire system only 115 volts is available. In a three-wire system there is 115 or 230 volts (or both) available.
4. The white wire in the house line system is the grounded wire. It is also known as the "common wire," or the "neutral wire."
5. Never replace a fuse, or return a circuit breaker to operation, until you have determined the cause of overload.
6. Appliances and lights in the home are connected in parallel so that they can be operated independently.

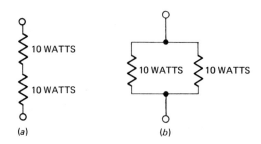

Figure 22-3. The total power in a circuit is determined by adding each power. (a) *The total power dissipated in the circuit is 20 watts.* (b) *The total power dissipated in the circuit is 20 watts.*

WHAT IS DEMAND FACTOR?

Before we discuss the meaning of demand factor, let us review how power is determined in an electric circuit. In a purely resistive circuit the power is determined by multiplying the rms voltage by the rms current.

Figure 22-3 shows how power is calculated for series and parallel circuits. In Fig. 22-3a two resistors are connected in series, each dissipating 10 watts. The power dissipated in this circuit is 20 watts. The resistors shown may actually represent any kind of purely resistive load in a house wiring system. For example, two 10-watt light bulbs may be shown as the resistors.

Figure 22-3b shows two resistors connected in parallel, each dissipating 10 watts. In this circuit the total power dissipated is also 20 watts. The important thing about this illustration is that it shows that the procedure for determining power is the same for series and parallel circuits. Remember the total power is the sum of the individual powers dissipated in either a series or a parallel circuit.

In calculating the total power required by a home, you could simply add all the power dissipated by every light fixture and appliance connected into the system. This assumes that all lights and appliances are on at the same time. However, this would give a total power demand that would be much greater than needed in normal needs. It is very unlikely that all the lights in the house and all the appliances would be turned on simultaneously.

The *demand factor* is the amount by which the total power demand can be reduced and still be within safe limits. When calculating the demand factor, the type of load must be taken into consideration. In lighting circuits there is no correction made for up to 3,000 watts. However, for over 3,000 watts of lighting, it is necessary to count only 35 percent of the portion above 3,000 watts. Certain appliances, such as clothes dryers and electric hot-water heaters, are counted at their full power rating. In other words, there is no demand factor applicable for these. For electric ranges the value of 8,000 watts is given as a rating for any range that has an actual value of power up to 12,000 watts. For a range that uses over 12,000 watts, you rate it as 8,000 watts plus 400 watts for each 1,000 watts above 12,000 watts.

The important thing to remember about demand factor is that it permits the calculation of power needed by a home to be lower than would be obtained if you added all the powers required by each light and appliance. It presumes that not all lights and appliances are on at the same time.

HOW ARE LOADS ON POWER LINES BALANCED?

In a three-wire line the appliances that require 230 volts are connected between the two hot wires. The 115-volt appliances are connected between ground and

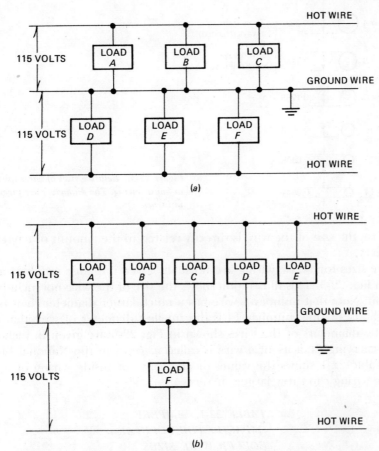

Figure 22-4. Comparison of a balanced and unbalanced power line. (a) *The loads are balanced on this three-wire system.* (b) *This line is not balanced.*

one of the hot wires as shown in Fig. 22-4. In Fig. 22-4*a* there is an equal number of loads connected between the two sides of the line. This is an ideal condition if it can be presumed that the loads are relatively equal. In other words, load *A* plus load *B* plus load *C* must be approximately equal to load *D* plus load *E* plus load *F* in order for the system to be balanced. If the line is balanced, then each of the hot wires carries approximately the same amount of power. When the lines are exactly balanced, there is no current flow through the center or common line. This is another important reason for having a balanced three-wire line.

Figure 22-4*b* shows the highly undesirable condition in which the loads are not balanced. Most of the load is connected between one of the hot wires and ground, and the other hot wire to ground connection consists only of load *F*.

When an electrician is called upon to add new outlets for appliances on the line, one of the things he must do is determine how the loads are distributed so he will not connect the appliances in such a way as to unbalance the line.

HOW IS WIRE SIZE RELATED TO CURRENT CAPACITY?

It stands to reason that the larger the diameter of the wire, the greater the amount of current—that is, electron flow per second—that can pass through it.

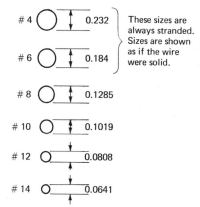

Figure 22-5. Relative sizes of wires used in house wiring. The diameters are given in inches.

Therefore, the size of the wire is directly related to the amount of current that it can carry.

Wire sizes for house wiring are normally given in gauge numbers as illustrated in Fig. 22-5. This illustration shows the size of the wire not including the insulation. Note that number 4 wire has a much larger diameter than number 14. For house wiring, number 14 is the smallest diameter of wire that can be used. The diameters of the wires shown in Fig. 22-5 are given in inches. The current-carrying capacity of a wire is called *ampacity* in the National Electrical Code. Table 22-1 shows the values of ampacity for inside wiring of a house. (Outside wiring can carry larger currents.)

TABLE 22-1. AMPERE CAPACITY FOR POPULAR COPPER WIRE SIZES

Wire Size	Current Capacity
4	70 amperes
6	55 amperes
8	40 amperes
10	30 amperes
12	20 amperes
14	15 amperes

The ampacity values in Table 22-1 are based on the presumption that the wire is made of copper. It the wire is made of aluminum, it is necessary to increase each by one wire size for each value in the table. In other words, number 4 aluminum wire could carry 55 amperes, number 6 aluminum wire could carry 40 amperes, number 8 aluminum wire could carry 30 amperes, etc.

Aluminum wire has not been used extensively in house wiring in the past. However, in recent years, because of the advantages of lower cost and lighter weight, it is being considered in certain wiring installations.

WHAT IS THE MINIMUM POWER REQUIREMENT?

You have learned that the demand factor is a method used to determine how much power is delivered to a home. The minimum power requirement for a

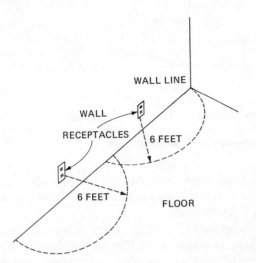

Figure 22-6. The requirement for spacing of receptacles is shown here.

home is the amount that the home must be wired to use. In other words, the minimum power requirement is the power that can be delivered to the home without regard to the demand factor.

A home should be wired in such a way that it can supply the power necessary for operating all the appliances that are in it. In addition to this it is necessary to wire the home for possible future increases in power requirements. The power requirement for a home is stated in the National Electrical Code, but may be modified by local codes.

As a general rule, the wiring for lighting and small appliances must be capable of delivering 3 watts per square foot of living space. Also, there must be two 20-ampere, 1,500-watt services for larger appliances. Wiring to the receptacles should be done with number 12 wire, and there should be no more than 10 receptacles connected to each circuit breaker or fuse circuit.

HOW MANY RECEPTACLES ARE NEEDED?

It has already been noted that no more than 10 receptacles are to be connected to each fuse circuit. If more receptacles are needed, then more fuse circuits are to be added.

The number of receptacles used in a room is set by the National Electrical Code. Figure 22-6 illustrates the requirement for the minimum number of receptacles in a room. According to the code, there is to be no point at the wall line that is further than 6 feet from a receptacle. This means that receptacles may be as much as 12 feet apart on a large wall. On a shorter wall line, the receptacles may be more closely spaced.

HOW DOES A THREE-WAY SWITCH WORK?

Occasionally it is desirable to install a switch which can be used to operate a light from two different locations. For example, it may be desirable to have a switch that will turn on the garage light from either the house or the garage. A *three-way switch* system is used for this purpose.

The wiring for the three-way switch is not complicated. It is illustrated in

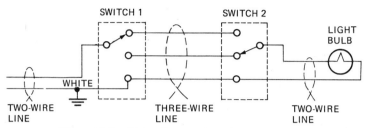

Figure 22-7. Wiring diagram for a three-way switch.

Fig. 22-7. The input power comes from a two-wire line which provides 115 volts for the circuit. Note that between switch 1 and switch 2 it is necessary to run three wires. One of these wires is a common.

In the position shown for the switches in Fig. 22-7, there is no complete path for current flow to the light bulb. You can check this by tracing the current path. Start at the ground side of the two-wire line, and trace a path through switch 1, through switch 2, and through the light bulb. You will notice that you cannot get back to the power source on the two-wire line. However, if you switch either of the two switches to its opposite position, the path will be complete. This means that you could turn the light on from either switch position. Furthermore, once the light is on, it can be turned off from either switch position.

As a practice exercise you could redraw this circuit with the switches illustrated in combinations of positions to show that the light can be turned on or off from both positions.

SUMMARY

1. Demand factor permits the power delivered to a house to be less than would be required if all lights and appliances were drawing power at the same time.
2. Demand factor is based on the principle that not all electrical systems would be in operation at the same time.
3. Before connecting an additional 115-volt circuit into a 230-volt line, the electrician must determine how the line is balanced.
4. In a balanced three-wire line the same amount of power is taken from each half, and there is no current flow in the common line.
5. The ampacity of a wire is a measure of its ability to pass a current. It means the *amp*ere cap*acity* of the wire.
6. A three-way switch system is used for operating a light (or any load) from two different positions.

Programmed Review Questions

(Instructions for using this programmed section are given in Chap. 1.)

We will now review the important concepts of this chapter. If you have understood the material, you will progress easily through this section. Do not skip this material, because some additional theory is presented.

1. You would expect a black wire in a house wiring system to be a
   A. hot wire. (Proceed to block 7.)
   B. ground wire. (Proceed to block 17.)

2. *Your answer to the question in block 7 is **B**. This answer is wrong. The national code should be considered a minimum standard for safe wiring practices.* Proceed to block 13.

3. *The correct answer to the question in block 10 is **A**. Since copper is a better conductor than aluminum, it is necessary to use a larger diameter of aluminum wire. A suggested rule is to increase the aluminum by at least one wire size. Only even numbers are used in the wire table, so increasing the wire size by one means that you take the next even number. For example, number 12 wire is the next larger size to number 14 wire.* Here is your next question.
   To provide for appliances that draw heavy currents, house wiring must provide at least
   A. two 20-ampere, 1,500-watt lines. (Proceed to block 9.)
   B. one 10-ampere, 1,000-watt line. (Proceed to block 19.)

4. *Your answer to the question in block 9 is **A**. This answer is wrong. Neither the mayor nor the city council would be ordinarily concerned with testing components. (They may, however, be directly or indirectly concerned with local wiring codes.)* Proceed to block 12.

5. *Your answer to the question in block 12 is **A**. This answer is wrong. A three-wire system can be used to obtain 115 volts, but it can also be used to obtain 230 volts.* Proceed to block 14.

6. *Your answer to the question in block 14 is **A**. This answer is wrong. You always add the powers in a circuit, regardless of whether they are in series or parallel.* Proceed to block 18.

7. *The correct answer to the question in block 1 is **A**. The white wire is the ground wire.* Here is your next question.
   As a general rule, local codes are
   A. more rigid than national codes. (Proceed to block 13.)
   B. designed to eliminate most of the strict requirements of the national code. (Proceed to block 2.)

8. *Your answer to the question in block 18 is **B**. This answer is wrong. As shown in Fig. 22-1, there is only 115 volts between one hot line and ground.* Proceed to block 11.

9. *The correct answer to the question in block 3 is **A**. This is a minimum requirement, and local codes may require a larger service.* Here is your next question.
   To be sure that the components he installs are safe for public use, the electrician chooses those components
   A. recommended by the city council or mayor in his area. (Proceed to block 4.)

B.   that are listed by the Underwriters' Laboratories. (Proceed to block 12.)

10. *The correct answer to the question in block 13 is **B**. It will be useless to replace the fuse until you have corrected the problem that caused the fuse to burn out.* Here is your next question.
Which of the following statements is correct?
A.   A number 10 copper wire can carry more current than a number 10 aluminum wire. (Proceed to block 3.)
B.   A number 10 aluminum wire can carry more current than a number 10 copper wire. (Proceed to block 16.)

11. *The correct answer to the question in block 18 is **A**. As shown in Fig. 22-1, there is 230 volts between the two hot lines.* Here is your next question.
The minimum size of wire that is allowed for house wiring is number _____. (Proceed to block 20.)

12. *The correct answer to the question in block 9 is **B**. The Underwriters' Laboratories (usually abbreviated UL) do not issue approvals for components. Instead, they list the components which pass the minimum requirements for safety. It is not correct to say "approved by Underwriters' Laboratories." Instead, you should say "listed by Underwriters' Laboratories."* Here is your next question.
A three-wire system delivers
A.   only 115 volts. (Proceed to block 5.)
B.   either 115 or 230 volts, or both. (Proceed to block 14.)

13. *The correct answer to the question in block 7 is **A**. The national code sets the minimum standards. This code is often modified by local codes that set more rigid standards.* Here is your next question.
Do not replace a fuse until
A.   you call the power company to make sure everything is OK. (Proceed to block 15.)
B.   you determine what caused the overload. (Proceed to block 10.)

14. *The correct answer to the question in block 12 is **B**. The method of obtaining 115 and/or 230 volts from a three-wire line is shown in Fig. 22-1.* Here is your next question.
If two 100-watt light bulbs are connected in parallel, the actual power being used is
A.   50 watts. (Proceed to block 6.)
B.   200 watts. (Proceed to block 18.)

15. *Your answer to the question in block 13 is **A**. This answer is wrong. there is no need to call the power company when a fuse burns out.* Proceed to block 10.

16. *Your answer to the question in block 10 is **B**. This answer is wrong. Remember that copper is a better conductor of electricity than aluminum.* Proceed to block 3.

17. *Your answer to the question in block 1 is* **B**. *This answer is wrong.* Proceed to block 7.

18. *The correct answer to the question in block 14 is* **B**. *The fact that the light bulbs are in parallel means that the voltage across each is the same. The total power in a series or parallel circuit is found by adding the power dissipated by each component.* Here is your next question.
A 230-volt load is connected between the
A. two hot lines. (Proceed to block 11.)
B. ground wire and one of the hot lines. (Proceed to block 8.)

19. *Your answer to the question in block 3 is* **B**. *This answer is wrong. A 10-ampere, 1,000-watt line would be too small to service most heavy appliances.* Proceed to block 9.

20. **(Number)** *14. This is the minumum size recommended, but local codes may state that the minimum size must be number 12.*
You have now completed the programmed questions. There is no "Experiments" section for this chapter. Proceed to the "Self-Test" that follows.

Self-Test with Answers

(Answers with discussions are given in the next section.)

1. For general lighting in the home, the code requires a service of (*a*) 3 watts per square foot of living space; (*b*) 3 watts per square yard of living space.
2. In the wiring system of the home, a white wire is used for (*a*) a ground wire; (*b*) a hot wire.
3. If you have three lamps in your front room, they will be connected (*a*) in parallel across the power line; (*b*) in series across the power line.
4. The minimum size of wire that can be used for house wiring is (*a*) number 10; (*b*) number 14.
5. To install a light in a garage that could be turned on or off from either the garage or the house, you would use (*a*) two-way switches; (*b*) three-way switches.
6. A number 14 wire can pass (*a*) more current than a number 10 wire; (*b*) less current than a number 10 wire.
7. On a large wall, receptacles may be spaced a maximum of (*a*) 12 feet apart along the wall line; (*b*) 6 feet apart along the wall line.
8. In the normal operation of electrical equipment in the home, it can be presumed that not all the equipment will be in operation at the same time. Thus, the power requirement of the home is lower than the sum of all individual powers. The amount by which the power rating can be reduced is called (*a*) reduction coefficient; (*b*) demand factor.
9. Which of the following is the correct way to say that a wire meets the

safety requirements of the Underwriters' Laboratories? (*a*) Approved by UL; (*b*) Listed by UL.

10. The ground (or common) wire for a house wiring system is (*a*) white; (*b*) black; (*c*) red; (*d*) green.

Answers to Self-Test

1. (*a*)
2. (*a*) — Another name for "ground wire" is "common wire."
3. (*a*) — By connecting them in parallel they can be operated individually. Also, the voltage across each would be the same.
4. (*b*) — Some local codes require a minimum size of number 2 or lower.
5. (*b*)
6. (*b*)
7. (*a*) — See Fig. 22-6.
8. (*b*)
9. (*b*)
10. (*a*)

A.
Safety

The experiments described in this book are done with very low voltages (6 volts and 12 volts) applied to the circuitry. This minimizes the chance of electric shock. Since you are reading this book, you may be considering electricity as a career. Therefore, this appendix contains information which you should know for even the most basic jobs in electricity. Safety should be one of your most important concerns on a job not only if you are working with electricity, but also if you are working on any job where accidents can be fatal.

Much work is being done in industry to reduce the job hazards. The first accurate statistics available on industrial accidents date to 1912 when it was reported that 35,000 people lost their lives in industrial accidents. These are not people that lost their lives working in electricity, but rather, people working in all the industrial trades. Today, the number of deaths by accidents in industry has been reduced to about one-third that number, but this is still considered to be too many by safety experts.

Of course, the most tragic part of an accidental death is the grief that it causes. Accidents that do not result in death can result in a serious impairment of physical ability and unnecessary cost and waste. Do not be surprised when you go to work in industry if a great amount of emphasis is placed on your safety and the safety of your fellow employees.

THE IMPORTANCE OF TRAINING

One of the first things to avoid is performing jobs for which you are not trained to do or are not familiar with. You might find yourself doing this even when working on simple laboratory experiments. If you catch yourself saying, "I wonder what would happen if I connected this wire to that point?" you are probably about to make a serious mistake.

If you understand the theory of a circuit (which means that you have been trained for that particular work), you will *know* what will happen if you "connect this wire to that point." In other words, *an important first step in safety is to learn as much as you can about the equipment that you are working with*. Do not experiment in electric circuits by touching wires just to see what would happen. Here is a good motto:

**STUDY FIRST
WORK LATER**

ELECTRICITY IS A SERIOUS BUSINESS

Horseplay and practical jokes are not funny in a laboratory. They *can* be fatal. What may seem like a practical joke can turn into a terrible disaster. If someone were to make a survey of the dumbest remarks in history, here is one that would certainly have to be included: "I didn't know it would hurt him."

DO NOT PAY ATTENTION TO OLD WIVES' TALES

To a person who has not studied electricity, it is a mysterious and very dangerous thing. This is probably one of the reasons that so many old wives' tales are told about electricity. One of the most dangerous of these is that "you can build up a resistance to electricity by taking repeated shocks of increasing intensity." Frankly, this just is not so. Not only is it *untrue*, it is an extremely dangerous thing to try. You can ruin your health—especially your heart—by repeated shocks with electricity.

Another old wives' tale which is just as dangerous as the one mentioned above is the idea that "a little bit of electric shock is good for you," or that it is fun to see who can take the biggest electric shock. The amount of electric shock that you can take is not in any way related to how strong you are or how big you are. There are many, many factors involved which do determine whether a shock will be fatal, but the final determining factor is *how much current flows through your body*. Note this important fact: It is *not* the *voltage* that kills you, but rather, it is the amount of *current* that passes through your body.

A current of only $1/1000$ ampere (or 1 milliampere) can produce a shock which you can feel. At about $1/100$ ampere (or 10 milliamperes) the shock is so severe that it paralyzes the muscles and you cannot release the conductor. If you see someone holding a conductor that he cannot let go of, do not be foolish enough to grab him and try to pull him away. If you do, you will receive a shock too. Instead, de-energize the circuit when possible. If you cannot do this, then use a dry board, or rope, or other insulating material to force him away.

At about $1/10$ ampere (100 milliamperes) shock may be fatal if it lasts for a second or more. These are extremely small values of current when you stop to think that the fuses in your house are rated for 15 amperes of current. In other words, *a shock can be fatal long before the fuse blows*.

If someone tries to tease you by saying, "You are not afraid of a little electricity, are you?" remember that the answer to that question should be *yes*. Airplane pilots have a favorite saying: "There are old pilots and there are bold pilots but there are no old, bold pilots." The same could be said of electrical workers.

People that play in electrical labs are eliminated from the profession in two ways. First, they are likely to make a serious mistake. Second, they will find it hard to get jobs because nobody wants to work around a clown on an electrical job.

Another old wives' tale that should be disregarded is the idea that you cannot be injured when working on an electric circuit if you keep one hand in your pocket. The reason for this idea is based on the fact that you are much more likely to be hurt with both hands across a hot circuit. However, it is not true that putting one hand in your pocket will keep you from getting hurt.

SAFETY FEATURES

There are safety features installed on equipment which should not be defeated. *Fuses* and *interlocks* are two examples.

You already know that a fuse will not protect you from a serious injury. However, it will open the circuit to the defective electrical equipment. When a fuse burns out, always determine the reason why. Do not simply replace the fuse or defeat the fuse circuit.

An interlock is a device that shuts the equipment off when its cabinet is open, or when it is partially disassembled. *Interlocks are for the protection of people.* In this way they are different from fuses, which are for the protection of equipment. In some cases, it is necessary to work on the equipment when it is energized. To do this, an experienced electrician can defeat the interlock—that is, he can bypass it so that the equipment will work even though the cabinet is open. This procedure is acceptable if you know exactly what you are doing in the circuit, but if you are not familiar with the equipment, you should *never* defeat the interlock.

TOOLS ARE IMPORTANT IN A GOOD SAFETY PROGRAM

There are a number of operations related to electricity which require safety precautions. Regardless of what type of work you are doing, you should keep your tools and equipment in top-notch order. Be sure that screwdrivers are sharp and tools are clean. Always wear safety glasses and safety gloves on jobs that require them, and before you work on any equipment, be sure that it is de-energized (whenever possible).

It is especially important to know where the main power switch is in the area where you are working. You *might* find it necessary to turn the power off quickly. If you find that it is necessary to work around high-voltage equipment, never work alone.

FIRE HAZARDS

You no doubt know that electricity can start fires. For example, it is not uncommon for a house with poor wiring to go up in flames. You should always keep in mind the fact that electricity can create fire hazards, and you should have a good understanding of how to deal with these hazards.

You may find it necessary to work with soldering irons or soldering guns, and you should remember that these guns or irons get hot enough to set fires. Furthermore, you should not get into the habit of flicking solder from an iron or gun to clean it. Flecks of solder can hit you or someone near you. This causes skin burn, or worse yet, the solder may go into an eye or ear.

When you are soldering, be especially careful not to breathe in smoke that is created by the soldering and the flux. As a matter of fact, you can make a general rule about smoke:

**IT IS NEVER GOOD TO INHALE SMOKE
REGARDLESS OF WHERE IT COMES FROM**

If electrical equipment catches on fire, you must be very careful *not* to throw water on the fire, and you must not use a foam-type fire extinguisher. Electricity can actually follow the water or foam and cause you a serious injury. Even if it were not for the possibility of injury, it would not be a good idea to use water or foam extinguishers on electrical fires, because both could destroy the equipment. What might have been a small burnout can end up to be a complete destruction of the equipment if water or foam is applied while the equipment is energized.

Some types of equipment use selenium rectifiers which produce a considerable amount of smoke when they burn out. This smoke can actually be injurious to your health, and again the above general statement can be repeated: *Never breathe in smoke of **any kind**, regardless of where you think it is coming from.*

Take a few seconds to investigate the fire extinguishers in your working area. There is usually an identification plate on the extinguisher which tells whether or not it can be used for an electrical fire. The kind usually recommended for electrical fires is the CO_2 type.

The following steps should be taken if a fire breaks out in an electrical lab.

1. De-energize the circuit immediately. You may have to do this by throwing the main switch.
2. Call the fire department or turn in a fire alarm. Remember that firemen are the experts in all kinds of fires, and you need their help as quickly as you can get it.
3. Using the correct fire extinguisher (a CO_2 type), direct the extinguisher at the *base* of the flame. Be careful not to let the fire get between you and the exit of the building.

KNOW YOUR FIRST AID

If a person receives a serious shock, his breathing may be impaired, and it is important to get his breathing started again *as quickly as possible*. Do not stop to loosen his clothes—let someone else do that. Instead, get him into the position shown in Fig. A-1 and follow the procedure for artificial resuscitation. You may have been taught a different method of administering artificial resuscitation in a health or first-aid class, and in that case you should use the method that you are familiar with.

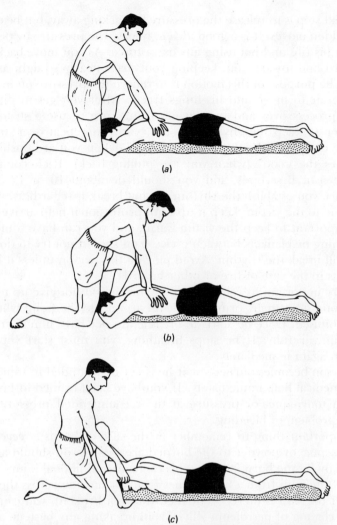

(a)

(b)

(c)

Figure A-1. Three steps in a cycle of artificial resuscitation. (a) The first step is to get the victim into this position. Place your hands on each side of his back as shown. (b) In this step pressure is exerted on the back to force the air out of the victim's lungs. Your arms should be straight and nearly vertical. (c) As you move back, lift the victim's arms and pull them forward as shown here. His arms should be raised enough to arch his back, and his chest should be barely lifted from the ground in order to ease the intake of fresh air into the victim's lungs.

Make sure that the person's mouth and nose are free from restrictions and that he does not have something in his mouth, like gum or candy, which he could choke on. Once you are in the position shown in Fig. A-1*a*, place your hands in the middle of his back—just below the shoulder blades—so that the fingers are spread downward and outward and the thumb tips are almost touching. Move forward until your arms are approximately vertical, and use the weight of your body to press down. Exert a slow, steady pressure on your hands until a firm resistance is met. The purpose of this action is to force the air out of his lungs. Figure A-1*b* shows your position for this operation.

The next step is to release the pressure by backing away, but be careful not to make sudden moves. It is a good idea to release the pressure by peeling your hands from his ribs and not using any extra force. As you move back, draw his arms upward and toward you, keeping your arms nearly straight as shown in Fig. A-1c. The purpose of this motion is to relieve the pressure on his chest and to allow the air to move into his lungs freely. After you get to this position, repeat the process over and over again, being sure to use a steady rhythm.

In order to keep the rhythm smooth, you can repeat over and over to yourself the expression, "out goes the bad air" (when you are pushing down), "and in goes the good" (when you are pulling back). Each complete cycle should take 5 or 6 seconds, and you should do about 10 or 12 cycles each minute. Once you establish the rhythm, do not break it regardless of what else is being done to the victim; keep it up until professional help arrives.

It is important to keep the victim warm, and you can have someone wrap him in clothing or blankets between cycles. You should not try to do this your-self, or it will break the rhythm. Avoid moving the person unless it is *absolutely* necessary (as in the case of fire or other hazard).

It is very important to remember that you should not give up too soon. It may take hours to actually bring the victim around to the point where he can breathe by himself. Once he starts to breathe, do not have him sit up. Be sure to watch him carefully. If he stops breathing, you must start your artificial resuscitation again immediately.

If a person becomes burned, or if he is cut or wounded in some way, you must seek medical help immediately. If you have been trained in first aid, you can apply a tourniquet or pressure at the recommended pressure points to reduce the problem of bleeding.

The important thing to remember in the case of burns is **never** to apply iodine, antiseptic, or powder to the burned area. Also, you should avoid using cotton directly on the burn, because this can cause additional injury.

Most of the pain from a burn comes from air moving across the area, and therefore, if the skin is merely reddened (but not broken), you can cover the area with a coating of petroleum jelly. Avoid breaking any blisters, and do not pull clothing away that is stuck to a burned area of skin. Instead, cut around it.

The information given here is not intended to substitute for first-aid training.

Here is a summary of some of the important safety precautions to be taken when working around electrical equipment.

Avoid loose clothing and jewelry that can be caught in electrical machinery.

Wear shoes with thick insulating soles; avoid the kinds of shoes that have cleats and metal hobnails.

Keep your tools clean and sharp, and make sure that they are properly insu-lated.

Avoid inhaling vapors and smoke of any kind.

Never operate defective electrical equipment, such as a power tool in which the three-way plug has been defeated so that it can fit into a two-way recep-tacle. Also equipment in which insulating mounts or knobs have been removed should never be operated.

Never use alcohol or carbon tetrachloride ("carbon tet") as a cleaning agent around electrical systems. Alcohol will burn if it gets too hot, and carbon tet will emit a poisonous gas when in contact with hot metals.

Do not rely on interlocks, fuses, and high-voltage safety relays to protect you in an electric circuit.

When at all possible, completely de-energize a circuit when working on it.

If you find it necessary to replace electric fuses, use an approved type of fuse puller. Do not reach into the box with your bare hands to remove cartridge fuses even though the circuit is off.

Never take *anything* for granted.

Do not work on electrical equipment that you are unfamiliar with. Take a little time to learn how a circuit works and what it is supposed to do. Your knowledge of electricity is one of your best safety devices.

Always know where the main switches are and where the fire extinguishers are in the area where you are working. Determine which types of fires the fire extinguishers are designed for.

Never put water, carbon tetrachloride, or foam extinguishers on electrical fires.

Avoid horseplay and make it clear that you do not appreciate practical jokes and horseplay.

Take some time to learn first-aid procedures, because they can help you (not only in the electrical lab, but also in many other life situations).

Never pull a line cord out by the wire. Always pull by the plug, being careful not to touch the conductor.

Programmed Review Questions

(Instructions for using this programmed section are given in Chap. 1.)

We will now review the important concepts of this appendix. If you have understood the material, you will progress easily through this section. Do not skip this material, because some additional theory is presented.

    1. Which of the following is true?
        A. If there is a fire extinguisher in an electrical lab, you can presume that it is safe for use on electrical fires. (Proceed to block 7.)
        B. Always take time to learn which fire extinguishers are safe to use *before* a fire starts. (Proceed to block 17.)

    2. *Your answer to the question in block 19 is **B**. This answer is wrong. In some cases a voltage of 70 volts can be fatal, but in other cases a voltage of 20,000 volts will not be fatal. Obviously, it is not the voltage that is the determining factor.* Proceed to block 22.

    3. *The correct answer to the question in block 18 is **B**. You will find that those marked "Safe for Electricial Fires" are CO_2 extinguishers.* Here is your next question.
    In case of an electrical fire, you should first
        A. turn in a fire alarm. (Proceed to block 9.)
        B. de-energize the circuit. (Proceed to block 16.)

    4. *Your answer to the question in block 21 is **A**. This answer is wrong. It can be dangerous and costly to "try a few things" in a circuit when you do not know anything about the equipment.* Proceed to block 13.

5. *Your answer to the question in block 13 is **A**. This answer is wrong. Although it is a good practice to keep one hand in your pocket, this is no guarantee against injury. Your best protection is to know what you are doing.* Proceed to block 19.

6. *Your answer to the question in block 18 is **A**. This answer is wrong. If you spray water on an electrical fire, there are two things that may happen. **First,** the water will probably ruin the circuit. (What might have been a small fire can turn into an expensive loss.) **Second,** electricity can follow the spray to your body, and this can be fatal.* Proceed to block 3.

7. *Your answer to the question in block 1 is **A**. This answer is wrong. There may be fire extinguishers placed in the lab for solvent or grease fires, or for paper or wood fires. You should **not** use these for electrical fires.* Proceed to block 17.

8. *The correct answer to the question in block 16 is **B**. Of course, it is important to loosen the clothing and get obstructions away from his nose and mouth, but let someone else do this. The breathing cycle must be started as soon as possible.* Here is your next question.
    Is this statement true or false? "If you plan to work around electrical equipment, you should build up your resistance to electricity by taking repeated shocks."
    A.   This statement is true. (Proceed to block 14.)
    B.   This statement is false. (Proceed to block 21.)

9. *Your answer to the question in block 3 is **A**. This answer is wrong. It is very important to get help, so you **should** turn in an alarm. However, you should **first** de-energize the circuit. The best way to do this is to turn off the main power switch or circuit breaker. Again, you should know where this switch is **before** a fire starts so you do not have to look for it in an emergency.* Proceed to block 16.

10. *The correct answer to the question in block 25 is **A**. Besides not putting iodine on burns, you should avoid the use of cotton swabs, and avoid the use of medicated powders.* Here is your next question.
    The complete cycle of artificial resuscitation should take about
    A.   20 seconds for each complete cycle, with a total of 10 to 12 complete cycles per minute. (Proceed to block 12.)
    B.   5 or 6 seconds for each complete cycle, with a total of 10 to 12 complete cycles per minute. (Proceed to block 18.)

11. *Your answer to the question in block 18 is **C**. This answer is wrong. Most foam-type extinguishers should never be used on electrical fires for the same reasons that water should not be used. (See block 6 for these reasons.)* Proceed to block 3.

12. *Your answer to the question in block 10 is **A**. This answer is wrong. A human could not live with such a little amount of air. Review the procedure described for Fig. A-1.* Then proceed to block 18.

13. *The correct answer to the question in block 21 is **B**. It is not safe, and it is not good practice, to experiment with electrical equipment. The correct procedure is to learn about the equipment, and **then** work with it.* Here is your next question.
Is this statement true or false? "When working in high-voltage circuits, you cannot be injured when working on electrical equipment as long as you remember to keep one hand in your pocket."
A. This statement is true. (Proceed to block 5.)
B. This statement is false. (Proceed to block 19.)

14. *Your answer to the question in block 8 is **A**. This answer is wrong. You cannot build up a resistance to electricity, and it is dangerous to your health to try it.* Proceed to block 21.

15. *Your answer to the question in block 17 is **A**. This answer is wrong. A fuse is used to protect the **equipment**, but not necessarily to protect the user.* Proceed to block 25.

16. *The correct answer to the question in block 3 is **B**. If you take time to de-energize the circuit, you may prevent additional fire hazards. In some cases, de-energizing the circuit will also put the fire out.* Here is your next question.
Which of the following steps should be taken first, if you are going to apply artificial resuscitation?
A. Loosen his shoes and belt. (Proceed to block 20.)
B. Get him into position as quickly as possible and start the rhythmic motion of artificial resuscitation. (Proceed to block 8.)

17. *The correct answer to the question in block 1 is **B**. Since it is possible that more than one type of extinguisher may be located in the lab, you should make it your own obligation to learn which are for electrical fires. A CO_2 extinguisher is usually used on electrical fires.* Here is your next question.
Which of the following statements is true?
A. If a circuit is properly fused, you cannot receive a fatal shock by touching it. (Proceed to block 15.)
B. Even though there is a fuse in the circuit, you could receive a fatal shock by touching it. (Proceed to block 25.)

18. *The correct answer to the question in block 10 is **B**. You should practice this rhythm until you can do it repeatedly.* Here is your next question.
For an electrical fire you should use
A. water. (Proceed to block 6.)
B. CO_2 extinguishers. (Proceed to block 3.)
C. foam-type extinguishers. (Proceed to block 11.)

19. *The correct answer to the question in block 13 is **B**. Keep one hand in your pocket as a safety precaution, but always be careful. Be sure that you are insulated from the electrical conductors, even though you believe the circuit to be de-energized.* Here is your next question.

Which of the following is the thing that actually determines if a shock is fatal?

A. The amount of current that flows through the body. (Proceed to block 22.)

B. The amount of voltage in the circuit. (Proceed to block 2.)

20. *Your answer to the question in block 16 is **A**. This answer is wrong. You should start **immediately** to get the breathing cycle started. Let someone else loosen his clothing. Proceed to block 8.*

21. *The correct answer to the question in block 8 is **B**. The rumor that a person can build up a resistance to electricity is a dangerous one. Here is your next question.*

Is this statement true or false? "If you are not familiar with a piece of electrical equipment, try a few things. After all, that's how you learn new things."

A. This statement is true. (Proceed to block 4.)

B. This statement is false. (Proceed to block 13.)

22. *The correct answer to the question in block 19 is **A**. A current of $^1/_{10}$ ampere could be fatal. Here is your next question.*

When you are working near high-voltage circuits, you should

A. work alone. (Proceed to block 24.)

B. work with someone. (Proceed to block 26.)

23. *Your answer to the question in block 25 is **B**. This answer is wrong. You should **never** put iodine on a burn. Proceed to block 10.*

24. *Your answer to the question in block 22 is **A**. This answer is wrong. It is never a good idea to work alone in any system where accidents may occur. Proceed to block 26.*

25. *The correct answer to the question in block 17 is **B**. Defective equipment **may** blow a fuse, and in this way you could receive some protection by the fuse being in the circuit. However, the fuse is mostly for protecting against fires and against destruction of electrical equipment. Remember that a 0.1-ampere shock can be fatal, and a 15-ampere fuse would be no protection at all. Here is your next question.*

Which of these statements is true?

A. Never put iodine on a burn. (Proceed to block 10.)

B. The best thing that you can put on a burn is iodine. (Proceed to block 23.)

26. *The correct answer to the question in block 22 is **B**. At the very least, a partner could go for help if an accident should occur.*

You have now completed the programmed questions. The next step is to take the "True-False Safety Test" that follows.

Self-Test with Answers

(Answers are given at the end of this appendix.)

1. Safety should be one of your most important concerns.
2. You should not work on equipment that you are not familiar with.
3. You should not experiment with expensive equipment.
4. You should not engage in horseplay in the lab.
5. You cannot build up a resistance to electricity.
6. It is not true that a little bit of electricity is good for you.
7. A current of only $\frac{1}{1000}$ ampere can produce a shock that you can feel.
8. With a current of $\frac{1}{100}$ ampere you cannot let go of the circuit.
9. A current of only $\frac{1}{10}$ ampere for about 1 second could be fatal.
10. A fuse in a circuit will not protect you from a serious shock.
11. Keeping one hand in your pocket when working around high voltages is a safe practice, but this is not an absolute guarantee of safety.
12. It is a good safety practice to keep tools sharp and clean.
13. You should find out where the main power switch and the fire extinguishers are when working in an area.
14. Never use water or foam-type fire extinguishers on electrical fires.
15. You should never inhale smoke or fumes of any kind.
16. When there is an electrical fire, you should de-energize the circuit.
17. You should never put iodine on a burn.
18. You should not use cotton to swab a burn.
19. Each cycle in artificial resuscitation should take about 5 or 6 seconds.
20. There should be about 10 or 12 cycles per minute when applying artificial resuscitation.

Answers to Self-Test

All answers are true.

B. Symbols and Vocabulary Words

AC GENERATOR
SYMBOL: OR
DEFINITION: A generator that produces an alternating current or alternating voltage.

ACTIVE CIRCUIT — A circuit that contains some type of voltage source such as a battery or generator.

ADJUSTABLE RESISTOR — See *Variable resistor.*

ALTERNATOR
SYMBOL: Same as for ac generator.
DEFINITION: See *Ac generator.*

AMMETER
SYMBOL:
DEFINITION: A meter used for measuring current. In general, an ammeter measures amperes, a milliammeter measures milliamperes, and a microammeter measures microamperes.

AMPERE
SYMBOL: A
DEFINITION: A unit of electric current flow. The current flow is one ampere if one coulomb of electricity flows past a point in the circuit in one second.

ANODE — 1. The positive terminal of a battery. 2. The terminal in an electrical or electronic device that is connected to the positive side of the voltage source.

APPLIED VOLTAGE

SYMBOL: E

DEFINITION: The voltage applied to a circuit. Some authors distinguish between applied voltage and voltage drop by representing the former with the letter symbol E and the latter with the letter symbol V.

ARMATURE — 1. The moving part of a relay. 2. The part of the generator in which the voltage is induced.

ARTIFICIAL MAGNET — A magnet that is produced by inducing a strong magnetic field in the material.

ATOM — The smallest division of an element that retains all the properties of that element.

AVERAGE — In alternating currents, the average value actually means the half-cycle average. For a pure sine wave the average value is 0.636 × the maximum value.

BATTERY

SYMBOL: ——|||——

DEFINITION: A combination of cells used for producing a voltage.

BRANCHES — The parts of a circuit into which current flows.

BRUSH — A conducting material, usually carbon, which is used to remove the voltage from the rotating part of a generator.

BUSBAR — A conductor used for carrying large currents. The busbar is sometimes used as a common ground line, and also as a return line to the generator.

CAPACITANCE

SYMBOL: C

DEFINITION: A unit of measurement that signifies the ability of a capacitor to store electricity. The unit of measurement for capacitance is the farad.

CAPACITOR

SYMBOL: ——|(——

DEFINITION: A component that is capable of storing an electrical charge.

CARBON-COMPOSITION RESISTOR

SYMBOL: ——/\/\/——

DEFINITION: (Usually called "carbon resistor") A resistor which is made of a carbon material. As a general rule it is less expensive than a wire-wound resistor.

CATHODE — 1. The negative terminal of a voltage source. 2. The terminal of a component that is connected to the negative terminal of the source.

CELL

SYMBOL: ——|———

DEFINITION: A voltage source that is capable of generating electricity by chemical means. It consists of two dissimilar conductors immersed in an acid or alkali solution. A number of cells in combination make a battery.

CELSIUS

SYMBOL: °C

DEFINITION: A system of measuring temperatures in which the temperature of freezing water is 0° and that of boiling water is 100°. This system used to be called "centigrade."

CENTIGRADE
SYMBOL: °C
DEFINITION: A system for measuring temperatures. See *Celsius.*

CHOKE
SYMBOL: —〰〰— OR 〰〰
SYMBOL WITH IRON CORE: 〰〰
DEFINITION: A component that opposes the flow of alternating current. Chokes are also called "inductors."

CIRCUIT BREAKER
SYMBOL: —⌒— OR —□—
DEFINITION: A device that automatically opens a circuit when the circuit current is excessive or when the circuit voltage is excessive. It serves the same purpose as a fuse. However, the circuit breaker is not destroyed when it opens a circuit. It can be reset either manually or electrically after the overload is removed.

CIRCULAR MIL—A unit of area measurement obtained by squaring the diameter of a circle. It is used for measuring the cross-sectional area of wire.

CLOSED CIRCUIT—A circuit in which there is a complete current path from the voltage source, through the circuit, and back to the voltage source.

COLOR CODE—A method of marking components so that their electrical values for a circuit can be determined. For example, the resistance value of a carbon-composition resistor can be determined by the colored stripes on its body.

COMMON
SYMBOL: ⏚ OR ▽
DEFINITION: A point in a circuit where a number of connections are made. In some circuits the ground (or earth) is used as a common connection for a number of branches in the circuit. The metal chassis in an electronic system is also used for a common connection. In an automobile the frame is used for returning electricity to a battery, and the frame is called the "common."

COMMUTATOR—A number of conductors, insulated from each other, which are connected to the rotating part of a generator. The generator voltage is taken from the commutator by the brushes.

COMPLETE CIRCUIT—See *Closed circuit.*

COMPOUND—A material that is made by combining two or more elements.

CONDENSER—Another term for "capacitor." In automotive ignition systems the term "condenser" is still popular.

CONDUCTOR—A material that will pass an electric current with very little opposition or resistance.

CONVENTIONAL CURRENT—An electric current that is presumed to flow from the positive terminal of the voltage source, through the circuit, and back to the negative terminal. Its direction of flow is opposite to the electron current flow.

COULOMB—A unit of electrical charge. It is equal to the combined charge of 6,240,000,000,000,000,000 electrons.

CURRENT

SYMBOL: *I*

DEFINITION: The flow of electrons (or charge carriers) through a circuit. Current flow is measured in amperes, but this unit may be too large for some applications, so milliamperes (thousandths of an ampere) and microamperes (millionths of an ampere) are also used for measurement.

CURRENT LAW — See *Kirchhoff's current law.*

CUTOUT — An automatic switch connected into the circuit of a dc generator to prevent it from turning like a motor when a dc voltage is applied to it.

D'ARSONVAL MOVEMENT — A name that is sometimes incorrectly used to refer to permanent-magnet, moving-coil meter movements.

DC GENERATOR

SYMBOL: OR

DEFINITION: An electromechanical device that produces a dc voltage.

DEGAUSSING — A term that means demagnetizing.

DIAMAGNETIC — The opposite of magnetic. A diamagnetic material will move away from a strong magnet, whereas a magnetic material will move toward it.

DIODE

SYMBOL:

DEFINITION: A component that will conduct electricity in only one direction.

EDDY CURRENTS — Circular currents induced in the iron frame and core of generators and transformers. These currents do not serve a useful purpose, but rather represent a loss in motor and transformer circuits.

EFFECTIVE VALUE

SYMBOL: rms

DEFINITION: The effective value, also called the "rms value," in an ac voltage or current is that value of dc voltage or current that could be used to replace the ac and still produce exactly the same heating effects in the circuit resistance. For a pure sine wave the rms value is $0.707 \times$ the maximum value.

ELECTRICAL CHARGE — An accumulation of electrons (negative charges) or a deficiency of electrons (positive charges) in a body.

ELECTRICAL FRICTION — A term that is sometimes used to describe circuit resistance.

ELECTRICAL PRESSURE — A term that is sometimes used to describe an applied voltage in a circuit. It is a convenience to think of the voltage as an electrical pressure that forces current to flow through the circuit.

ELECTROLYTE — A solution that conducts an electric current.

ELECTROMAGNET — A magnet that is produced by causing a current to flow through a coil.

ELECTROMOTIVE FORCE

SYMBOL: emf

DEFINITION: An older term, which is now going out of style, that has been used to mean voltage.

ELECTRON — A small negative charge in an atom. It is the charge carrier for electric current in metal conductors.

ELECTRON CURRENT — A current that leaves the negative terminal of a source and flows through the circuit back to the positive terminal of the source. Electron current is presumed to be a flow of electrons within the circuit.

ELECTROSCOPE—A simple instrument used to detect the presence of an electrical charge.

ELECTROSTATIC—A term used to refer to electrical charges at rest.

ELECTROSTATIC GENERATORS—A type of voltage generator that produces a large difference in potential as a result of friction. An example is the Van de Graaff generator.

ELEMENTS—The basic building materials of the universe. There are 92 elements in the universe which are found in their natural state. From these elements, all other materials are made.

ENERGY—The capacity to do work.

FARAD—A unit of measurement for capacitance. It is too large for most practical applications, so microfarads and picofarads are preferred.

FERROMAGNETIC—Materials that are ferromagnetic are strongly attracted to a magnet.

FIELD—A region of influence. Thus, the field of a magnet is the region around the magnet where a ferromagnetic material will be attracted to the magnet.

FIXED RESISTOR—A resistor with an unvarying resistance value.

FREQUENCY
SYMBOL: f
DEFINITION: A measurement of how rapidly an alternating current changes direction. It is measured in hertz.

FUSE
SYMBOL: —⌒⌣— OR ⊏▭⊐
DEFINITION: A circuit protective device. When the circuit voltage or current becomes excessive, the fuse burns out and opens the circuit path.

GALVANOMETER—An instrument for measuring very small currents.

GROUND
SYMBOL: ⏚
DEFINITION: A return path to the generator through the earth, or through a common connection.

HENRY—The unit of measurement of inductance.

HERTZ
SYMBOL: hz
DEFINITION: The unit of measurement for frequency. One hertz is equal to one cycle per second.

HYSTERESIS—A type of loss in iron-core electromagnetic circuits. It is due to the fact that some magnetism remains in the iron after each cycle of current, and energy must be used to remove it before the magnetic material can be magnetized in the reverse direction.

IMPEDANCE
SYMBOL: Z
DEFINITION: Opposition to the flow of alternating current. It is measured in ohms.

INDUCTANCE
SYMBOL: L
DEFINITION: The property of a coil that causes it to oppose the flow of alternating current. Inductance is measured in henrys.

INDUCTOR — Another name for choke.

INERTIA — The property of a body that makes it tend to stay at rest when a moving force is applied, or tends to make it continue in motion after the moving force has been removed.

INSULATOR — A material that will not readily conduct electricity.

ION — An atom that has an electrical charge. The charge is due to the fact that the atom has gained or lost an electron.

JEWELED MOVEMENT — A type of meter movement in which the moving element turns on a jewel and pivot.

KILO

SYMBOL: k

DEFINITION: A prefix meaning thousand. Thus, a kilohertz is one thousand hertz.

KINETIC ENERGY — The energy that a body has because of its motion.

KIRCHHOFF'S CURRENT LAW — A law of current behavior in networks. It states that the sum of the currents entering any junction equals the sum of the currents leaving that junction.

KIRCHHOFF'S VOLTAGE LAW — A law of voltage relationships in networks. It states that the sum of the voltage drops around any closed circuit must be equal to the sum of the voltage rises in that circuit.

LAMINATIONS — Thin layers of iron in iron-core chokes and transformers. Cores are made in laminations in order to reduce eddy currents.

LEADS — Connections to a circuit, or to an instrument.

LOAD — 1. Technically, the "load" of a circuit refers to the amount of current delivered to that circuit. A circuit with a light load is one in which very little current flows. 2. The term "load" is sometimes used to mean load resistance.

LODESTONE — A natural magnet.

MAGNETIC FIELD — The region of magnetic influence around a magnet.

MAGNETIC POLE — The place where the magnetic flux lines leave or enter a magnet.

MEGA

SYMBOL: M

DEFINITION: A prefix meaning million. Thus, a megohm is a million ohms.

MICRO

SYMBOL: μ

DEFINITION: A prefix meaning millionth. Thus, a microampere is a millionth of an ampere.

MICROAMMETER — See *Ammeter*.

MIL — A thousandth of an inch.

MILLI

SYMBOL: m

DEFINITION: A prefix meaning one-thousandth. Thus, a milliampere is a thousandth of an ampere.

MILLIAMMETER — See *Ammeter*.

MOLECULE — The smallest possible division of a compound that has all the properties of that compound.

MOTOR
SYMBOL: —(MOT)—
DEFINITION: An electrical component that converts electricity to rotation.

NANO
SYMBOL: n
DEFINITION: A prefix meaning a thousandth of a millionth. Thus, a nanosecond is 0.000 000 001 second.

NONMAGNETIC—See *Paramagnetic*.

NUCLEUS—The center of an atom.

OHM
SYMBOL: Ω
DEFINITION: A unit of measurement for resistance.

OHM'S LAW—A law that expresses the relationship between current, voltage, and resistance. It states that the current (in amperes) is equal to the voltage (in volts) divided by the resistance (in ohms). Mathematically, $I = E/R$.

OHMMETER —(Ω)—
SYMBOL:
DEFINITION: An instrument used for measuring resistance.

PARALLEL CIRCUIT—A circuit in which there is more than one path for current to flow.

PARAMAGNETIC MATERIAL—A material that is not noticeably affected by the presence of a magnetic field.

PASSIVE CIRCUIT—A circuit in which there is no source of voltage.

PERMANENT MAGNET—A magnet that retains its magnetism over a long period of time.

PERMANENT-MAGNET, MOVING-COIL METER MOVEMENT—A meter in which the current being measured flows through a movable coil that is located in the field of a permanent magnet. The magnetic field of the coil causes it to turn and move a meter needle up scale.

PHASE ANGLE
SYMBOL: ϕ
DEFINITION: The angle between the voltage and the current in an ac circuit.

PHOTOCELL
SYMBOL:
DEFINITION: A component that generates a voltage when light falls on it. It is important to note that the letter lambda (λ) is used on the schematic symbols for other components that are light-operated.

PICO
SYMBOL: p
DEFINITION: A prefix meaning millionth of a millionth. Thus, a picofarad is a millionth of a millionth of a farad, or a millionth of a microfarad.

PIEZOELECTRICITY—Electricity generated by exerting pressure on certain crystalline materials.

POLES—The points on a permanent magnet where the magnetic flux lines are presumed to leave or re-enter the magnet.

POTENTIAL ENERGY—The energy that a body has because of its position.

POTENTIOMETER

SYMBOL:

DEFINITION: A variable resistor connected in such a way that it can be used to vary the amount of voltage to a component or circuit.

POWER—A measure of the rate at which energy is expended or at which work is done. In electric circuits, power is measured in watts.

POWER FACTOR

SYMBOL: P.F.

DEFINITION: A measure of how nearly in phase the voltage and current are in an ac circuit. A power factor of 1.0, or 100 percent, means that the current and voltage are in phase. A power factor of 0, or 0 percent, means that the current and voltage are 90° out of phase.

POWER SUPPLY—A source of voltage or current used to supply electricity to a circuit. In some circuits the power supply simply converts the electricity from one form (such as ac) to another form (such as dc).

PRECISION RESISTOR—A resistor that is manufactured to a very close tolerance so that its measured resistance is very nearly equal to its rated value of resistance.

PRIMARY CELLS—Cells that cannot be recharged.

PROTON—A positive particle in an atom.

Q—1. The symbol for quantity of charge as measured in coulombs. 2. A measure of how sharply a circuit will turn. The higher the Q, the narrower the band of frequencies that will pass through the circuit. 3. A comparison of the reactance and resistance of a coil. Mathematically, $Q = X_L/R$.

REACTANCE

SYMBOL: X

DEFINITION: A measure of the opposition that an inductor or capacitor offers to the flow of alternating current. It is measured in ohms. The symbol for inductive reactance is X_L, and the symbol for capacitive reactance is X_C.

RECTIFIER—A component that allows current to flow through it in only one direction. It is used for converting alternating current to direct current.

RELAY

SYMBOL: 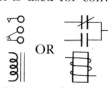 OR

DEFINITION: An electrically-operated switch.

RELUCTANCE

SYMBOL: $\mathscr{R}$

DEFINITION: A measure of the opposition that a material offers to the establishment of magnetic flux lines.

RESISTANCE

SYMBOL: R

DEFINITION: A measure of the opposition that a component or circuit offers to the flow of an electric current. Resistance is measured in ohms.

RESISTOR

SYMBOL: OR

DEFINITION: A component used for introducing resistance into a circuit.

RHEOSTAT

SYMBOL: ─\/\/\─OR─\/\/─

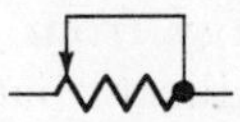

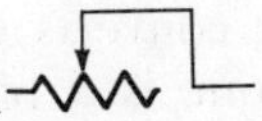

DEFINITION: A variable resistor connected in such a way that varying the resistance changes the amount of circuit current.

RMS—Abbreviation for "root mean square." This is another way of saying "effective value." If the rms value of an ac voltage is 100 volts, and this voltage is placed across a 100-ohm resistor, then a 100-volt dc battery across the same resistor will produce exactly the same amount of heat dissipation.

SATURATION—In the iron core of a choke or transformer, saturation occurs when the maximum number of flux lines possible are established in the core.

SCHEMATIC DRAWING—A circuit drawing using symbols to represent the components.

SECOND

SYMBOL: s

DEFINITION: The standard unit of time for scientific work.

SECONDARY CELL—A cell that can be recharged.

SEMICONDUCTOR—A material that is neither a good conductor nor a good insulator.

SERIES—A connection of components arranged so that the same current flows through each.

SHUNT—A parallel connection in a meter that prevents an excessive current from flowing through the meter movement.

SOFT IRON—Iron that cannot retain magnetism.

SWITCH

SYMBOL: ─o╱ o─

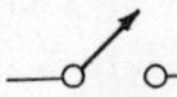

DEFINITION: A component that opens or closes a circuit path.

TAUT BAND—A rugged method of suspending a meter movement. The taut band provides the restoring force to return the meter needle to zero after a measurement has been made.

TEMPERATURE COEFFICIENT—A method of rating a component by how much its value will change with a change in temperature. If a component has a positive temperature coefficient, its value will increase when the temperature increases. With a negative temperature coefficient, the value of the component goes down when the temperature goes up.

TERMINAL VOLTAGE—The voltage measured at the terminals of a voltage source.

THERMOCOUPLE—A junction of two dissimilar metals that generates a voltage when heated.

TOLERANCE—The allowable variation between the rated value of a component and the actual measured value of that component.

TOTAL CURRENT

SYMBOL: I_T

DEFINITION: The current delivered by the power supply to a circuit.

TOTAL RESISTANCE

SYMBOL: R_T

DEFINITION: The total opposition that a circuit offers to the flow of current, measured in ohms.

TRACING THE CIRCUIT—The process of following the path of current flow in a circuit.

TRANSDUCER—A component that converts energy from one form to another. For example, a microphone is a transducer that converts sound energy into electrical impulses.

TRANSFORMER
SYMBOL:
SYMBOL FOR AN IRON-CORE TRANSFORMER:
DEFINITION: A component that will pass ac, but will not pass dc. It can be used to step an ac voltage up or down, or to step an alternating current up or down.

VAN DE GRAAFF GENERATOR—A static-electricity generator capable of generating very high voltages.

VARIABLE RESISTOR
SYMBOL: —⋀⋀⋀— OR ∘—⋀⋀⋀—∘
DEFINITION: A resistor with a resistance value that can be varied. It is used for a rheostat or potentiometer.

VOLT
SYMBOL: V
DEFINITION: A unit of measurement for the voltage rise or voltage drop in a circuit.

VOLTAGE DROP—The voltage across a resistor that results when a current flows through the resistor. When you are tracing the path of electron current flow, a voltage drop occurs when you enter the negative side of a voltage and leave the positive side of that voltage.

VOLTAGE LAW—See *Kirchhoff's voltage law.*

VOLTAGE RISE—The applied voltage in a circuit. When you are tracing the path of electron current flow, a voltage rise occurs when you enter the positive side of a voltage and leave the negative side of that voltage.

VOLTMETER
SYMBOL: —(V)—
DEFINITION: An instrument used for measuring voltages.

WATT—The unit of measurement for electric power.

WIRE-WOUND RESISTORS—Resistors that are made by winding a resistance wire on an insulator. This type of resistor can be made with a high power rating and an accurate resistance value.

WORK—Work is performed when a force is exerted through a distance.

C. Circuit Board for Experiments

The experiments described in this book can be performed on a simple circuit board like the one shown in Fig. C-1. The arrangement of parts is optional, and you may wish to mount only those parts needed for each experiment as you perform it. Here is a parts list of all the components that you will use in the experiments:

AC POWER LEAD

ALUMINUM FOIL

BATTERIES, 6-volt (2)

CAPACITOR, 25-microfarad (not electrolytic), 200-volt

CAPACITOR, electrolytic, 25 microfarads, 50 WVDC

COIL, reed relay or 150 turns of number 20 wire on cardboard form

COMB

COMPASS

CONNECTORS, Fahnestock type or equivalent

COPPER, flat strips (2)

COPPER WIRE, number 20, insulated

COPPER WIRE, number 8 or 10, bare

DC MOTOR, permanent-magnet, slot-car type (2)

DIODE, silicon, 2-ampere, 150-volt (2)

FLEXIBLE TUBING, to fit over motor shafts

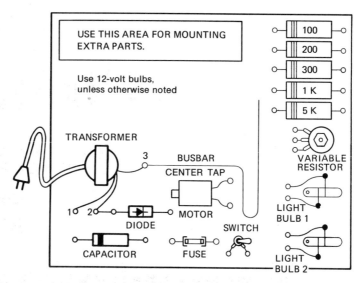

Figure C-1. The experiment board.

FUSE, ¼-ampere, cartridge type
INSULATED WIRE
JAR, or glass
LIGHT BULBS, 6-volt and 12-volt (3 of each)
LIGHT SOCKETS (3)
METAL, from tin can, 3-inch square
NAIL, not aluminum
PAPER CLIPS
RESISTORS, 2-watts minimum
     100-ohm
     200-ohm
     300-ohm
     1-kilohm
     5-kilohm (3), wire wound
RHEOSTAT, 0–1,000-ohm, 25-watt
STRING
SWITCH, single-pole, single-throw (2)
TEST LEADS (2)
TRANSFORMER, 117-volt, 12.6-volt center tapped secondary
VOLT-OHM-MILLIAMMETER
WIRE, 1-foot length from coat hanger

Index